ADVANCED METROLOGY

ADVANCED METROLOGY

Freeform Surfaces

X. JANE JIANG
Professor, Dame, EPSRC Hub for Future Metrology,
Centre of Precision Technologies,
University of Huddersfield, Huddersfield, United Kingdom

PAUL J. SCOTT
Professor, EPSRC Hub for Future Metrology,
Centre of Precision Technologies,
University of Huddersfield, Huddersfield, United Kingdom

ELSEVIER

ACADEMIC PRESS
An imprint of Elsevier

Academic Press is an imprint of Elsevier
125 London Wall, London EC2Y 5AS, United Kingdom
525 B Street, Suite 1650, San Diego, CA 92101, United States
50 Hampshire Street, 5th Floor, Cambridge, MA 02139, United States
The Boulevard, Langford Lane, Kidlington, Oxford OX5 1GB, United Kingdom

Notices
Knowledge and best practice in this field are constantly changing. As new research and experience broaden our understanding, changes in research methods, professional practices, or medical treatment may become necessary.

Practitioners and researchers must always rely on their own experience and knowledge in evaluating and using any information, methods, compounds, or experiments described herein. In using such information or methods they should be mindful of their own safety and the safety of others, including parties for whom they have a professional responsibility.

To the fullest extent of the law, neither the Publisher nor the authors, contributors, or editors, assume any liability for any injury and/or damage to persons or property as a matter of products liability, negligence or otherwise, or from any use or operation of any methods, products, instructions, or ideas contained in the material herein.

Library of Congress Cataloging-in-Publication Data
A catalog record for this book is available from the Library of Congress

British Library Cataloguing-in-Publication Data
A catalogue record for this book is available from the British Library

ISBN 978-0-12-821815-0

For information on all Academic Press publications
visit our website at https://www.elsevier.com/books-and-journals

Publisher: Matthew Deans
Acquisitions Editor: Brian Guerin
Editorial Project Manager: Peter Adamson
Production Project Manager: Sojan P. Pazhayattil
Cover Designer: Miles Hitchen

Typeset by SPi Global, India

Contents

Preface

Free-form surfaces are now becoming ubiquitous in industrial products and are widely used in optics, astronomy, aerospace, automotive, semiconductors, telecommunications, and medical devices. As well as being able to provide aesthetical benefits, they can improve the functional performance of a geometrical product while at the same time reducing size, weight, and cost. As a result, high added-value products can have almost any designed geometry, including free-form shapes, which demand much more complicated geometries than traditional products. Further the specification of allowable geometric variability for free-form surfaces is also very important for the control of manufacture and function.

The book introduces discrete/continuous surface representations, decomposition to solve the nonuniqueness problem of free-form shape, and filtration based on diffusion equations (PDEs), morphology (alpha shape), wavelet decomposition and numeric characterization. The book discusses methods able to handle the geometry of free-form surfaces and bridge the knowledge gap between research and practical industrial applications. To be ready for future smart manufacturing, the book also introduces a machine-readable and machine-interpretable semantic system that deals with data/information involved in manufacturing.

The book provides a systematic guide for engineering designers and manufacturers interested in exploring the benefits of this technology. The inclusion of industrial case studies and examples will help readers to implement these techniques that are being developed across different industries as they offer improvements to the functional performance of products and reduce weight and cost.

Acknowledgments

Jane Jiang and Paul Scott would like to express deepest thanks to all contributors, our current, and former research fellows who contributed to the book on free-form surfaces.

Hussein Abdul-Rahman
Philip Cooper
Shan Lou
Luca Pagani
Qunfen Qi
Wenhan Zeng
Xiangchao Zhang
Jian Wang

Jane Jiang and Paul Scott also gratefully acknowledge the UK Royal Society under a Wolfson-Royal Society Research Merit Award, the European Research Council under ERC Advanced Investigator Grant, and the UK Engineering and Physical Sciences Research Council for strategic funding for manufacturing metrology and Fellowships.

CHAPTER 1

Introduction

1.1 Introduction

Free-form surface is becoming ubiquitous in advanced industrial products, either for functional or aesthetical reasons. Free-form surfaces have been widely used in optics, astronomy, aerospace, automotive, semiconductors, telecommunicating, and medical devices. Many high-added value products can have almost any designed geometry, a free-form shape, which demands much more complicated geometries than previous conventionally specified geometries either for function, cost, size, or aesthetical reasons [1, 2].

Leading-edge manufacturing has allowed the production of free form to emerge in modern optics, bringing the advantages of excellent optical performance and system integration [3]. Free-form optics are gradually maturing, with broad uses in imaging systems for green energy, illumination in automotive, aerospace, and biomedical engineering. Compared with traditional optical components, free-form optics have an increased range of manufacturable surfaces, giving optical designers more flexibility and scope for innovation, enhancing the optical system's performance to the maximum extent. For instance, free-form optics enable optical performance that is otherwise impossible, such as simultaneously correcting aberrations, increasing depth of field, and expanding the field of view. It can also simplify system structure with fewer surfaces, lower mass, lower cost, smaller package size, and reduced stray light, allowing easy system integration and reducing difficulties in assembly. In addition, multiple optical surfaces can be made on one free-form element. Although free-form optics have many advantages, the one disadvantage is that tolerances tend to be smaller than in spherical optics and there is less room for correcting manufacturing errors during the setup than with spherical optics.

Other types of free-form surfaces are often called sculptured or curved surfaces, for example, in automotive (car bodies, headlights, panel displays, 3D cams, gears, etc.) and in aerospace (turbine blades, fan blades, impellers, etc.). Geometrical deviations from manufacturing can cause inefficiencies that give poor aesthetics and ineffective functionality and waste large quantities of energy. Design, engineering, and the testing of parts for functional surfaces are key activities for developing products with better performances due to their free-form geometry. For large components with complex shapes in aerospace and automotive, computer-aided design (CAD)/finite element method (FEM) is the most common tool for design and optimization [4].

Advanced Metrology
https://doi.org/10.1016/B978-0-12-821815-0.00001-0

1.2 Geometrical variability

The specification and characterization of free-form surfaces is immature; both optical and many CAD packages only allow the nominal free-form geometry to be specified. To control manufacturing and function, there needs to be a specification of allowable geometrical variability as currently international standards for geometrical products specify allowable geometry through geometrical tolerancing. At the moment, geometrical tolerancing has proved to be a very blunt instrument with many cases, particularly in the aerospace and biomedical industries, failing in the function of free-form surfaces due to inadequate specification.

To qualify geometrical variability a geometrical product will normally proceed through four conceptual realms: the design realm, where an artifact as imagined by the designer according to product functional requirement is specified; the realm of manufacture, where the artifact is created; the realm of verification, where the artifact is measured for conformance to specification; and the realm of product life, where the artifact is used.

An analogy between the chain of four realms and music is used here to illustrate the basic idea of specification, manufacture, verification, and comparison: Beethoven specifies some music (composes). This is manufactured (by playing); see Fig. 1.1.

How do we know it is manufactured to specification? The played music is measured by listening to it and the notes written down on a musical scale to produce a musical score; see Fig. 1.2. This score can be then compared with the original specification to check for conformity.

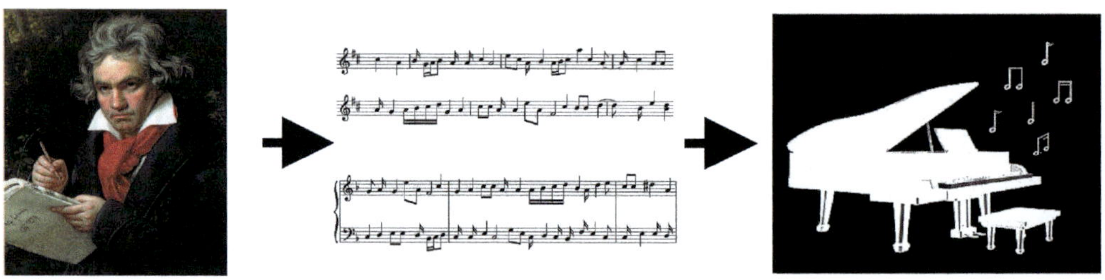

Fig. 1.1 Beethoven composes some music that is played.

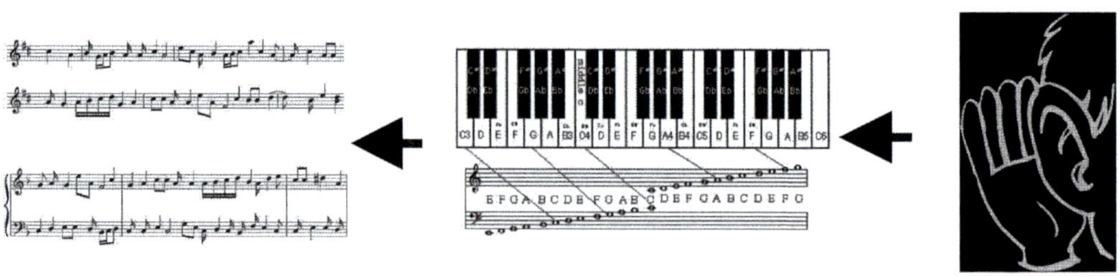

Fig. 1.2 Identifying the individual notes on a musical scale to produce a musical score.

1.3 Classification of geometrical surfaces

A useful approach to the classification of surfaces, used in geometrical products, is provided by the invariance class concept, based on technologically and topologically related surface (TTRS) theory (see Table 1.1) [5, 6]. Each class is defined by the ridged motions (translations and rotations) that leave the geometrical surface unchanged (invariant).

True free-form geometries fall into the complex classification of invariant geometries in that the surface shape has no continuous translational or rotational symmetry around any axes. Many types of surfaces are also defined as free form by industry, for example, off-centered rotational optics and toroidal. These are mathematically not true free-form surfaces; they do have an axis, but this axis does not pass through the functional surface. In this book, these surfaces will be called quasifree form surfaces, and what follows in the subsequent chapters is also applicable to quasifree form surfaces.

1.4 Basic classification of geometrical surface modifications

It was proposed by [7] that free-form geometrical surfaces be classified into three kinds according to modifications made to the basic form geometry by superimposing (or not) additional surface features (Table 1.2).

1.5 The ISO system for geometrical products

At the international standards level, geometrical products are the raison d'être of ISO/TC 213 *Geometrical Product Specification and Verification* (GPS). ISO/TC 213 started as a technical committee in 1996 in Paris and was the coming together of several ISO technical committees for the sole purpose of harmonizing the two realms of design and verification for geometrical products. The ISO/TC developed a common language, a common system, and a common understanding of geometrical products. ISO/TC 213 has had a very great influence on the other two realms of manufacture and the world of product life in particular through close cooperation with other ISO technical committees.

A new ISO/TC 213 inherited a very ad hoc group of disparate ISO documents. The first task was to identify and resolve the contradictions within these documents, and the second task was to identify missing documents. This was achieved through the concept of "the GPS matrix" (the current GPS matrix is to be found in [8]). The GPS matrix consists of a series of chains of standards for different aspects of the GPS system, for example, size, orientation, roundness, and profile roughness. Each chain of standards currently consists of seven links of which the first three (A, B, and C) are for specification and cover: (A) symbols and indications; (B) feature requirements; and (C) feature properties.

Table 1.1 The seven invariance classes.

Type	Nature of invariance	Example
1. Spherical	Three rotations around a point	
2. Planar	One rotation around a straight line and two translations in a plane perpendicular to the straight line	
3. Cylindrical	One translation along and one rotation around a straight line	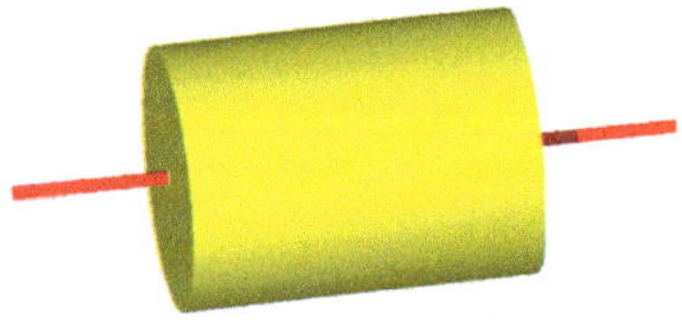
4. Helical	One translation along and one rotation combined, at a fixed ratio, around a straight line	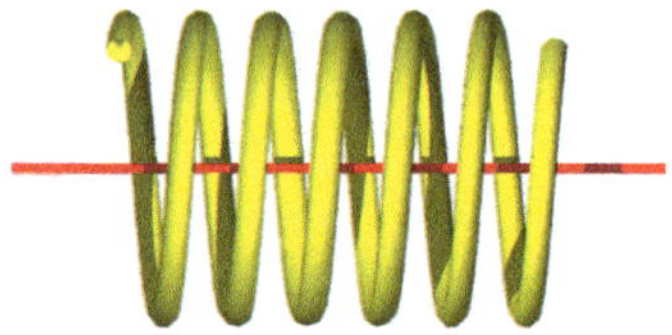
5. Revolute	Rotation around a straight line	
6. Prismatic	Translation along a straight line	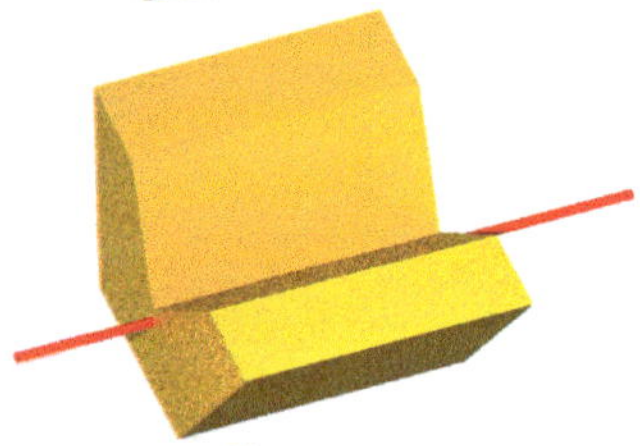
7. Complex	No rotational or translational invariant movement possible	

Table 1.2 Additional surface feature classes.

<table>
<tr><td>Class 1</td><td>Surfaces that include steps, edges, and facets superimposed on to the basic geometry. The classic nonfree form illustrative example is the Fresnel lens in Fig. 1.3 that is used to magnify the light in lighthouses</td><td>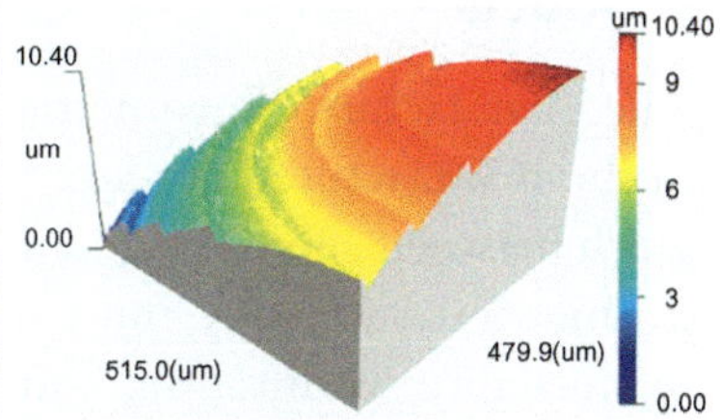

Fig. 1.3 Fresnel lens.</td></tr>
<tr><td>Class 2</td><td>Surfaces that have patterns of features superimposed on to the basic geometry. Nonfree form examples include the golf ball and the 3M TRIZAC abrasive surfaces (in Fig. 1.4), which consist of an array of triangular-based micron-sized pyramids</td><td>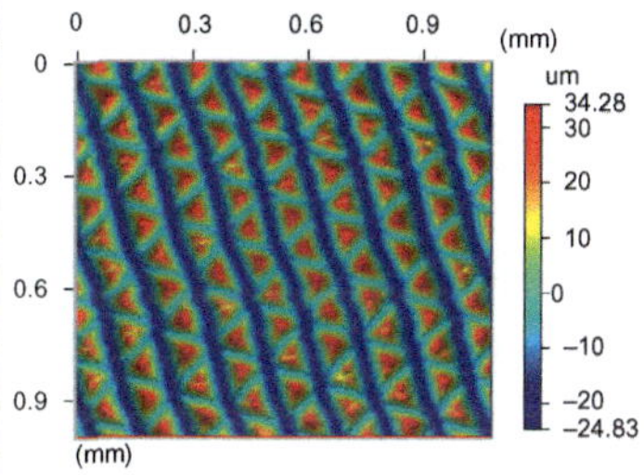

Fig. 1.4 3M TRIZAC abrasive surface.</td></tr>
<tr><td>Class 3</td><td>Smooth surfaces, that is, surfaces designed with no steps, edges, or patterns but rely purely on the global geometry. Fig. 1.5 is a free-form surface example from the optics industry, which requires submicroform error and nanosurface topography</td><td>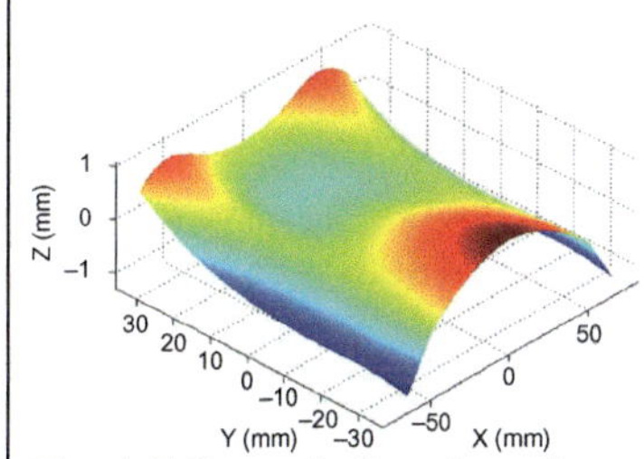

Fig. 1.5 Smooth free-form lens.</td></tr>
</table>

The last three links (E, F, and G) are for verification and cover: (E) measurement; (F) measuring equipment; and (G) traceable calibration. The middle link (D) covers conformance and nonconformance; see [8] for details.

The structure of the chain of links also reflects the very important GPS principle, namely, the **duality principle** [9], where it is stated as follows:

(1) A specification defines a specification operator independent of any verification procedure or measurement equipment.

(2) The specification operator is realized in a verification operator that is independent of the specification itself but is intended to mirror the GPS specification operator.

This means that the specification procedure and the specification procedure alone defines the measurands (the quantity to be measured [10]). The verification procedure mirrors the specification procedure; any deviation introduces specification uncertainty [11] that contributes to the measurement uncertainty [12].

1.6 A brief history of the specification and metrology of geometrical products

The manufacture of geometrical products is inevitably imprecise, resulting in variations in the manufactured product from the designer's specification. A mechanism to distinguish between allowable and unacceptable product variances is required. The control of geometrical variability must be taken into account in all of the four realms: design, manufacture, verification, and product life. This is more necessary when specifying a product's geometry and managing the manufacturing process.

As mentioned earlier, there are two types of geometrical applications: general geometrical surfaces and geometrical optics; each has its own design, manufacturing, and measurement technologies. We will begin with the former, the specification and measurement of general geometrical surfaces, and then move on to the specification measurement of optical surfaces.

Historically, various approaches have been taken to control geometrical variability. Early mechanisms to control geometrical variability include the following:

- (Pre 1800s) The use of a master component with which the manufactured part could be compared and then adjusted as required. Here the specification is the master component, and the comparison measurement was often carried out with calipers or by physically placing the master component on top of the manufactured component and compared by eye or by rubbing a fingernail along the edge to determine the difference.
- (Early 1800s) The use of hard gauging, which enabled parts to be manufactured accurately enough to be assembled interchangeably. Hard gauges are physical devices with specified features of size and length with which manufactured components can be compared to check the size/length of the manufactured object. The basic mathematical theory behind hard gauging was not fully developed until Taylor's principle [13].
- (c.1850) Dimensional drawings: gauging technology allows dimensional drawings, rather than representational drawings or models, to be used for design specifications. Although not directly concerned with geometrical variability, dimensional drawings are used as a mathematical nominal model of the desired geometry, a precursor to dimensioning tolerancing.
- (Early 1900s) Dimensional tolerancing: a system for specifying explicit limits on dimensional requirements for dimensional drawings to control workpiece variability (plus/minus tolerancing). In 1905 William Taylor formulated the Taylor principle in Patent No. 6900 for improvements in gauges for screws [13]. The most important principle associated with the design of hard gauges is as follows: a go gauge should check the full feature (maximum material limit) of an object, while a no-go gauge should check each individual feature or dimension of that object.

- (1940s) Geometrical tolerancing: a system of specifying tolerance zones in engineering drawings to control workpiece variability. Geometrical tolerancing was developed to improve the weakness of previous tolerance systems and to handle imperfect form and ambiguous references. Geometrical product specification and verification is defined by the International Organization for Standardization (ISO), and as such, geometrical tolerancing is still at the core of today's tolerancing systems.

  Traditional metrology is essentially analogue, using hard gauges and hand gauges to verify that an artifact meets its specification. The mathematics is encapsulated within the mechanical setup (jigs and fixtures) and the relationships of the gauges (Fig. 1.6). Postmeasurement calculation (analytics) is relatively simple and, traditionally, carried out by hand. Traditional metrology was restricted mainly to simple geometrical shapes (planes, spheres, cylinders, cones, etc.) that described the idealized geometry of artifacts of the day [14], which is to say invariance classes 1–6.

  Free-form surfaces (invariance class 7) are more difficult to apply to traditional metrology since it is not easy to mechanically set up a datum system. One approach is to create fiducial surfaces, in addition to the free-form surface, to establish a precision datum system. Alternatively a mechanical device (a jig or a fixture) can lock the six degrees of freedom uniquely, from the free-form surface, with repeatable precision to establish the datum system. One such mechanical device is a plate with six spheres, in which each sphere is designed to touch the free-form surface in a unique position. Once the datum system is determined, then a measurement using this datum system can be applied.

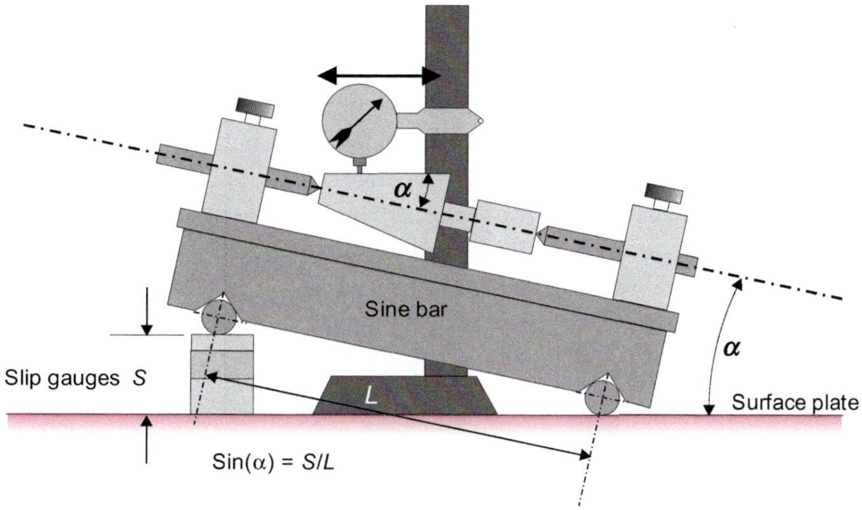

Fig. 1.6 Traditional metrology setup to measure cones.

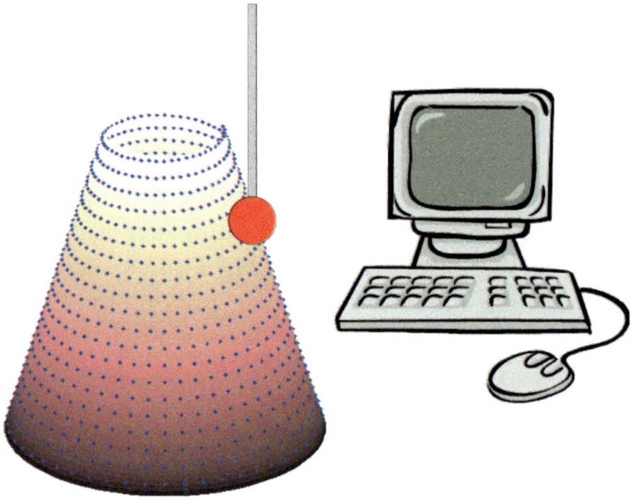

Fig. 1.7 Example of digital metrology.

- (1960s) Digital metrology, through sampling in a coordinate system, provides a digital representation of an artifact, enabling computational techniques to be applied (Fig. 1.7). The mathematics is encapsulated in the postmeasurement calculations or analytics captured through software implementing algorithms that became increasingly complex. When digital metrology first appeared, standards and methodologies still reflected the language and concepts of traditional metrology, but now, digital metrology extends far beyond the traditional metrology domain.
- Savio et al. [2] surveyed metrology for general free-form surfaces, ranging from car body parts to optics. For optical surfaces, metrology requirements extend down to the nanometer range. Both contact (tactile) and noncontact measuring techniques are used, either offline or on machine. Another way to discriminate the different measurement possibilities is between measurements with respect to a datum and those without any datum (e.g., for inspecting the form of a turbine blade without reference to the root profile). Commercial instruments usually operate offline. The results contain systematic errors due to the machine error motions. Many instruments have a small measuring range or a long measuring time. On-machine metrology can avoid the errors caused by moving and repositioning the workpiece. On-machine measurements can use the machine axes to extend the measuring range and improve the measuring efficiency. In this case, however, the systematic errors of the machine may be hidden.

1.7 Remarks

Currently the GPS matrix does not have a specific chain of standards for free-form geometries. Part of the intention of this book is to contribute to the formation of a chain of links for free-form geometries for the GPS system.

The book provides the emerging methodology for handling the geometry of free-form surfaces for industrial applications, including free-form data representation (Chapter 2), sampling for free-form surfaces (Chapter 3), free-form surface fitting (Chapter 4), free-form surface reconstruction (Chapter 5), filtration including linear and nonlinear diffusion filters (Chapter 6), morphological filtration (Chapter 7), segmentation (Chapter 8), wavelets and multiscale decomposition (Chapter 9), characterization of free-form surfaces (Chapter 10), and characterization of free-form structured surfaces (Chapter 11). Finally the book will initiate a smart free-form design and metrology platform using category semantics (Chapter 12).

References

[1] Jiang X, Whitehouse DJ. Technological shifts in surface metrology. CIRP Annals 2012;61(2):815–36.
[2] Savio E, De Chiffre L, Schmitt R. Metrology of freeform shaped parts. CIRP Annals 2007;56(2):810–35.
[3] Fang FZ, Zhang XD, Weckenmann A, Zhang GX, Evans C. Manufacturing and measurement of freeform optics. CIRP Annals 2013;62(2):823–46.
[4] Hirz M, Dietrich W, Gfrerrer A, Lang J. Integrated computer-aided design in automotive development. Heidelberg: Springer; 2013. ISBN 978-3-642-11939-3. https://doi.org/10.1007/978-3-642-11940-8.
[5] Clement A, Desrochers A, Riviere A. Theory and practice of 3-D tolerancing for assembly. In: Tire de: CIRP international working seminar on computer-aide tolerancing, Penn State University, University Park, PA, May 16–17; 1991. p. 26–56.
[6] ISO 17450-1:2011. Geometrical product specifications (GPS)—general concepts—part 1: model for geometrical specification and verification. Geneva: ISO; 2011.
[7] Jiang X, Scott P, Whitehouse D. Freeform surface characterisation-a fresh strategy. CIRP Annals 2007;56(1):553–6.
[8] ISO 14638:2015. Geometrical product specifications (GPS)—matrix model. Geneva: ISO; 2015.
[9] ISO 8015:2011. Geometrical product specifications (GPS)—fundamentals—concepts, principles and rules. Geneva: ISO; 2011.
[10] ISO/IEC Guide 99:2007. International vocabulary of metrology—basic and general concepts and associated terms (VIM). Geneva: ISO; 2007.
[11] ISO 17450-2:2012. Geometrical product specifications (GPS)—general concepts—part 2: basic tenets, specifications, operators, uncertainties and ambiguities. Geneva: ISO; 2012.
[12] ISO/IEC Guide 98-1:2009. Uncertainty of measurement—part 1: introduction to the expression of uncertainty in measurement. Geneva: ISO; 2009.
[13] Taylor W. Improvements in gauges for screws. British Patent no. 6900; 1905.
[14] Rolt FH. Gauges and fine measurements. 2 vols. London: MacMillan and Co. Ltd; 1928.

CHAPTER 2

Fundaments for free-form surfaces

2.1 Introduction

The theory of measuring and characterizing ordinary simple surfaces such as planes, spheres, and cylinders has been developed and is now mature [1–5]. Indeed, many research papers and industrial standards have been published to describe the measurement and characterization of such surfaces [3–13]. However, as stated in Chapter 1, with the development of science and technology, an increasing number of complex surfaces are being produced that, unlike the conventional surfaces, have **no axes of rotation** and **no translational symmetry** and could have any shape or design; such complex surfaces are called free-form surfaces.

In the case of simple planar geometry, surfaces are regarded as a continuous function that defines a height value over a planar domain. This planar domain is Euclidean in nature, as the form has zero Gaussian curvature. In this case, many well-established methods for studying the surface data are available, for example, Fourier analysis and Gaussian filters. However, in non-Euclidean free-form surfaces, the underlying domain is no longer planar. Instead, apart from the developable (Euclidean) free-form surfaces, which have zero Gaussian curvature everywhere and can be mapped onto planes without distortion (see Section 2.2.4.2), free-form surfaces will have positive or negative surface curvatures according to the surface's geometry. Thus the underlying domain is non-Euclidean in nature.

Characterization and parameterization of the surface texture of such free-form geometries is challenging and requires the rethinking of each characterization step for simple surfaces. Traditionally the characterization and parameterization of surface texture is carried out using three major steps, namely, surface sampling and representation; decomposition, filtration, and reconstruction; and characterization and parameterization as shown in Fig. 2.1.

Moving from simple geometries to non-Euclidean free-form geometries, many of the traditional techniques start to fail when used to perform any of the tasks shown in Fig. 2.1. Therefore new theories and tools are required that can cope with the new emerging surfaces. For example, surface decomposition is an essential step of the texture characterization system. During the last decade, decomposition and filtration techniques for simple Euclidean surfaces have been comprehensively investigated, and many algorithms based on Fourier, Gaussian, spline, morphological, and wavelet techniques were proposed and became the industrial filtration standards [6–12]. Unfortunately, all of these techniques

Advanced Metrology
https://doi.org/10.1016/B978-0-12-821815-0.00002-2

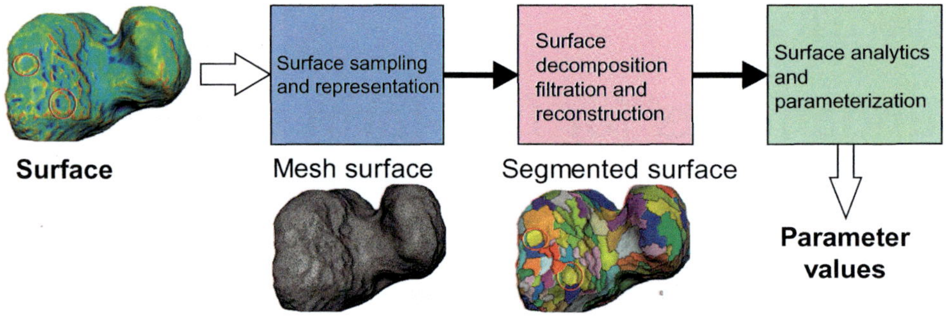

Fig. 2.1 The three main steps for surface characterization.

are designed to decompose and filter Euclidean surfaces, so most of these techniques fail to filter non-Euclidean free-form surfaces.

As mentioned in Chapter 1, part of the intention of this book is to contribute to the formation of a chain of links for free-form geometries for the ISO GPS system. In this chapter an overview of the three main stages for free-form surfaces is introduced, the challenges will be elucidated, and solutions are presented that have industrial utility; this should be considered a roadmap into the rest of this book. As a first step, free-form surface representation is considered in detail.

2.2 Free-form surface representation

The representation of any surface can be defined as the way that this surface is displayed, stored, and manipulated using digital computers. The traditional method that has been widely used to represent different types of surfaces in the field of surface metrology is when the surface is sampled over a regular grid. This means that the surface is represented as height values defined over a planar domain. Each point of this domain defines a single height value of the surface. This planar domain has zero Gaussian curvature everywhere; this is called the Euclidean domain, and consequently, such simple surfaces are called Euclidean surfaces. These Euclidean surfaces can be projected onto a plane without causing any distortion or loss of surface information, which validates the representation of such areal surfaces as height values over a regular 2D grid.

According to Gauss's **Theorema Egregium** [14] in differential geometry:

> *The Gaussian Curvature of a surface can be determined entirely by measuring angles, distances and their rates on the surface itself.*

One consequence of this theorem is that:

> *If a curved surface is developed upon any other surface whatsoever, the measure of curvature in each point remains unchanged.*

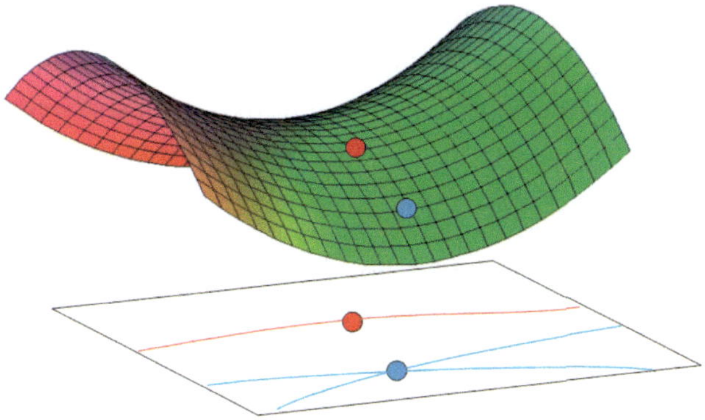

Fig. 2.2 A non-Euclidean geometry.

In other words, surfaces with the same Gaussian curvature can be mapped into each other without any distortion. For example, surfaces with zero curvature like planes, cylinders, cones, and developable free-form surfaces can be transformed into each other without any distortion. In contrast, surfaces with different Gaussian curvature cannot be mapped into each other without distortion. For example, with a free-form surface measured by a lattice method, there can be numerous (or zero) geodesics through a point that are parallel to one another (instead of just one in Playfield's version of Euclid's fifth postulate), thus invalidating Euclid's original fifth postulate [15] (i.e., it is a non-Euclidean geometry). The invalidity of Playfield's postulate for non-Euclidean geometries is illustrated in Fig. 2.2.

For non-Euclidean free-form surfaces, the underlying domain is no longer a plane and contains points that have nonzero curvature. Subsequently, non-Euclidean free-form surfaces cannot be projected onto a plane without distortion or loss of some surface information. As a result, non-Euclidean free-form surfaces can no longer be represented as height values over a two-dimensional grid without distortion. For example, if attempts are made to represent a non-Euclidean hemisphere surface over a 2D grid, this will cause a distortion on the surface of at least one of the following attributes: the geodesic distances, the surface angles, or the areas [16]. The geodesic distance between two points on a given surface is defined as the shortest path between the two points on that surface. Another example showing the invalidity of representing free-form non-Euclidean surfaces using 2D grid is the lack of such a method to represent closed surfaces such as a sphere. For the aforementioned reasons a new method is needed to represent free-form surfaces.

Fortunately the representation of free-form surfaces has been one of the core aspects of computer graphics, and many different methods have been constructed over the years to tackle this problem. Before we investigate characterization methods, it is important to define a set of requirements that the surface representation method has to meet to accurately represent free-form surfaces.

2.2.1 Free-form surface requirements

Any representation technique for free-form surface manufacturing has to meet minimum requirements that should be adopted to represent free-form surfaces with high accuracy and speed. Although there are many other requirements that are well defined in the field of computer graphics for visualization, modeling, and editing free-form surfaces [17], this book only considers the main requirements for free-form representation as follows:

1. to be a universally representative method that can be applied to different surfaces with different geometrical complexities, such as closed surfaces or surfaces with boundaries, and also to be able to support local variation in the curvature of the surface;
2. to provide a good approximation of the original surface that minimizes the digitalization errors and distortions;
3. to calculate different surface information and geometrical properties, such as distance and curvature,
4. to have simplicity to build and store the representation of the surface in linear complexity or in order of $O(n \cdot log(n))$; and
5. to represent the surface at different resolutions.

2.2.2 Surface representation models

According to the defined requirements given in the preceding text, various representation models are discussed in this section with consideration of the merits and disadvantages of each [17–22]. **Discrete representation** is a relatively simple way to give a good approximation of the free-form surfaces. Two main types of discrete representation for a free-form surface can be found: the **basic cloud of points**, the representation that does not give any surface connectivity information between the points, and **a faceted surface polygon mesh**, constructed from the basic cloud of points. **Continuous models** for free-form surface include RBF, B-spline, and NURBS. For "gentle" free-form geometries, particularly used in optics, orthogonal polynomials or free-form geometries modified by orthogonal polynomial and/or other features can also be used (see Chapter 4). Such techniques can accurately describe the free-form surface using mathematical equations; therefore more mathematical operations can be carried out on the surface. Continuous representation methods are the best choice for representing surfaces with simple geometries. However, finding the mathematical description using B-spline, RBF, or any other method for non-Euclidean free-form surfaces is not a trivial task. **Other representation methods** present a surface using iterative operations that are carried out on a base surface or base mesh. The base mesh contains only a limited number of points, and at each iteration, new points are generated, and more detail appears in the surface. Such techniques are usually referred to as subdivision representation. Generally, these methods are suitable for computer-generated surfaces where the symmetry on the surface is very high and are not particularly suitable for surface metrology.

2.2.3 Discrete surface representations

There are two main distinct types of discrete point representation: the basic points cloud and a faceted surface polygon mesh [17, 19]. Table 2.1 gives the key advantages and disadvantages for both.

The basic point set does not really provide any useful information other than to provide an input into mesh generation model. As an alternative to the polygon mesh representation, there are few methods for acquiring geometrical information from a basic point set, the most widely used of which are triangular meshes. Usually the Delaunay triangulation provides a simple way of triangulating points in a plane or giving a tetrahedral mesh for a 3D point cloud, but to obtain a sheet or skin surface of triangles over a 3D point cloud, more sophisticated algorithms such as the alpha hull or alpha shape are required (see Chapter 7).

Different data structures exist for polygon meshes [19]. These data structures can be roughly classified into two main categories, face-based and edge-based data structures. Face-based methods define the surface as a group of different faces grouped together, whereas the edge-based methods define the surface as a group of connected edges that together form the surface. Although edge-based methods can provide more flexibility

Table 2.1 Advantages and disadvantages of the discrete representation models of free-form surfaces.

Representation models	Advantages	Disadvantages
Points cloud	Very simple	Provides no quantitative description of the surface (e.g., no connectivity information)
	Requires no calculations as it is immediately available from the measurement data	It is not possible to derive geometrical properties or quantities from a basic point set
	Low storage size compared with mesh	
Polygon mesh	Provides a topology (connectivity)	Requires complex algorithms to generate a mesh from the basic point set (e.g., Delaunay and alpha hull)
	Mesh information can be used to calculate approximate geometric quantities (e.g., when the mesh is triangular, the discrete mean and Gaussian curvature can be calculated approximately)	When calculating certain geometric quantities, the mesh may also require a parameterization
	Provides a piecewise linear interpolation of the surface as each polygon in the mesh is planar	Far higher storage size required to store the connectivity information between data points

and represent the mesh connectivity, these methods usually require more sophisticated data structure and more memory to represent the surface. In contrast the face-based methods require less memory, and they are easier to construct, but they provide no mesh connectivity information. For this reason, it is common to refer to the face-based methods as triangle soup (or polygon soup if the facets are not triangles).

Two main structures are used to store the facets information for the face-based methods. The first structure represents the facets by their vertex position. This method will result in redundant vertices that are repeated many times over, as many facets use the same vertices. To overcome this problem the indexed face structure is used, where the facets are represented by the indices of the vertices rather than the vertices themselves.

Because the face-based method is simple and efficient in storing and representing different type of surfaces, this structure is used in many file formats such as STL, OFF, OBJ, and VRML. The STL file format is one of the simplest data structures where not only vertices and facets but also the facet normal vectors could be stored.

2.2.4 Continuous surface representations

The main types of continuous surfaces used in surface analytics are NURBS [17, 18, 20, 22], B-spline [17, 18, 20], and radial basis function (RBF) [21] surfaces. Other types of spline function are available, for example, those based on Bézier polynomials, but they give very similar advantages and disadvantages to the B-spline. The key advantages and disadvantages of the use of each type are summarized in Table 2.2.

Table 2.2 Advantages and disadvantages of the continuous representation models of free-form surfaces.

Representation models	Advantages	Disadvantages
B-spline	Flexible approximations B-spline basis functions are numerically stable Compact support When knots are fixed the approximation form is linear in the control point parameters Easy to calculate derivative values of any order B-spline basis functions easy to calculate due to recurrence relation definition Compact support ensures that alteration of an individual control point only affects the surface locally	Usually requires parametric fitting Fitting usually needs data to be on a mesh Need for a knot selection algorithm

Table 2.2 Advantages and disadvantages of the continuous representation models of free-form surfaces—cont'd

Representation models	Advantages	Disadvantages
NURBS	Utilizes B-spline basis functions, so all positive numerical properties for B-splines hold for NURBS also	Requires parametric fitting
	Nonlinearity of form allows more flexible approximations than B-splines	Fitting usually needs data to be on a mesh
	Interactive local modification of surface can be achieved in an intuitive manner via control point weight modification (without modification of the control net itself)	Need for a knot selection algorithm
	Infinitely differentiable inside patch boundaries, finitely differentiable at knots	Nonlinearity of form vastly complicates the fitting process (linear fitting methods are unsuitable, and general optimization algorithms are necessary)
	Linear fitting methods can be used if control point weights are fixed (comes at the expense of flexibility and a method of assigning suitable weight values)	
RBF	Linear approximation form when centers are fixed	Center selection algorithms needed in addition to the fitting process itself
	Very good for approximation of scattered data	RBF approximation theory not as well developed as for RBF interpolation
	Easy to generalize to higher dimensional approximation problems as basic functions are functions of a single parameter (norm of distance from data points to centers)	Numerically unstable when centers are close together
	No need for parameterization	No clear methods for "good" center selection, particularly on or close to data boundaries
	If the set of centers is a subset of the data points, the observation matrix satisfies the Haar condition	Derivative calculation less straight forward than for NURBS and B-splines

In general, B-spline surfaces have been found to be the most useful for data fitting purposes. RBF surface approximations have some nice properties and are very useful when it comes to the approximation of scattered data, but they can exhibit poor approximation quality at data boundaries. Only careful center selection in the region of the boundary can overcome this. Another downside of the RBF surface fitting process is the numerical instability of the observation matrix when centers are close together. In addition, the benefit of RBFs when it comes to multidimensional approximation is not particularly relevant to our applications as we will only ever be interested in the fitting of surfaces in three dimensions.

Given the nature of surface metrology, all data are acquired from measurement instruments, and as such, we have an element of control over the spacing of the measurements. This allows for data measurements to be made over a mesh, and for data of this type approximation, using B-spline surfaces is excessively difficult. The use of NURBS surface allows greater flexibility in the approximation form, but the highly nonlinear nature of the form means that the fitting process is far more complicated than for B-spline surfaces. Higher accuracy of approximation can be obtained using B-spline surfaces by inserting extra knots and control points until a satisfactory error norm is attained, but such operations are considered difficult for complex surfaces.

2.2.4.1 Subdivision surfaces

The subdivision surface is a continuous surface that is not necessarily given as a closed form, but can be built up using convolutions of a discrete point set. At each subdivision a new point is generated as a linear combination of some neighboring points (such as spline subdivision). In one sense, it behaves as a discrete surface representation as it not only is always a finite number of points that are generated at each step but also has elements of continuity as the limiting surface of the subdivision process is continuous and some properties of the limiting surface may be known. Subdivision falls into two categories: interpolation subdivision and approximation subdivision [18, 23]. The advantages and disadvantages of the subdivision methods are given in Table 2.3.

2.2.4.2 Ruled surfaces

One important class of surfaces are the ruled surfaces [24]. These are constructed from a series of straight lines that form the ruled surface (see Fig. 2.3 for an example). Mathematically, they take a parameterization of the form

$$\underline{q}(u, v) = \underline{p}(u) + v \cdot \underline{e}(u) \quad \text{with} \quad \underline{e}(u) \neq [0, 0, 0]^T$$

The line $\underline{p}(u) = \underline{q}(u, 0)$ is called the directing curve. The vectors $e(u)$ are called the generators and give the direction of the one parameter set of straight lines.

Generalized cylinders and cones are particular types of ruled surfaces.

Table 2.3 Advantages and disadvantages of the other continuous representation models of free-form surfaces.

Representation models	Advantages	Disadvantages
Subdivision surfaces	No fitting methods required—new data points created via discrete convolution of data points with weighting vector (mask)	No explicit continuous functional form to work with
	Properties about the limiting surface may be known (such as levels of continuity)	No derivative information—derivatives will need to be calculated using numerical approximations
	Can interpolate or approximate the initial set of data points	For approximating subdivision surfaces, there is no given statistical distribution for the errors (e.g., as opposed to least squares or Chebyshev approximation)
Rule surfaces	Very simple mathematical structure	Limited range of ruled surface types
	Compact support	Not all free-form surfaces can be well approximated by ruled surfaces
	Developable surfaces can be mapped onto planar surfaces without distortion	
	Developable surfaces can be constructed through "bending" a plane without stretching	
	A developable surface has straight-line contours with respect to any central or parallel projection (see [24]). This is a way to recognize developable surfaces	

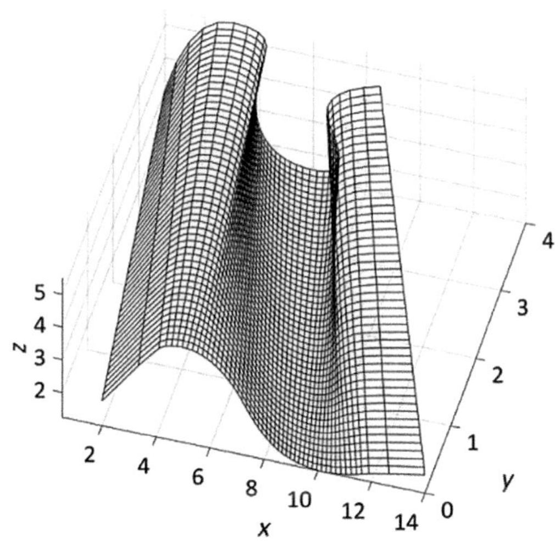

Fig. 2.3 Example of a ruled surface.

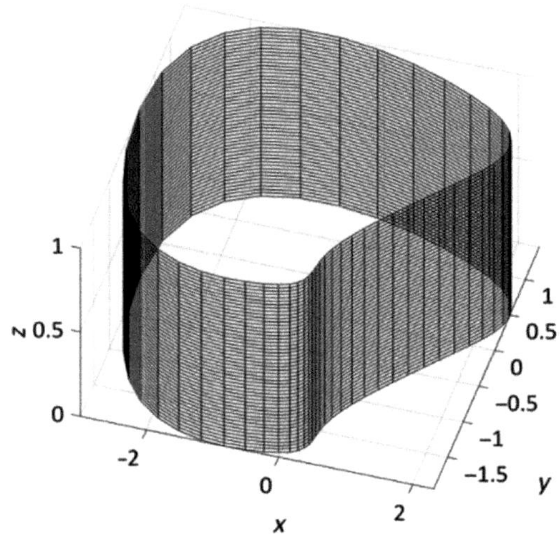

Fig. 2.4 Example of a generalized cylinder.

A generalized cylinder has a constant vector $e(u) = e$ as its generator and is defined by e and its directing curve (see Fig. 2.4 for an example).

In a generalized cone the generators run through a given point in space s and is defined by s and its directing curve (see Fig. 2.5 for an example).

An important subset of ruled surfaces are the developable surfaces [24]. These are surfaces that have zero Gaussian curvature and as a result of Gauss's **Theorema Egregium** can be mapped onto a planar surface without distortion. There are only three types of developable surfaces (all ruled surfaces), namely,

(i) generalized cylindrical surfaces
(ii) generalized cones
(iii) tangent surfaces to directing curves (see Fig. 2.6)

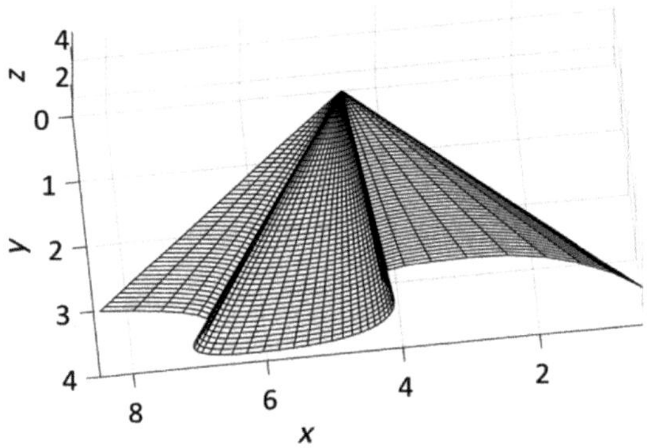

Fig. 2.5 Example of a generalized cone.

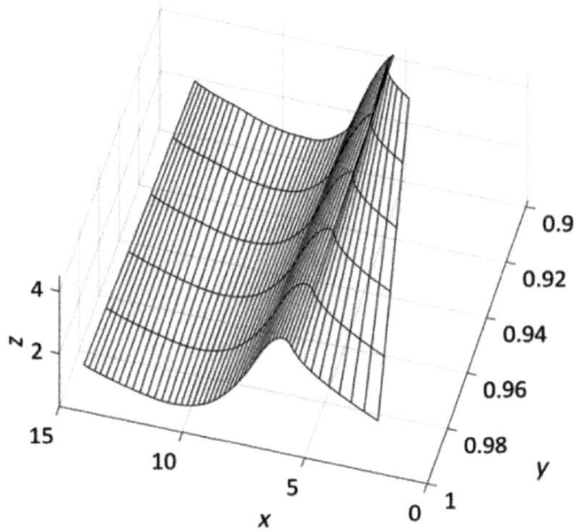

Fig. 2.6 Example of a tangent surface to a directing curve.

The first two have been mentioned previously. Tangent surfaces are defined when the generators take the special form:

$$\underline{e}(u) = \dot{\underline{p}} = \frac{d\underline{p}(u)}{du}$$

That is to say that the generator vectors are the first differential of the directing curve.

For more information on ruled and developable surfaces, including some classical results, see reference [24].

2.2.5 Other surfaces methods

This section finishes with a selection of other surfaces commonly used in free-form design. This selection is neither complete nor very detailed. Furthermore, new methods are constantly being developed for different applications. This section is here to illustrate some other approaches to the specification of free-form surfaces.

2.2.5.1 Skinned surfaces or multisection surfaces

A surface that contains a given set of profile curves or cross sections, $p1(u)$, $p2(u)$, ..., $pn(u)$, is called a skinned surface [24, 25]. The shape of the desired surface is defined by the given cross sections (see Fig. 2.7). Since a skinned surface contains all of its profile curves, it interpolates these profile curves into the design. The construction of a skinned surface is not unique with different interpretation methods and different orderings of the cross sections giving different skinned surfaces.

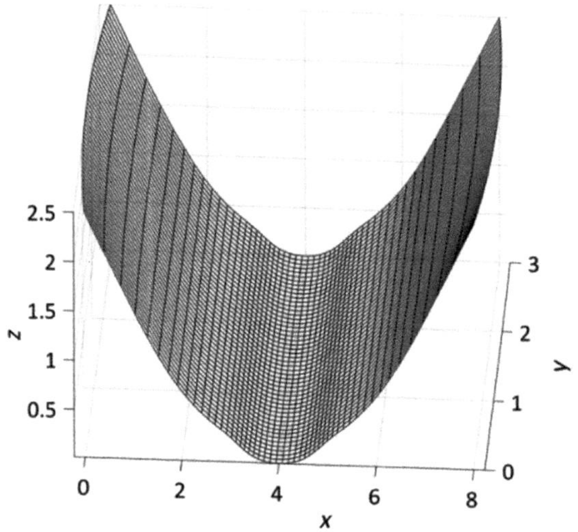

Fig. 2.7 Example of a skinned surface.

2.2.5.2 Swept surfaces

A swept surface is the surface generated by moving a curve $p1(v)$ (the profile curve) along a second curve $p2(u)$, the trajectory curve (see Fig. 2.8) [24, 25]. The profile curve may be rotated and scaled depending on the value of u in the trajectory curve $p2(u)$.

2.2.5.3 Swung surfaces

A swung curve is a generalization of a surface of revolution. Here a generatrix (profile curve) is rotated a full 360 degrees around an axis to generate the revolute surface. The generatrix is never scaled. In a swung surface, there is a second trajectory curve that

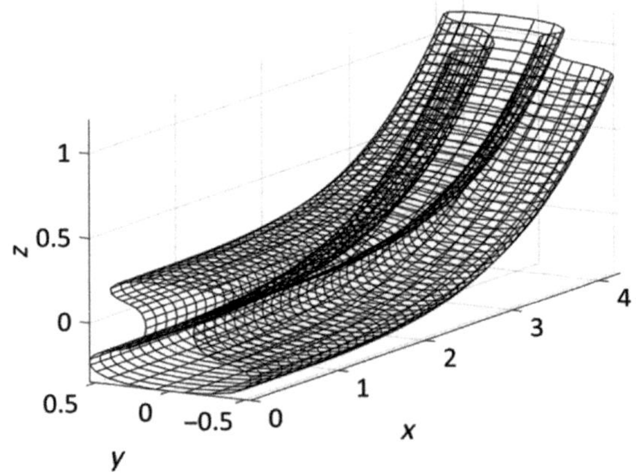

Fig. 2.8 Example of a swept surface.

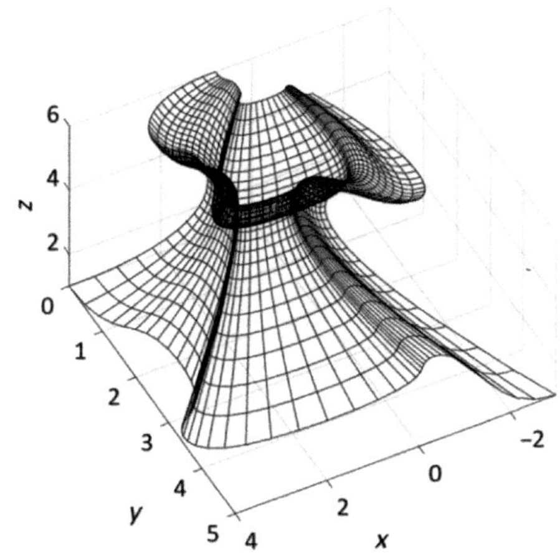

Fig. 2.9 Example of a swung surface.

guides the generatrix's scale as it rotates about the axis of rotation (see Fig. 2.9) [25]. In addition, the rotation does not have to be full 360 degrees but just a partial rotation.

2.3 Free-form analysis

Measurement of geometrical products can be separated into sampling and characterization. Sampling converts real-world continuous signals into discretized data points. Following this, characterization extracts out interesting attributes from the data points for evaluation. In past decades, characterization techniques have been developed at a fast pace, such as feature-based methods, fractal analysis, wavelet, and morphological filtration [4, 26]. Simultaneously, sampling techniques have progressed relatively slowly.

2.3.1 Sampling and reconstruction

Finding optimal sampling strategies is a central problem in surface representation and in surface reconstruction. Since Chapter 3 of this book is concerned with sampling for free-form surfaces, this section will limit itself to comments and observations of a general nature.

Sampling and reconstruction for planar (Euclidian) surfaces are well developed with mathematical foundations and many approaches to practical sampling. All of these techniques can still be used on developable (Euclidean) free-form surfaces that is to say that those free-form surfaces that can be mapped onto a plane without distortion (see Section 2.2.4.2 for more details). For non-Euclidean free-form surfaces, these techniques are generally not applicable, and other sampling approaches need to be used. In particular, it is not possible to use equally spaced sampling on a general non-Euclidean surface, even for simple geometry such as a sphere [27].

There are a number of different approaches to sampling and reconstruction of non-Euclidean free-form surfaces. The following contains an outline of the most popular general approaches that have utility. Again, this is not a complete list, and other approaches do exist (see Chapter 3 for more details).

2.3.1.1 Feature- or attribute-based sampling

Here the density of the sampled points is dependent on the type and size of local features contained in the nominal or specified designed surface (e.g., hill, dale, saddle point, ridge, and course line) or the local attribute of the surface (e.g., the local curvature of the surface). The smaller the feature or the larger the curvature, the denser the sampling. A practical approach to achieve this type of sampling is to start with course sampling and then zoom into the identified features of interest. Some more complicated sampling may use a combination of features and/or attributes to determine the local density of sampling.

2.3.1.2 Sampling based on orthogonal functions

Here the nominal free-form shape or more commonly the residuals from the nominal free-form shape are modeled using a set of complete orthogonal functions (e.g., Zernike polynomials, 3D Zernike polynomials, wavelets, and eigenfunctions of a smoothing linear PDE such as the Laplace-Beltrami operator). The sampling points then consist of the local maxima or local minima of all the functions up to a given specified order of function depending on the sampling resolution required for the application. The extrema are used since they give the best signal-to-noise ratio for the fitting.

2.3.1.3 Simplification of meshes

Simplification of meshes describes a class of techniques to transform one mesh into another mesh, usually with fewer points. There are many reasons for carrying out mesh simplification: comparison of meshes with differing point densities, limitations in computational power, etc. There are two basic approaches:

(i) Reducing the number of mesh points (with appropriate readjustments of the triangular edges).

(ii) When the required new mesh points (such as measured points) do not align with the original mesh points, using a suitable reconstruction, the position of the new mesh point can be determined.

There are a number of approaches to reconstruction, from simple planar facets to splines and polynomials, which allow various degrees of smoothness along the edges of the mesh (see Chapter 3 for more details).

2.3.2 Free-form form fitting

Unfortunately, some subsequent analytics only work with the residual vector field from a defined reference surface. There are two main approaches: when the nominal form is known and when the nominal form is not known. The following two sections are a brief introduction to these two approaches.

2.3.2.1 Nominal form is known

If the nominal form is known and fixed, then the reference surface can be calculated from a best fit surface based on the nominal form (e.g., a best fit total least squares reference surface, a best fit total absolute fit reference surface, and a best fit Chebyshev reference surface) with the residual vectors calculated normal to the reference surface. If the free-form surface contains features of size, then for stability the size of these features should be allowed to vary.

The best fit algorithm is usually one that works in two stages: the first stage is a course approximate fit, and the second stage is a more refined optimized fit. This is because the more refined fit is usually an iterative algorithm and requires a good starting value to iterate into the optimized solution. A good starting solution is provided by the first-stage course approximate fit. More details of this approach can be found in Chapter 4.

2.3.2.2 Nominal form is not known

If the fixed nominal form is not known, then a filtration operator that does not require a reference surface can be used to determine the reference surface (e.g., morphological closing filter with spherical structuring element). Fig. 2.10A shows an additively manufactured strut whose nominal form is not known. Fig. 2.10B shows the same surface after a morphological alternating sequence rolling ball filter has been applied to determine the form. Fig. 2.10C shows the original surface with the determined form as a reference surface.

Alternatively, fitting a free-form continuous model to the data, such as NURBS, where the parameterization of the model is not fixed but allowed to float, can also work. Again, more details of all these approaches can be found in Chapter 4.

2.3.3 Free-form filtration and multiscale decomposition

Decomposing the surface into different scales is one of the key operations required to analyze a surface and study its characterization. Traditionally, simple Euclidean surfaces were decomposed into three major components: the roughness, waviness, and the form. Various mathematical tools often carry out the decomposition of such surfaces such as Fourier transforms, Gaussian filtration, and wavelets [6–12]. Non-Euclidean free-form surfaces should also be decomposed into various scales by extending some of the mathematical tools to suit the nature of non-Euclidean free-form surface. Jiang et al. proposed

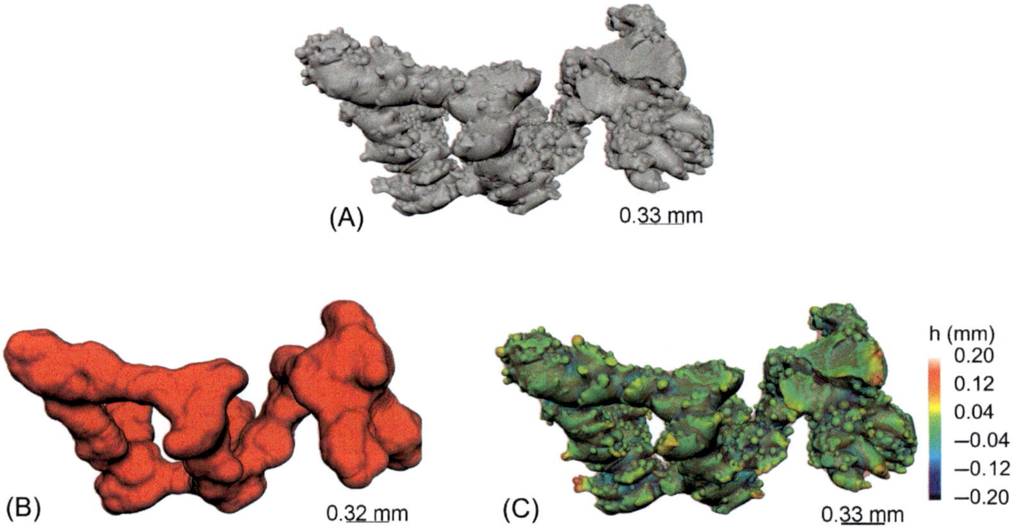

Fig. 2.10 Additively manufactured strut: (A) measured data, (B) form determined from morphological ball filter, and (C) measured data and form data.

decomposing the free-form surface into three scale-limited surfaces, namely, the short-scale surface, the middle-scale surface, and the long-scale surface [3, 4, 28]. Here, we will use the alternative classification for this decomposition: texture, shape, and form.

The basic concept of decomposing a surface into various scales is given in ISO/16610 part 1 [27]. The nesting index λ indicates the scale of decomposition for a particular filter, with lower nesting index values corresponding to smaller scales and higher nesting indices corresponding to larger scales.

In this section, we will explore various decomposition models that are used to decompose free-form surfaces represented by triangular meshes as explained in the previous section. The decomposition models that will be discussed in this section are based on partial differential equations (manifold harmonics), morphological models, and, finally, the lifting wavelets. All can be used to decompose free-form surfaces into large-scale (smooth) and small-scale components. Alternatively, they can be used for multiscale decomposition that partitions free-form surfaces into a series of components with a disjointed and exhaustive range of feature scales.

2.3.3.1 Diffusion filtering

The Fourier decomposition is the classical tool for decomposing a given surface or more generally a signal into its basic components. Using Fourier decomposition the surface is broken down into different scales using a set of sine and cosine functions with different frequencies. However, Fourier decomposition is not available for non-Euclidean free-form surfaces, and therefore a generalization of the Fourier theory to non-Euclidean free-form surfaces is required. The missing link is provided by an important observation

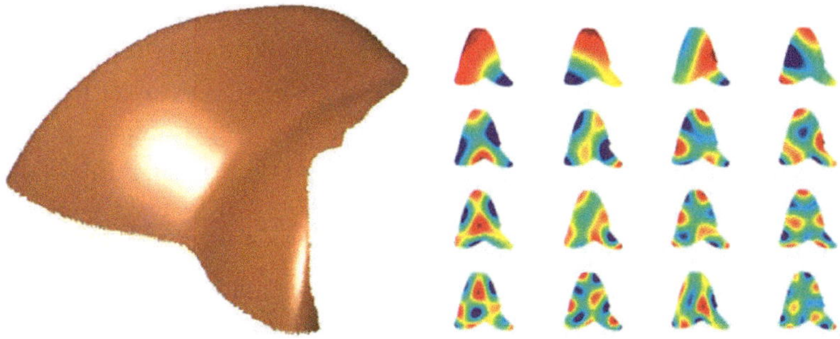

Fig. 2.11 Decomposition of measured knee joint into the first 16 eigenfunctions using the manifold harmonic decomposition technique.

that sine and cosine functions are eigenfunctions of the Laplace operator (heat equation) for the Euclidean cases. In the case of general surfaces, it is therefore natural to choose the eigenfunctions of the Laplace-Beltrami operator (LBO; generalized heat equation) as a generalized basis function to decompose that given surface [29, 30].

Fig. 2.11 shows an example of decomposing a measured knee joint, represented by 3D triangular mesh into the different scales using the first 16 eigenfunctions from the Laplace-Beltrami operator of this surface. As can be seen from the figure, the eigenfunctions contain more details as further decomposition has taken place. The first scale at the top left is more homogeneous and contains fewer details than the last scale, the bottom right in the figure.

In this decomposition model the nesting index λ corresponds to the square root of the eigenvalues associated with the eigenfunctions. Different scales can be obtained from the surface by reconstructing the surface from a particular subset of the eigenfunctions.

The eigenfunctions and eigenvalues can also be used for spectral analytics of the measured free-form surface that is a generalization of Fourier analytics (spectrogram) of areal surfaces. As can be seen in Fig. 2.4, the eigenfunctions of the Laplace-Beltrami operator come in pairs, just like sine and cosine functions in Fourier analytics, with each pair having the same (or very similar) eigenvalues.

2.3.3.2 Morphological filtering

Morphological operations are very important functional methods for surface texture assessment. They were derived from the traditional envelope filtration system proposed by Von Weingraber [31], which is obtained by rolling a ball with the selected radius over the surface. With the introduction of mathematical morphology, morphological filters offer more tools and capabilities than its predecessor [32, 33]. They are conducted by performing morphological operations on the input surface with spherical or flat structuring elements. Over the last decade, morphological filters have found many practical applications and were accepted by ISO 16610 [34, 35] as a useful part of the filtration toolbox.

Surface decomposition using morphological operations is considered as a type of sieving technique, which could date back to the morphological granulometry. It is similar to the process of sieving small solid particles with a series of sieves with increasing mesh openings. The sieve with the smallest opening is used first. The grains that are bigger than the mesh opening are kept and counted. The bigger sieve then sifts the remaining grains, and this process continues until all the sieves are used. In this way, grains are classified according to the size of mesh openings.

Similar to the sieving process, the surface could be decomposed into various scales using the morphological operation by using different sizes of the structuring elements, that is, by choosing a different radius for the ball. In this case the nesting index λ corresponds to the sizes of the structuring elements, the radius of the ball, and the different subspaces $\mathcal{V}_\lambda$, the scales, correspond to the results of applying the structuring element with size λ over the original surface.

2.3.3.3 Segmentation

Segmentation is a technique of dividing a surface into a number of nonintersecting regions such that each region is homogeneous, but the union of any two adjacent regions is not homogeneous [36, 37]. Segmentation, using the watershed algorithm with Wolf pruning, was introduced to areal surface texture by Scott [38–40] and later standardized in ISO 25178-2 [13] and ISO 16610-85 [12]. Others also introduce variants of this approach for areal surface texture [41].

An important fact is that the watershed segmentation is applied to a scalar function. The choice of the scalar function depends on specific applications. They can be image gradient, surface curvature, or any value of interest. The basic algorithm for watershed segmentation is independent of the scalar function. Scott's method, based on the Maxwell's theory and Pfaltz graph [39], is adopted in this work; its computation strategy conforms to the water flowing principle and is theoretically independent of data structure (topology). Either it is lattice data or mesh data. In terms of mesh data, the exact positions of mesh vertices are not directly related to the computation. It is the topological connection of the vertices (being based on local connections, the approach is independent of the topology of the mesh) and the values of the chosen scalar function that determine the watershed segmentation. Another reason for choosing Scott's method relies on its stability. Small changes to the input profile will not cause a significant change in the analysis of the results, with "relevant" features appearing or disappearing almost randomly [39]. This property is of great importance, particularly for the field of engineering, for example, precision metrology. More details on segmentation on non-Euclidian free-form surfaces can be found in Chapter 8.

Finally an example of segmentation on a free-form surface is illustrated in Fig. 2.12. This is the triangular mesh data of 67P/Churyumov-Gerasimenko comet obtained from the Rosetta mission. The data (STL file to produce an additive model) can be downloaded from the European Space Agency's website [42]. The figure illustrates the initial

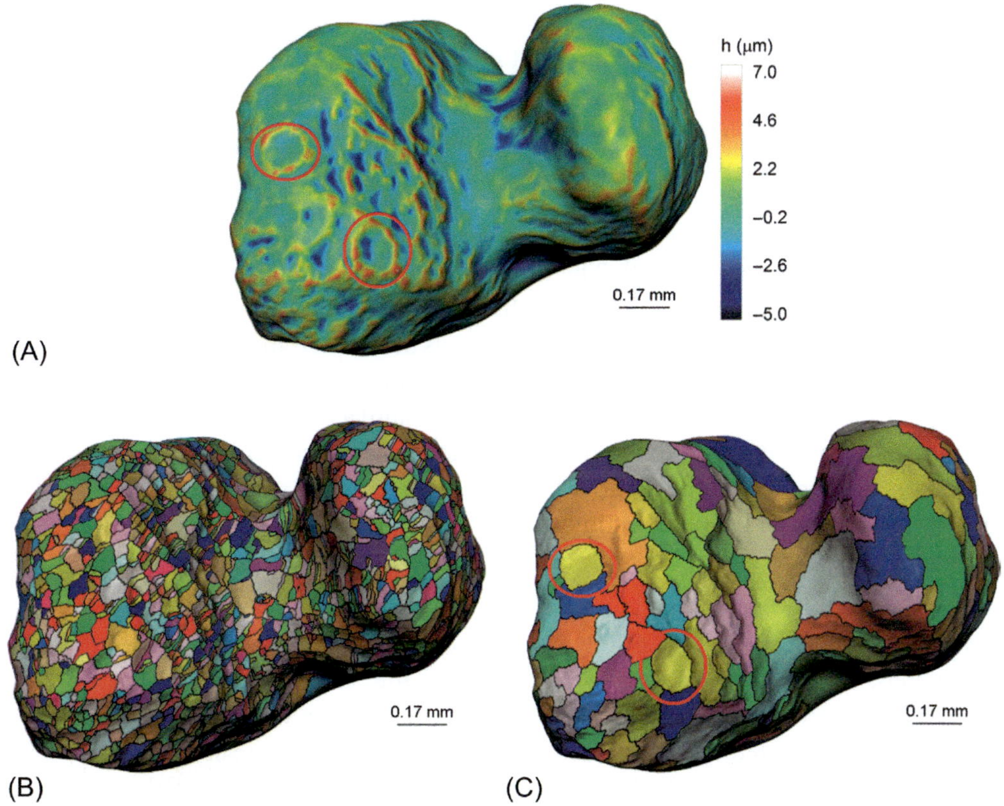

Fig. 2.12 3D Watershed segmentation: (A) comet surface with a color map indicating the heights of surface texture, (B) watershed segmentation without Wolf pruning (6101 regions), and (C) watershed segmentation with 10% Wolf pruning (244 regions). *Note* that the scale is for the STL model and not the comet itself.

watershed oversegmentation (6101 regions) and the watershed segmentation with 10% Wolf pruning (244 regions), identifying the two crater regions indicated in the first figure.

2.3.3.4 Wavelets

Wavelets are a very powerful tool for representing and simplifying general functions, curves, surfaces, or any type of data set into their basic components. The power of wavelets derives from representing the input data sets using time-scale space and with different levels of resolution. Wavelet analysis is defined from the translation and dilation of one particular function called the mother wavelet. In 1995 Swelden proposed the lifting scheme algorithm, which can be applied to both regular and irregular data sets [43–45]. The lifting scheme also extended the wavelet analysis to include free-form surfaces and made the extension of wavelet multiscale decomposition possible for all types of meshes [46, 47].

2.3.4 Free-form analytics

Free-form analytics or characterization follows the spirit of the standardized GPS verification characterization as contained in the ISO GPS documents, in that the free-form surface is decomposed into form, shape, and texture parameters [13]. These parameters can then be compared with the free-form specification for conformance.

The characterization of the form is carried out on the unaltered mesh. The characterization of the shape and texture parameters is carried out on the residual vector field after the form has been removed. The difference between shape and texture is that, for the shape vector field, the residual surface has only a lower limit in scale Lc, whereas the texture vector field has both upper and lower limits in scale Uc and Lc, respectively. Currently, it is very common that Lc and Uc have the same scale value. However, for some modern manufacturing processes such as additive manufacturing, further decomposition of the shape and texture vector fields can yield useful analytics for control of the manufacturing processes or functional prediction, with each individual decomposition having its own set of characterization parameters.

2.3.4.1 Form parameters

The parameterization of the fitted form representation model determines the form parameters, usually in a one-to-one mapping of the standard attributes of the form representation model. Nevertheless, this is not always the case with an alternative parameterization being used to determine the form parameters. Also, in many cases, the parameterization of the form can be unstable, with many different values of the parameters giving the same form to within a very small tolerance. This leads to the measured parameter values jumping randomly, leading to great difficulties in comparing the measured parameter values and the nominal parameter values. In these cases the best approach is to characterize the residual differences between the nominal form and the fitted form in a similar manner to shape characterization (see next section).

The following three sections give a brief overview of shape and texture parameters. A fuller description is given in Chapter 10.

2.3.4.2 Shape parameters

Shape parameters are associated with geometrical tolerance in that they characterize the deviation from nominal form through the shapes residual surface. Again, following the spirt of standardized GPS documents, there are four different types of shape parameters based on the following deviations (see e.g., [48]):

- **(i) peak-to-valley deviation:** value of the largest positive local deviation added to the absolute value of the largest negative local deviation,
- **(ii) peak-to-reference surface deviation:** value of the largest positive local deviation,

(iii) valley-to-reference surface deviation: value of the largest negative local deviation, and

(iv) root-mean-square deviation: square root of the sum of the squares of the local deviations from the least squares reference surface.

Where the reference surface is the associated free-form surface, the reference free-form surface is the surface from which deviations from free form are referred. The deviation is negative if from the reference surface the point lies in the direction of the material and is normal to the local reference surface. The mathematics for calculation is the same as for the surface texture field parameters given in the next section.

2.3.4.3 Surface texture field parameters

It is possible to redefine many standard planar field parameters when the residuals are a vector field on the reference surface. The new definitions give the same results as the standard planar parameters when they are applied to planar surfaces [49]. They also work when reentrant features are present on the free-form surface. Free-form "areal field" parameters can be computed as the integral of a vector (or scalar) field (represented by the scale-limited surface) on the form surface and take the structure

$$\iint_{\sum \text{form}} f(u,v) \cdot \eta d\sigma_{\text{form}}$$

where $d\sigma_{form}$is the infinitesimal surface element and η is the surface normal. For more details, see Chapter 9.

Examples of surface parameters are given in free-form Table 2.4.

2.3.4.4 Surface texture feature parameters

Surface texture feature parameters are derived from a segmented free-form surface (see Section 2.3.3.3). The choice of the scalar function for the segmentation algorithm

Table 2.4 Comparison of areal and free-form surface texture parameters.

	ISO 25178-2		Proposed
A	$(x_1 - x_0) \cdot (y_1 - y_0)$	$\rightarrow$	$\iint_{\Sigma_{\text{form}}} d\sigma_{\text{form}}$
Sa	$\frac{1}{A}\int_{x_0}^{x_1}\int_{y_0}^{y_1} \lvert z(x,y)\rvert dxdy$	$\rightarrow$	$\frac{1}{A}\iint_{\Sigma_{\text{form}}} \lvert \boldsymbol{r}_{sl}(u,v) \cdot \boldsymbol{n}_{\text{form}}(u,v)\rvert d\sigma_{\text{form}}$
Sq	$\sqrt{\frac{1}{A}\int_{x_0}^{x_1}\int_{y_0}^{y_1} z^2(x,y)dxdy}$	$\rightarrow$	$\sqrt{\frac{1}{A}\iint_{\Sigma_{\text{form}}} (\boldsymbol{r}_{sl}(u,v) \cdot \boldsymbol{n}_{\text{form}}(u,v))^2 d\sigma_{\text{form}}}$
Ssk	$\frac{1}{ASq^3}\int_{x_0}^{x_1}\int_{y_0}^{y_1} z^3(x,y)dxdy$	$\rightarrow$	$\frac{1}{ASq^3}\iint_{\Sigma_{\text{form}}} (\boldsymbol{r}_{sl}(u,v) \cdot \boldsymbol{n}_{\text{form}}(u,v))^3 d\sigma_{\text{form}}$
Sku	$\frac{1}{ASq^4}\int_{x_0}^{x_1}\int_{y_0}^{y_1} z^4(x,y)dxdy$	$\rightarrow$	$\frac{1}{ASq^4}\iint_{\Sigma_{\text{form}}} (\boldsymbol{r}_{sl}(u,v) \cdot \boldsymbol{n}_{\text{form}}(u,v))^4 d\sigma_{\text{form}}$

depends on specific applications. They can be surface height normal to the local reference surface, surface gradient, surface curvature, or any other value of interest.

Since it is the topological connection of the vertices and the values of the chosen scalar function that determine the watershed segmentation, the calculation of the surface texture feature parameters from the segmented surface is exactly the same as for areal feature parameters. The same five tables, which appear in ISO 25178-2 [13] for the feature characterization process, can also be used for the definition of the required feature parameter, namely,

(1) selection of the type of texture feature
(2) segmentation
(3) determining significant features
(4) selection of feature attributes
(5) quantification of feature attribute statistics

Surface texture feature parameters are a relatively new approach to characterizing surface texture, particular surfaces with deterministic patterns and features. The feature parameter approach is seen as complementary to traditional surface texture field parameter characterization. It is strongly believed that surface texture feature parameters will open up many new applications for free-form surface texture characterization (such as structured features on free-form surfaces) for free-form surface texture characterization.

2.4 Summary

In this chapter an overview of the foundations of free-form surfaces is given, beginning with the three main steps for surface characterization: surface representation, surface decomposition, and surface analytics. Using Gauss's Theorema Egregium the main differences between Euclidean and non-Euclidean surfaces are explained. Given the requirements for free-form surfaces, the different ways to represent free-form surfaces are given for both discrete and continuous representations. Sampling strategies (extraction) of free-form surfaces and its dual reconstruction from sampled points of free-form surface are introduced using feature-based sampling, orthogonal function-based sampling, and mesh simplification. Free-form surface decomposition is discussed, which divides the free-form surface into various scales, starting with determining a reference form surface, through either optimized fitting if the nominal form is known, or using special filters if the nominal form is not known. Decomposition of the free-form surface into elements at differing scales using filters defined on non-Euclidean surface is then described: diffusion filters using a PDE, Laplace-Beltrami operator, morphological filters, segmentation, and wavelet filters. The scales of interest can then be reconstructed to create the surfaces of interest for further analysis. Finally, free-form analytics is introduced that includes form characterization, shape characterization, and texture characterization, the latter being subdivided into surface field parameters and surface feature parameters as in areal surface texture.

References

[1] Abbott EJ, Firestone FA. Specifying surface quality: a method based on accurate measurement and comparison. J Mech Eng 1933;55:569–72.
[2] Greenwood JA. A unified theory of surface roughness. Proc R Soc A 1933;**393**:133–57. https://doi.org/10.1098/rspa.1984.0050.
[3] Jiang X, Scott PJ, Whitehouse DJ, Blunt L. Paradigm shifts in surface metrology. Part I, historical philosophy. Proc R Soc A 2007;**463**:2049–70. https://doi.org/10.1098/rspa.2007.1874.
[4] Jiang X, Scott PJ, Whitehouse DJ, Blunt L. Paradigm shifts in surface metrology. Part II, the current shift. Proc R Soc A 2007;**463**:2071–99. https://doi.org/10.1098/rspa.2007.1873.
[5] Whitehouse DJ. Surfaces and their measurement. Hermes Penton Ltd; 2002.
[6] ISO 16610-21:2010. Geometrical product specifications (GPS)—filtration—part 21: linear profile filters: Gaussian filters. Geneva: ISO; 2010.
[7] ISO 16610-22:2010. Geometrical product specifications (GPS)—filtration—part 22: linear profile filters: spline filters. Geneva: ISO; 2010.
[8] ISO 16610-29:2019. Geometrical product specifications (GPS)—filtration—part 29: linear profile filters: wavelets. Geneva: ISO; 2019.
[9] ISO 16610-41:2015. Geometrical product specifications (GPS)—filtration—part 41: morphological profile filters: disk and horizontal line-segment filters. Geneva: ISO; 2015.
[10] ISO 16610-49:2015. Geometrical product specifications (GPS)—filtration—part 49: morphological profile filters: scale space techniques. Geneva: ISO; 2015.
[11] ISO 16610-61:2010. Geometrical product specifications (GPS)—filtration—part 61: linear areal filters: Gaussian filters. Geneva: ISO; 2010.
[12] ISO 16610-85:2013. Geometrical product specifications (GPS)—filtration—part 85: morphological areal filters: segmentation. Geneva: ISO; 2013.
[13] ISO 25178-2:2010. Geometrical product specification (GPS)—surface texture: areal—part 2: terms, definitions and surface texture parameters. Geneva: ISO; 2010.
[14] Gauss CF. Disquisitiones generales circa superficies curvas. Typis Dieterichianis; 1828, https://archive.org/details/disquisitionesg00gausgoog.
[15] Hartshorne R. Geometry: Euclid and beyond. Springer; 2000. ISBN-10: 0-387-98650-2.
[16] https://en.wikipedia.org/wiki/Map_projection.
[17] Hubeli A, Gross M. A survey of surface representations for geometric modelling technical, report 335 ETH Zurich. Computer Science Department; 2000.
[18] Iske A, Quak E, Floater MS. Tutorials on multiresolution in geometric modelling. Berlin, Heidelberg, New York: Springer-Verlag; 2002, ISBN:3-540-43639-1.
[19] Botsch M, Kobbelt L, Pauly M, Alliez P, Levy B. Polygon mesh processing. AK Peters Ltd; 2010, ISBN:978-1-56881-426-1.
[20] Prautzsch H, Boehm W, Paluszny M. Bezier and B-spline techniques. Berlin, Heidelberg: Springer-Verlag; 2002, ISBN:3-540-43761-4.
[21] Zhang X. Freeform surface fitting for precision coordinate metrology [Ph.D. thesis]. University of Huddersfield; 2009.
[22] Kuhnel W. Differential geometry: curves-surfaces-manifolds. 2nd ed. New York: American Mathematical Society; 2006, ISBN:0-8218-3988-8.
[23] Schroder P, Sweldens W. Building your own wavelets at home. ACM SIGGRAPH course notes; 1996.
[24] Hirz M, Dietrich W, Gfrerrer A, Lang J. Integrated computer-aided design in automotive development. Heidelberg: Springer; 2013. ISBN: 978-3-642-11939-3. https://doi.org/10.1007/978-3-642-11940-8.
[25] Piegl L, Tiller W. The NURBS book. 2nd ed. New York: Springer; 1997, ISBN:3-540-61545-8.
[26] Jiang XJ, Whitehouse DJ. Technological shifts in surface metrology. CIRP Ann Manuf Technol 2012. https://doi.org/10.1016/j.cirp.2012.05.009.
[27] ISO 16610-1:2015. Geometrical product specifications (GPS)—filtration—part 1: overview and basic terminology, ISO, Geneva.
[28] Saff EB, Kuijlaars ABJ. Distributing many points on a sphere. Math Intell 1997;19(1):6–11. Springer.
[29] Jiang X, Cooper P, Scott PJ. Freeform surface filtering using the diffusion equation. Proc R Soc A 2011;467(2127):841–59. https://doi.org/10.1098/rspa.2010.0307.

[30] Scott PJ, Jiang X. Freeform surface characterisation: theory and practice. 14th international conference on metrology and properties of engineering surfaces. J Phys Conf Ser 2014;**483.** https://doi.org/10.1088/1742-6596/483/1/012005.

[31] Von Weingraber H. Über die Eignung des Hüllprofils als Bezugslinie für die Messung der Rauheit. Ann CIRP 1956;5:116–28.

[32] Srinivasan V. Discrete morphological filters for metrology. In: Proceedings 6th ISMQC symposium on metrology for quality control in production; 1998. p. 203–9.

[33] Scott PJ. Scale-space techniques. In: Proceedings of the X international colloquium on surfaces; 2000. p. 153–61.

[34] Lou S, Jiang X, Scott PJ. Correlation motif analysis and morphological filters for surface texture analysis. Measurement 2013;46(2):993–1001. ISSN 0263-2241.

[35] Dietzsch M, Gerlach M, Groger S. Back to the envelope system with morphological operations for the evaluation of surfaces. Wear 2008;264:411–5.

[36] Pal NR, Pal SK. A review on image segmentation techniques. Pattern Recognit 1993;26:1277–94.

[37] Cheng HD, Jiang XH, Sun Y, Wang JL. Color image segmentation: advances and prospects. Pattern Recognit 2001;34:2259–81.

[38] Scott PJ. The mathematics of motif combination and their use for functional simulation. Int J Mach Tools Manuf 1992;32:69–73.

[39] Scott PJ. Pattern analysis and metrology: the extraction of stable features from observable measurements. Proc R Soc A 2004;460:2845–64.

[40] Scott PJ. Feature parameters. Wear 2009;266:548–51.

[41] Barré F, Lopez J. Watershed lines and catchment basins: a new 3D-motif method. Int J Mach Tools Manuf 2000;40:1171–84.

[42] http://blogs.esa.int/rosetta/2015/11/30/new-comet-shape-model/.

[43] Sweldens W. The lifting scheme: a new philosophy in biorthogonal wavelets reconstructions. In: Wavelet applications in signal and image processing III; 1995. p. 68–79.

[44] Sweldens W. Wavelets and the lifting scheme: a 5-min tour. Zeitschrift Fur Angewandte Mathematik Und Mechanik 1996;76(2):4–7.

[45] Sweldens W. The lifting scheme: a construction of second generation wavelets. SIAM J Math Anal 1998;29(2):511–46.

[46] Guskov I, Sweldens W, Schroder P. Multiresolution signal processing for meshes. In: Proceedings of ACM SIGGRAPH; 1999. p. 325–34.

[47] Eck M, DeRose T, Duchamp T, Hoppe H, Lounsbery M, Stuetzle W. Multiresolution analysis of arbitrary meshes. In: SIGGRAPH 95 conference proceedings, ACM SIGGRAPH; 1995. p. 173–82.

[48] ISO 12181-1:2011. Geometrical product specifications (GPS)—roundness part 1: vocabulary and parameters of roundness. Geneva: ISO; 2011.

[49] Pagani L, Qi Q, Jiang X, Scott PJ. Towards a new definition of areal surface texture parameters on freeform surface. Measurement 2017;109:281–91.

CHAPTER 3

Free-form surface sampling

3.1 Introduction

Modern measurement of geometrical products can be separated into sampling, decomposition, and characterization. Sampling reduces real-world continuous signals into discretized data points following which characterization extracts out interesting attributes from the data points for evaluation. Fig. 3.1 presents the main processes in a complete modern measurement flow. In the past decades, characterization techniques have been developed at a fast speed, such as feature-based methods, fractal analysis, wavelet, and morphological filtration [1,2]. Simultaneously, sampling techniques were progressed relatively slow.

Sampling [3] is the process of discrete representation of a continuous signal (which is referred to as "source signal" in this chapter). As is commonly recognized, if the size of a sample is infinitely large, the sample set approaches the source signal. If the sample size is small, the sample set can only represent the source signal at an approximation level. In real-world measurement, sample size is usually limited. Reconstruction of a sample set in its continuous form produces a substitute signal (geometry, surface, etc.) that usually has deviation (errors) from the source signal. Design of optimal sampling strategies such as sample density and positions, which are able to best approximate the source signal using less sample points, has been a tough research topic for decades. Much research work has been carried out on sample design for surface measurement in the past three decades, with the development of computer-aided inspection techniques [4–6]. However, the development of sampling techniques and their application in surface measurement is immature; thus conventional uniform sampling is still the prevailing method.

Sampling cannot be implemented without a sensing tool and a scanning plan. Given a sample design, measurement with a specific sensor needs specialized design of scanning routes. Comprehensive concerns are needed in scan planning, including scanning cost, occlusion accessing, collision avoidance, and accuracy. For point touch sensing probes, measurement can be accomplished by sequentially moving a touch probe to every designed sampling position via an optimized route [7–9]. For profile (raster scan) and areal scanning sensors, which obtain a point set in a scanning process, one can choose an appropriate sampling rate or objective to scan several profiles or areas which can completely cover the required sample points [10], following which a sub-sampling process is taken

* For more examples and practical implementation, check the website: www.digitalsurf.com/freeform.

Advanced Metrology
https://doi.org/10.1016/B978-0-12-821815-0.00003-4

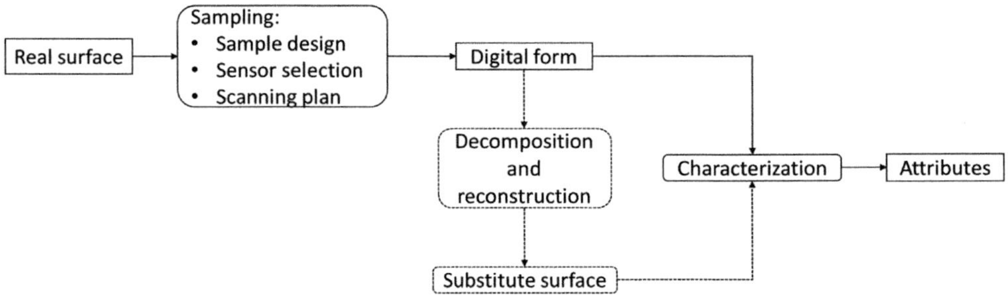

Fig. 3.1 A complete measurement flow in computer-aided inspection.

to reserve only the required data points. The next phases of sample design, scan planning and sensor selection are excluded from this report, only the design of the sampling strategies will be of concern.

Design of the sampling strategy for a specific measurement task aims to find an optimal density (or size), optimal interval, and optimal positions. Hence a source surface can be represented by its digital form with the lowest distortion if reconstruction is applied. In mathematics, sampling and reconstruction are a pair of inverse problems, and they must be synergistically considered, such as in the Shannon's theorem [3,11]. However, in engineering, sampling and reconstruction are separately investigated [4,5,12–14]. In this separate research framework, neither sampling nor reconstruction can be completely immune from uncertainty. The reasons behind this may be that ideal theorems are hard to apply to practical signals and uncertainty are allowed in engineering. For example, measurement of flatness of a plane may need a huge amount of sample points for uncertainty-free evaluation. This high-cost sampling behavior is unacceptable in engineering and usually replaced by sampling with only tens of sample points combined with the consideration for uncertainty.

Many sampling strategy designs [4–6] for surface measurement, in particular free-form surface measurement, have been developed in the past two to three decades. A review of the state of the art of sampling strategies is given in Section 3.2. Among the diverse strategies, adaptive sampling [15] has been found to be the most promising method, in terms of sampling inefficiency and accuracy. Section 3.3 introduces some reconstruction methods. In Section 3.4 a model-based sampling strategy based on the curvature is analyzed, while in Section 3.5 an adaptive sampling strategy for raster scan-based sampling tools is introduced. Adaptive sampling acquires sample points in a time sequence, and future sample positions are redirected by previous sample data. Finally, in Section 3.6, sampling method for triangular meshes is investigated.

3.2 The state of the art

Sampling techniques in geometrical measurement can be classified into categories according to the object type to be measured, for example, sampling for primitive geometries and sampling for free-form geometries. The primitive geometries here include planes, spheres, and cones. Generally, measurement of primitive geometries uses simple uniform sampling, due to the fact that such surfaces are homogeneous. Design of proper sample sizes or densities is more meaningful instead of sampling patterns or positions for primitive geometry measurement. For free-form surfaces with heterogeneous feature distributions, design of optimized sampling patterns and locations is usually more effective. Some literature reviews on the research of sampling techniques for surface measurement have been summarized, for example, in the Refs. [4,5,16] based on Table 3.1 that reports the historical efforts made on sampling techniques in a chronological order. In this table, research on scan planning, scanning methods, and probing path generation are excluded because they are regarded as the next work after sampling design.

3.2.1 Primitive surfaces

Historical research on sampling optimization for primitive surface measurement can be divided into two categories: optimization of sampling strategies (including sampling positions, size, and intervals) and optimization of sample size (or interval) only.

Table 3.1 Literature reported on sampling design.

	For primitive surfaces	**For free-form surfaces**	
Optimization of sample strategy (positions and size)	Woo and Liang [17] and Woo et al. [18] Lee et al. [19] Capello and Semeraro [20] Kim and Raman [21] Fang and Wang [22] Cho et al. [23,24] Barari et al. [25, 26] Barini et al. [27]	Model adapted	Cho and Seo [8], Cho and Kim [9], and Cho et al. [23,24] Pahk et al. [28] Ainsworth et al. [7] Elkott et al. [5] and Elkott and Veldhuis [29] Shih et al. [30] Wang et al. [6, 31] Pagani and Scott [32]
		Self-adaptive	Edgeworth and Wilhelm [33] Li and Liu [34] Barari et al. [25] Huang and Qian [35] Wang et al. [6,31]
Optimization of sample size (density and interval) only	Menq et al. [36] and Yau and Menq [37] Lin [38] and Lin et al. [39] Zhang et al. [40] Jiang and Chiu [41]		Menq et al. [36] and Yau and Menq [37]

Regarding sample strategy optimization, some low-discrepancy sampling patterns [14] originated from statistics were introduced for surface measurement in the 1990s by researchers such as Woo, Lee, Kim, and Cho et al. [17–19,21,23,24]. These methods have been found to dramatically reduce sample size compared with uniform sampling in terms of the evaluation of surface roughness, for example, *Sa*. For other indicators such as the RMS deviations and flatness, however, the low-discrepancy patterns have no advantage over uniform methods [6,18]. In 2007, Barari et al. [25] developed an adaptive strategy to improve sampling efficiency based on analyzing the probability density function (PDF) of historical sample values. This PDF-based method can also be used for free-form surface measurement with parametric fitting. This adaptive sampling was found to be effective on the minimum deviation zone (MDZ) estimation because it searches for critical sampling positions that contribute to MDZ evaluation. For other indicators, such as the RMS deviations, the effectiveness of this method is unknown.

Most real-world primitive surfaces have homogeneous random manufacturing error distributions, such as machined plane surfaces. Optimization of sample size or spacing (interval) only with uniform or simple random sampling instead of optimization of whole sampling strategies may be more effective and reasonable. Lin et al. [38,39] developed a systematic procedure to find the optimal sample size and interval for surface topography measurement, based on Fourier analysis. Statistical analysis is also widely used in the determination of sample size. Menq et al. [36,37] initially developed a sample size determination procedure based on the ratio of design tolerance to manufacturing uncertainty (or process capability). If the design tolerance is larger and manufacturing uncertainty is smaller, less sample points are needed and vice versa. This design tolerance and manufacturing uncertainty-based sample size design can be used for both primitive surfaces and free-form surfaces. Jiang and Chiu [41] also developed a variance analysis-based procedure for different primitive surfaces. In addition to statistical analysis, other mathematical tools were also used for sample size determination, such as artificial neural networks studied by Zhang et al. [40].

3.2.2 Free-form surfaces

Design of sampling strategies for free-form surface measurement is usually adapted to surface characteristics, for example, local curvatures, which are usually heterogeneous. For example, denser sample points are needed to be allocated on the areas with complex features, hence local details will not be missed out. There are plenty of adaptive sampling strategies that have been developed in the past two decades, and they can usually be divided into two categories: model-adapted sampling and self-adaptive sampling. Model-adapted methods design sampling positions based on a given CAD model. Sample points are adaptively distributed on the CAD model according to its local geometrical characteristics. Self-adaptive sampling methods design sample points by analyzing its earlier samples without referring to its nominal design model. Table 3.1 presents the historical research carried out from the two approaches.

3.2.2.1 Model-adapted sampling strategies

Some free-form surfaces can be decomposed into multiple primitive patches. Therefore a simple model-adaptive solution is to apply the sampling strategies developed for primitive surfaces by segmenting a free-form surface into different primitive patches. For example, Cho et al. [8,9,23,24] developed an automated inspection system by extracting primitive patches from a free-form surface and then applying low-discrepancy sampling strategies for measurement.

For complex free-form surfaces that are represented in parametric spaces, such as NURBS, efforts have been made to optimize sampling positions in the parametric spaces. For example, Pahk et al. [28] proposed parametric uniform sampling, normal curvature-adapted sampling, and a combination of the two methods in his research in 1995. Ainsworth et al. [7] developed an iterative subdivision sampling strategy for NURBS surfaces with consideration for multiple criteria, such as minimum sample density, uniform sampling in u and v space, and curvature. In Elkott et al. research work [5,29], model-adapted sampling strategies for NURBS surfaces were further extended.

For more general free-form CAD models, sampling positions are usually optimized by minimizing an objective function, such as the RMS deviation or other types of errors with a surface reconstruction technique. Shih et al. [30] developed three types of sampling methods for general CAD model-based measurements: direct sampling (including triangular and rectangular subdivision sampling), indirect sampling (smart section profiling), and local adjustment sampling. These methods have been shown to be advantageous for most general cases in terms of sample size saving or accuracy improvement.

Model-adapted sampling methods have difficulties in practical measurement due to the lack of consideration of unexpected defects from manufacturing and the errors that come with matching real parts to the design coordinate system.

3.2.2.2 Self-adaptive sampling strategies

Self-adaptive sampling strategies do not rely on a nominal design model and have the ability to adjust the sampling points in real time. Self-adaptive sampling methods normally rely on a fitting model to reconstruct sample data in a specific space. Evaluation of the properties of the reconstructed geometry, such as uncertainty, is then used to guide the positions of future sample points. Self-adaptive sampling is usually iteratively executed until one or multiple criteria are achieved, such as maximum iteration cycles, maximum measuring time or sample points, and maximum reconstruction error.

For example, Edgeworth and Wilhelm [33] proposed a position and normal measurement-based adaptive sampling method. This method iteratively looks for next sample points by comparing the interpolated error curves between each pair of consecutive points with a preset threshold. The iterative sampling stops until no further error curves exceed the allowed threshold in magnitude. In the research by Li [34] and Huang [35], an adaptive sampling method was developed by analyzing the covariance matrix of the fitting results from historical sample data, because the covariance matrix corresponds to the uncertainty in the evaluation of a fitting model. Wang et al. [6,31] have also developed a self-adaptive strategy based on compression of sample points in sectional profiling in which evaluation of the error of reconstructed curves from a nominal curve is iteratively implemented.

Other self-adaptive sampling techniques include the research from Barari et al. [25] in which a constructed probability density function from historical sample data is used as a guide for future sample collection.

Self-adaptive sampling strategies can easily be applied without the exact association of a real part to its design or manufacturing coordinate system. These techniques are able to effectively collect information to identify unexpected errors or defects from the manufacturing process. However, self-adaptive strategies are usually sensitive to initial conditions, such as initial sample positions.

3.3 Surface reconstruction

Surface reconstruction aims to obtain a continuous surface that best illustrates a given discrete data point set. In the majority of occasions, reconstructed surfaces have deviations from the original surface without sampling. Therefore reconstructed surfaces are usually named as substitute surfaces. Many common methods and models have been investigated for the construction of a substitute surface from regular lattice data and scattered data [16,42].

3.3.1 Tensor product B-spline reconstruction

The tensor product method has been widely used for reconstruction of regular lattice data (e.g., the uniform sampling result or partially regularly latticed data). This is because of the method's high numerical stability and computing efficiency. The tensor product method presents a surface as a tensor product of two bases, for example,

$$\begin{cases} \varphi_{\boldsymbol{a}}(x) = \sum_{k=1}^{n_a} a_k \varphi_k(x) \\ \phi_{\boldsymbol{b}}(y) = \sum_{l=1}^{n_b} b_l \phi_l(y) \end{cases}$$

in the x and y directions independently. Thus the surface can be expressed as

$$z = \varphi_{\boldsymbol{a}}(x) \times \phi_{\boldsymbol{b}}(y) = \sum_{k=1}^{n_a} \sum_{l=1}^{n_b} a_k b_l \varphi_k(x) \phi_l(y) = \sum_{k=1}^{n_a} \sum_{l=1}^{n_b} c_{k,l} \psi_{k,l}(x, y)$$

where φ_k and ϕ_l are preset base functions and the coefficient vectors $\boldsymbol{a}$ and $\boldsymbol{b}$ should be calculated from the data. Chebyshev polynomials, polynomial splines, and B-splines provide base functions for the tensor product surface reconstruction [42]. Considering the smoothness and computing stabilities, second- (linear) and fourth-order (cubic) B-splines [43] basis functions are adopted here.

3.3.2 Delaunay triangulation reconstruction

Intelligent sampling always results in nonregular latticed data that sometimes cannot be solved using the tensor product method. Triangulation-based methods are a simple and stable substitution. For example, Delaunay triangulation [44, 45] establishes neighborhood connections among the data points with the Delaunay triangulation algorithm, which neglects all the nonneighboring points in the Voronoi diagram of the given points and avoids poorly shaped triangles. Following this structuring process, regional reconstructions [45] (linear or cubic) within each triangular patch can be carried out. These methods are able to guarantee a reconstruction to a high degree of accuracy if the sample points are dense enough, which provides the theoretical foundation for developing new reconstruction techniques.

Considering that the amount of sample points for surface measurement is usually large, radial basis function (RBF)-based interpolations or fits are not generally suggested. For example, it has been stated elsewhere [46] that RBF-based reconstructions may be very unstable, computationally complex, and memory consuming; the RBF-based reconstructions are only employed when data points are no more than several thousand.

Typical examples using the tensor product reconstruction methods and Delaunay triangulation method are presented in Fig. 3.2 in which the sample result shown in Fig. 3.10A is tested and the performance differences of the reconstruction results are clearly shown. Selecting an appropriate method for reconstruction usually depends on many conditions, such as the surface complexity, distribution of the sample points, accuracy, and efficiency requirements. In this study, all the potential methods are tested, and the best one (with minimum residual errors) is selected for the performance validation in the next stage. Further reconstruction methods will be discussed in Chapter 5.

3.4 Curvature based sampling

Curvature-based sampling has been widely discussed in Elkott et al. [5], Hernández-Mederos and Estrada-Sarlabous [47], and Pagani and Scott [32]. In this section the sampling of parametric curves and surfaces is discussed.

3.4.1 Curve sampling

A parametric is a curve γ whose support is a function of a parameter t: $\gamma = \boldsymbol{r}(t) \in \mathbb{R}^p$, $t_0 \leq t \leq t_1$. The first step of the sampling procedure is the reparametrization of the curve

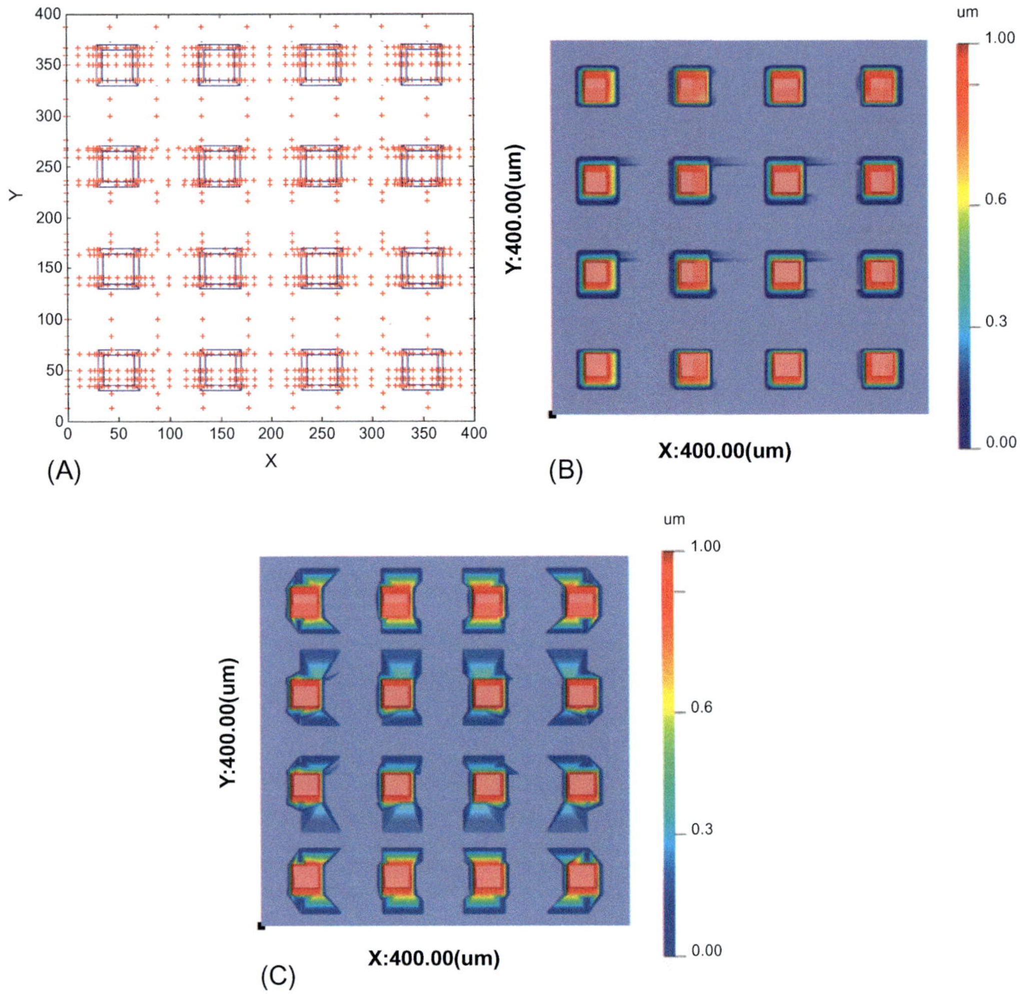

Fig. 3.2 Reconstruction of a sampled surface using tensor product B-splines and Delaunay triangulation. (A) Example surface and sampling, (B) Tensor product second-order B-spline reconstruction, and (C) Delaunay triangulation reconstruction (linear).

according to a measure that take into consideration the arc length and the curvature; then uniform sampling can be performed in the parametric space. The arc length parametrization of a generic parametric curve can be computed as [48]

$$l(t) = \int_{t_0}^{t} ||\mathbf{r}'(u)|| du$$

where $||\bullet||$ is the l^2 norm and $\mathbf{r}'(t) = \frac{d\mathbf{r}(t)}{dt}$. It is also called uniform parametrization because if a curve is reparameterized according to $l(t)$, its speed is unitary. On the contrary, if the curvature is a local geometric property that is commonly used to measure the complexity of a curve, it is computed as

$$k(t) = \frac{\|\mathbf{r}'(t) \times \mathbf{r}''(t)\|}{\|\mathbf{r}'(t)\|^3}$$

Since the curvature is a property of the curve, a parametrization according to the curvature can be computed as

$$k_p(t) = \int_{t_0}^{t} k(u)dl(u) = \int_{t_0}^{t} k(u) \cdot \|\mathbf{r}'(u)\| du$$

The integral is computed according to the arc length, so the parametrization does not depend on the specific parameter t chosen.

A reparametrization of the curve based on a mixture between the arc length (uniformity) and the curvature (complexity) can be computed as

$$p(t) = \alpha \frac{l(t)}{l(t_1)} + (1-\alpha) \frac{k_p(t)}{k_p(t_1)}$$

where $\alpha \in [0, 1]$ is a parameter that balances the contribution of the arc length and the curvature; in the following examples, it is set to 0.5.

An example of the reparametrization and the uniform sampling of 50 samples is shown in Fig. 3.3. The mixed parametrization allows more points to be added when the curvature is high, that is, where the curve needs more points for the reconstruction, but the samples span the whole curve. Using only the curvature information, there is an oversampling in some portions of the curve, while in the flat portions there are few points.

A profile extracted from a measured surface, along with 100 samples using the mixed parametrization, is shown in Fig. 3.4A. Four sampling strategies were tested: uniform, arch length parametrization (arc_length), the mixed parametrization (al_curv), and the Latin hypercube sampling [49] (lhs) and the method proposed by Estrada-Sarlabous [47] (al_curv2). The performance of the sampling points in the reconstruction using the Akima splines [50] is shown in Fig. 3.4B. The reconstruction based on mixed parametrization clearly outperforms the others.

3.4.2 Surface sampling

Sampling can be easily extended to parametric surfaces using the tensor product of the points the parametrized space. Let $\mathbf{r}(u, v)$, $u_0 \leq u \leq u_1$, $v_0 \leq v \leq v_1$ be a parametric surface; the extension of the arc length parametrization may be a parametrization based on the marginal cumulative area along the u and v directions.

The marginal area can be computed as

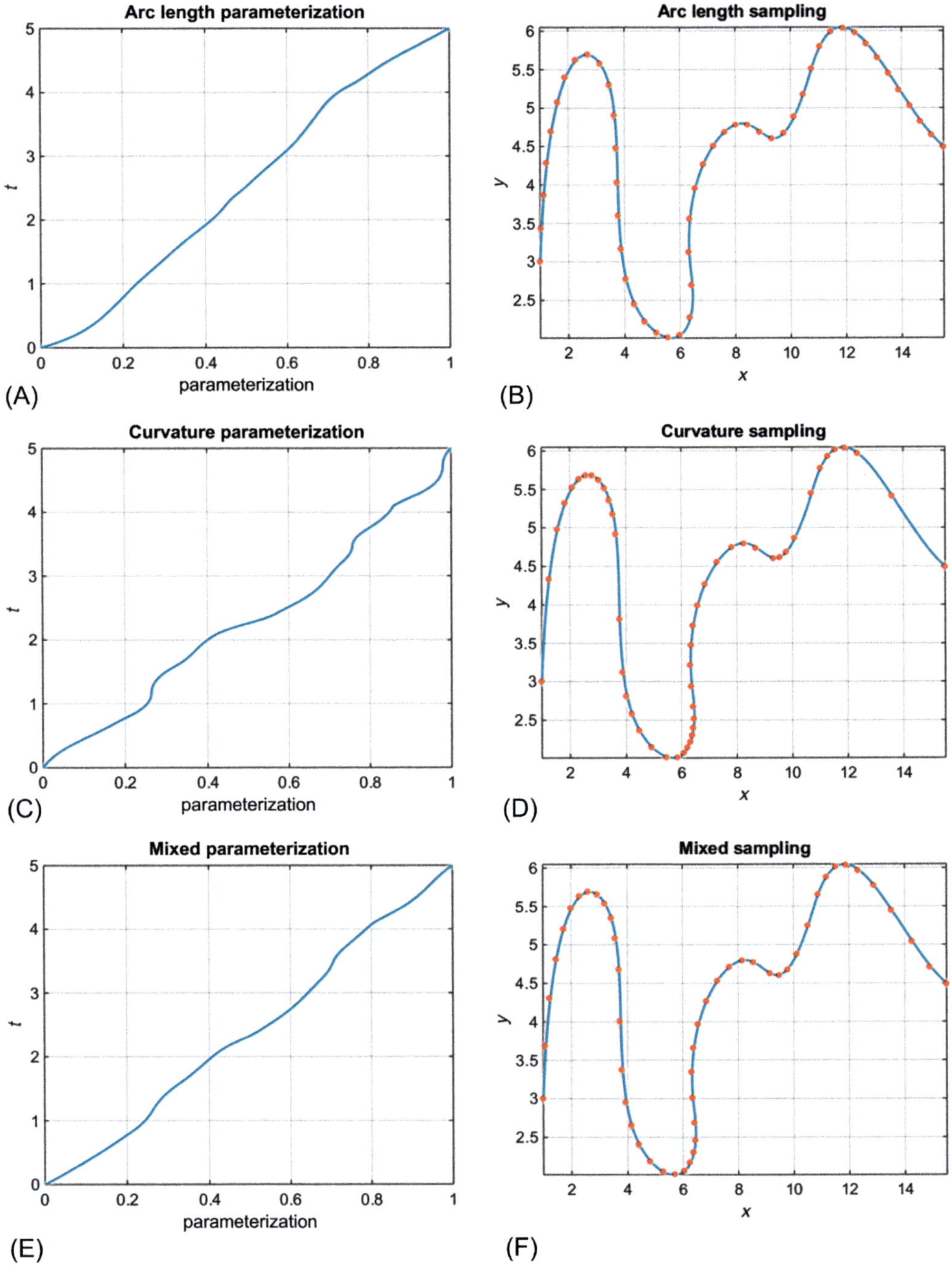

Fig. 3.3 Parametrization and sampling: arc length (A) and (B), curvature (C) and (D), and mixed (E) and (F). *(Credit: https://www.sciencedirect.com/science/article/pii/S0167839617301462.)*

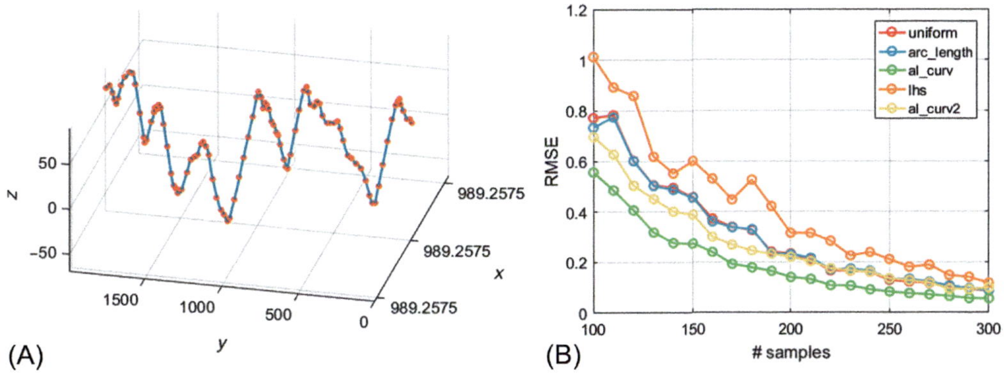

Fig. 3.4 Measured profile and sampled points (A) and reconstruction errors (B), values in μm. *(Credit: https://www.sciencedirect.com/science/article/pii/S0167839617301462.)*

$$A_u(u) = \int_{u_0}^{u} ds \int_{v_0}^{v_1} \|\boldsymbol{r}_s(s,v) \times \boldsymbol{r}_v(s,v)\| \, dv$$

along the u direction and

$$A_v(v) = \int_{v_0}^{v} ds \int_{u_0}^{u_1} \|\boldsymbol{r}_u(u,s) \times \boldsymbol{r}_s(u,s)\| \, du$$

in v direction, where $\boldsymbol{r}_u(u,v) = \frac{\partial r(u,v)}{\partial u}$ and $\boldsymbol{r}_v(u,v) = \frac{\partial r(u,v)}{\partial v}$ are the first-order partial derivatives in u and v direction, respectively. The sampling can be performed by applying a uniform sampling of $n_u(n_v)$ samples in u (v) direction; the final set of points can be performed by the tensor product of the set of points sampled in u and v directions. As for the curve an index of the complexity of the surface has to be chosen; in Pagani and Scott [32] the authors proposed to use the mean curvature:

$$k(u,v) = \frac{k_1(u,v) + k_2(u,v)}{2}$$

where $k_1(u,v)$ and $k_2(u,v)$ are the principal curvatures of the surface [48]. The mean curvature was chosen instead of the Gaussian curvature (the product of the principal curvatures) because the Gaussian curvature vanishes to 0 if one of the two principal curvatures is null. Using the Gaussian curvature is therefore not possible for taking into account the complexity of the surface if one of the principal curvatures is null.

As in the curve scenario, the marginal curvature parametrizations can be computed as

$$k_u(u) = \int_{u_0}^{u} ds \int_{v_0}^{v_1} k(s,v) \cdot \|\boldsymbol{r}_s(s,v) \times \boldsymbol{r}_v(s,v)\| \, dv$$

$$k_v(v) = \int_{v_0}^{v} ds \int_{u_0}^{u_1} k(u, s) \cdot \|\mathbf{r}_u(u, s) \times \mathbf{r}_s(u, s)\| \, du$$

The marginal mixed parametrizations can be computed as

$$p_u(u) = \alpha_u \frac{A_u(u)}{A_u(u_1)} + (1 - \alpha_u) \frac{k_u(u)}{k_u(u_1)}$$

$$p_v(v) = \alpha_v \frac{A_v(v)}{A_v(v_1)} + (1 - \alpha_v) \frac{k_v(v)}{k_v(v_1)}$$

where $\alpha_u \in [0, 1]$ and $\alpha_v = [0, 1]$. In the following test cases, both α_u and α_v are set to 0.5, that is, equal weight is given to the area and the mean curvature. Fig. 3.5 shows a free-form surface with the sampling of four different sampling strategies and the error map after the reconstruction: uniform sampling on the parametric space (Fig. 3.5B), uniform sampling based on the marginal area parametrization (Fig. 3.5C), and uniform sampling based on the mixed parametrization (Fig. 3.5D). The sampling based on the parameter

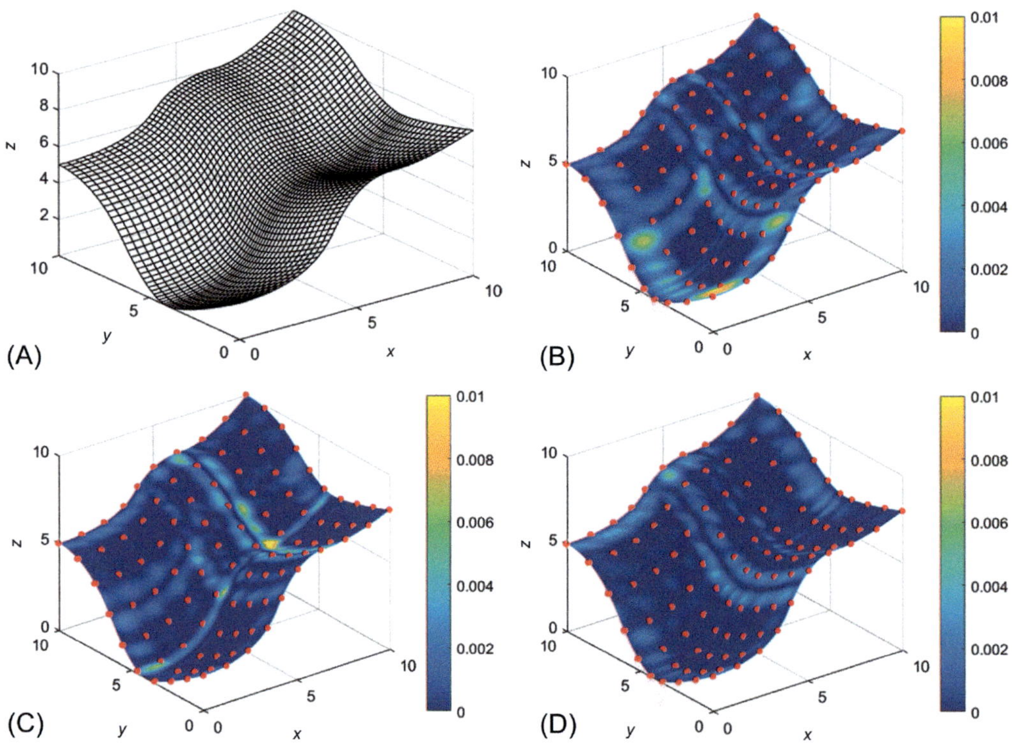

Fig. 3.5 Example of surface with different sampling strategies and error maps: surface (A), uniform sampling in the parameters' space (B), sampling based on the area (C), and sampling based on the mixed parametrization (D). *(Credit: https://www.sciencedirect.com/science/article/pii/S0167839617301462.)*

space does not take into account any geometrical property of the surface and varies if there is a transformation of one of the parameters, for example, if it is applied to a transformation like $u^{\star} = g(u)$, the surface does not change, but the sampling does. In the example, there are some errors on the bottom left of the surface due to the high space between the sampled points that can be improved by a parametrization based on the area. The area-based parametrization cannot perform an adaptive parametrization; that is, it cannot "see" that changing speed of the normal vector, which is measured by the curvature. The mixed parametrization allows a better reconstruction in the analyzed example.

Fig. 3.6 shows the sampling of 10×10 points of a simulated surface (top row) and the sampling of 30×30 points from a measured structured surface; the RMSE are shown as a function of the square root of the number of samples. The compared sampling strategies were the following: uniform in the parameters' space (uniform), uniform in the marginal cumulative areas function (area), according to the mixed parametrization (area_curv), Latin hypercube sampling (lhs), Halton [51], Hammersley

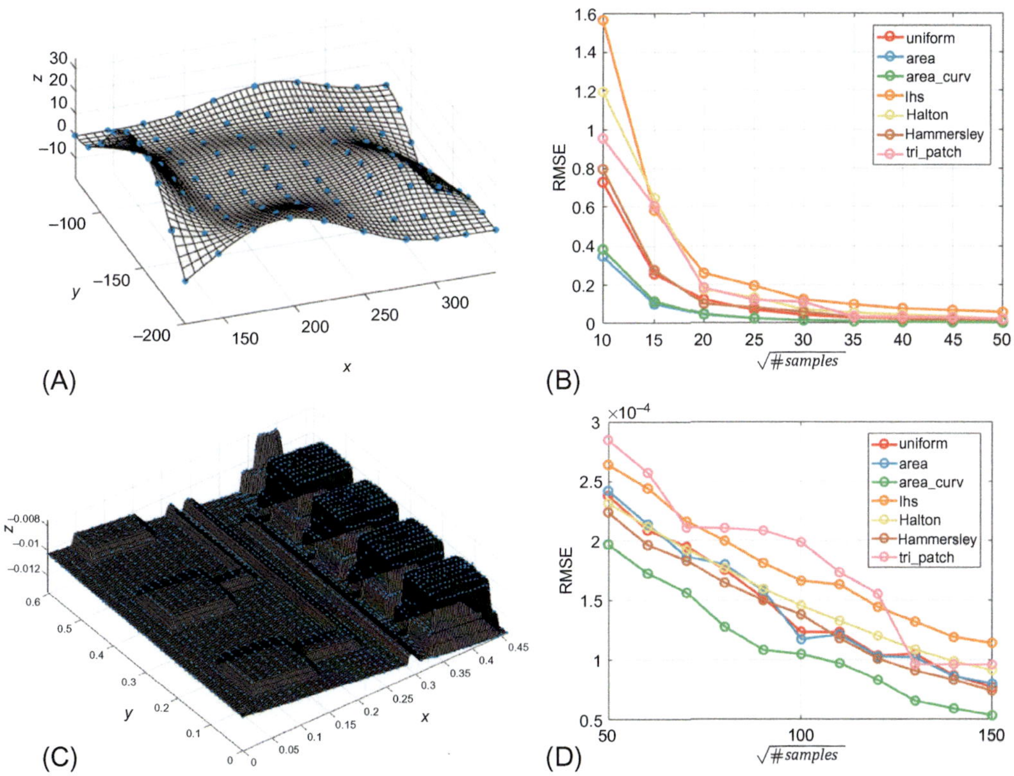

Fig. 3.6 Nominal surface and sampling using the mixed parametrization (A, C) and sampling method performance comparisons (B, D), values in mm. *(Credit: https://www.sciencedirect.com/science/article/pii/S0167839617301462.)*

[51], and triangle subdivision adaptive sampling (tri_patch) [30]. The values of the RMSE index show that the sampling based on the mixed parametrization allows a lower reconstruction error.

There is another class of free-form surface that can be generated by two or more curves. These surfaces can be classified into several groups: prism surface, ruled surface, surface of revolution, swung surface, skinned surface, swept surface, and coons surface; for a better explanation regarding their construction, see Piegl and Tiller [43]. During the design phase the generatrices are known; therefore it is possible to sample the points using the information provided by the curves. In this section, only swept and coons surfaces are presented; the sampling of the other classes were analyzed by Pagani and Scott [32].

A swept surface is generated by sweeping a section curve along an arbitrary trajectory. It is defined as

$$\boldsymbol{r}(u, v) = \boldsymbol{r}_2(u) + \boldsymbol{M}(v) \cdot \boldsymbol{r}_1(v)$$

where $\boldsymbol{r}_1(u)$ is the section curve, $\boldsymbol{r}_2(v)$ is the trajectory, and $\boldsymbol{M}(v) \in \mathbb{R}^{3\times 3}$ is a rotation and scaling matrix. A sampling strategy for swept surfaces can be computed as follows:

1. Compute n_u samples along the section curve using the mixed sampling curve method.
2. Compute n_v samples along the trajectory curve using the mixed sampling curve method.
3. Compute the samples on the surface as the cross product of the samples along the curves.

A swept surface and the reconstruction performance are shown in the first row of Fig. 3.7. The points represent sampling of 20×20 points along the section and the trajectory curves. The reconstruction using the area, the mixed parametrization, and the sampling based on the generatrices achieve better results.

Another surface based on four boundary curves is called coons surface. Let $\boldsymbol{r}_{1,\,u}(u)$ and $\boldsymbol{r}_{2,\,u}(u)$ be the two boundaries curves of the u parameter space and $\boldsymbol{r}_{1,\,v}(v)$ and $\boldsymbol{r}_{2,\,v}(v)$ be the two boundaries curves of the v parameter space, the bilinear blended coons surface is defined as

$$\boldsymbol{r}(u, v) = \boldsymbol{r}_1(u, v) + \boldsymbol{r}_2(u, v) - \boldsymbol{r}_3(u, v)$$

where $\boldsymbol{r}_1(u, v)$ and $\boldsymbol{r}_2(u, v)$ are the ruled surfaces between, respectively; the curves $\boldsymbol{r}_{1,u}(u)$ and $\boldsymbol{r}_{2,\,u}(u)$ and the curves $\boldsymbol{r}_{1,\,u}(u)$ and $\boldsymbol{r}_{2,\,u}(u)$ and $\boldsymbol{r}_3(u, v)$ is a bilinear tensor product surface (plane) defined as

$$r_3(u, v) = (1 \;\; u) \begin{bmatrix} p_{0,0} & p_{0,1} \\ p_{1,0} & p_{1,1} \end{bmatrix} \begin{pmatrix} 1 \\ v \end{pmatrix}$$

$$p_{0,0} = r_{1,u}(0) = r_{1,v}(0)$$
$$p_{1,0} = r_{1,u}(1) = r_{2,v}(0)$$
$$p_{0,1} = r_{2,u}(0) = r_{1,v}(1)$$
$$p_{0,0} = r_{2,u}(1) = r_{2,v}(1)$$

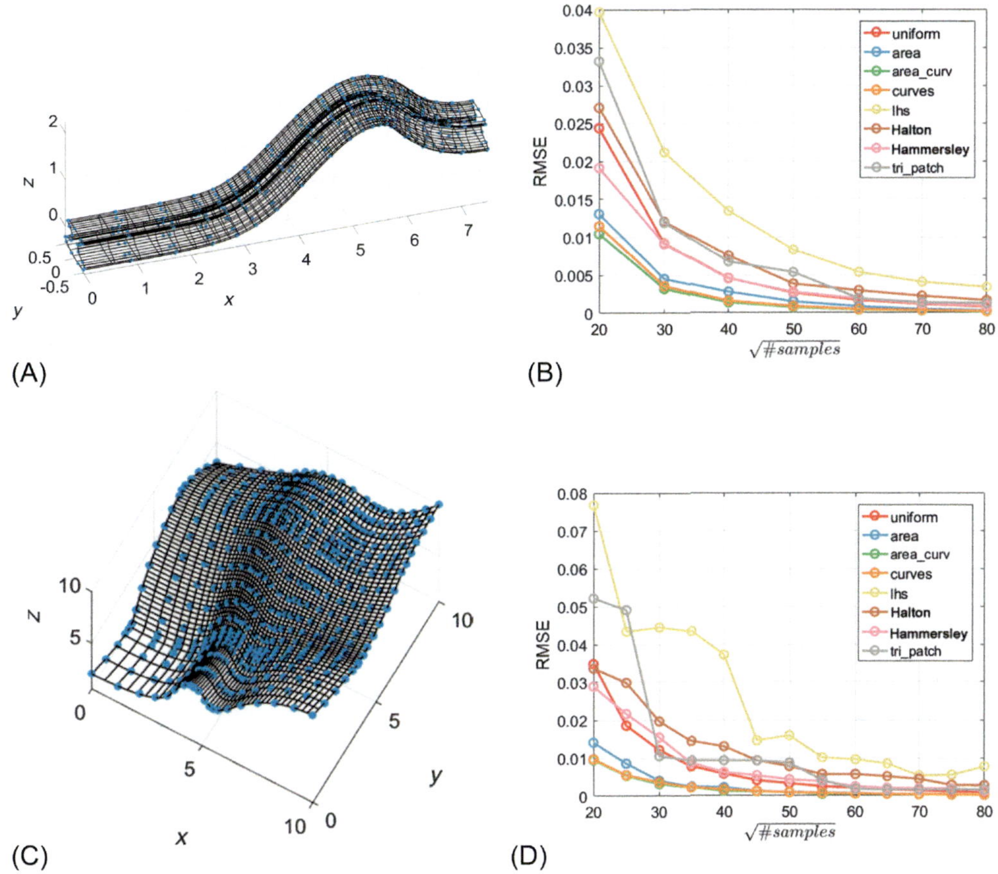

Fig. 3.7 Example of swept surface (A, B) and coons surface (C, D), values in mm. *(Credit: https://www.sciencedirect.com/science/article/pii/S0167839617301462.)*

Since a Coons surface is a sum of three surfaces that are linear in at least one direction, a possible sampling strategy is as follows:

1. Compute n_u samples along the two generatrices in u direction using the mixed parametrization method.
2. For each of the couple of n_u samples, perform an interpolation in the $u - v$ space.
3. Compute n_v samples along the two generatrices in u direction using the mixed parametrization method.
4. For each of the couple of n_v samples, perform an interpolation in the $u - v$ space.

An example of a coons surface and the sampled points are shown in Fig. 3.7 (second row); the sampling based on the mixed parametrization and on the curves achieve the best performances.

3.5 Adaptive sampling strategy

For a specific sensor (a stylus tip or areal microscope), sampling needs to be realized using an optimized scanning strategy. Sometimes, scanning and sampling can be designed in a united form; thus the design of the scanning process can be omitted. An adaptive sampling (and scanning) strategy called sequential-profiling adaptive sampling [6,31] was developed for raster scan-based sensing tools by the authors. Suitable reconstruction algorithms were reported following a validation of the performance of the developed method and other sampling strategies that was presented. The proposed strategy is found to be effective for free-form surfaces with structured surface patterns.

3.5.1 Method description

The sequential profiling method was developed based on the idea of so-called indirect sampling [30]. Considering the raster scanning mechanism used in stylus profilometry, this methodology comprises a two-stage algorithm: profile adaptive compression sampling and areal adaptive scanning.

3.5.1.1 Profile adaptive compression sampling

The core of the profile adaptive compression sampling method is a compression algorithm that is proposed to prune out the unnecessary samples for a given uniform sampling result. This method does not aim to reduce the sampling duration; rather, it reduces the sample size by maintaining necessary reconstruction accuracy. After this process, those key samples that have significant influence on minimizing the reconstruction error are retained. A simulation result of this method is shown in Fig. 3.8, and it can be seen that dense samples are retained near the high curvature regions. The method proceeds as follows:

1. For a given surface profile, obtain its digital measuring result using the instrument-permitted dense sampling setting (blue lines in Fig. 3.8).
2. Divide the digital profile at inflection points, if any, into several segments that are solely concave or convex.
3. For each segment, evaluate the approximation error (usually the residual error between the original profile and the approximated/interpolated profile) of each segment.
4. If the error exceeds an initially set threshold (usually a fraction of the initial error from step 3), an extra sampling point is inserted on the profile curve at the midpoint. Otherwise, stop.
5. For each subinterval formed by insertion of a new point, repeat steps 3 and 4 until the approximation error is smaller than the threshold value.

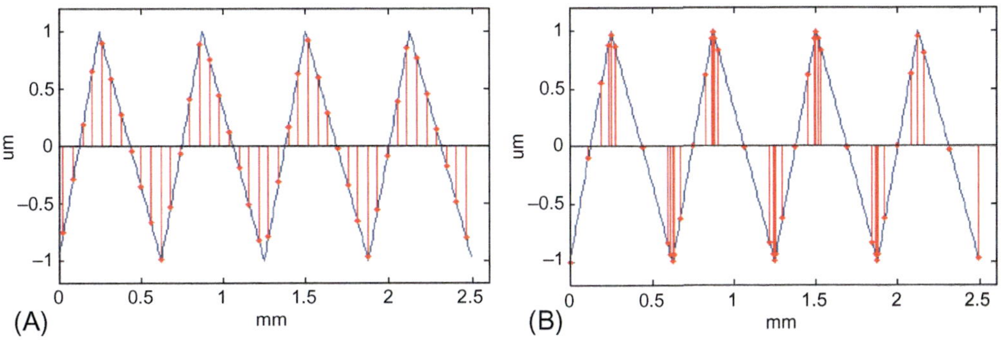

Fig. 3.8 A sample design from (A) uniform sampling and (B) the profile adaptive sampling.

3.5.1.2 Areal adaptive scanning

With the assistance of the profile adaptive compression sampling, the sequential-profiling adaptive sampling can be implemented. For a given surface the methodology searches the key sample positions in a perpendicular direction to the main measuring axis and then implements the profile adaptive compression for all the main axis profile measurements at the key positions. Here the main axis is the measuring axis of an instrument with the highest measuring repeatability. The technique comprises five steps as follows and as illustrated in Fig. 3.9.

1. Randomly (or uniformly) select n (usually ten) profiles parallel to the main measuring axis (x-axis in Fig. 3.9).
2. Implement profile adaptive compression sampling for each profile. Thus the key positions can be found.
3. Resort all the pruned key samples in accordance with their positions along the measurement axis (y-axis in Fig. 3.9).
4. Edit the key sample list produced in step 3 by the factor N to prune out samples that are too dense.
5. For the revised key sample positions, implement the profile adaptive compression sampling for each profile on the main axis direction.

An improvement in measurement efficiency has been presented by the adaptive sampling result shown as the red sample points in Fig. 3.9C. Dense sampling points are arranged near the high curvature areas (edges of the square step structures), while the low-curved regions have a sparse allocation of samples. Using this method the measuring size and duration can be effectively reduced, and simultaneously reconstruction errors such as the residual root mean square (RMS) error and the error of the dimensional parameter evaluation can be minimized.

Fig. 3.10 presents a top-view of the sample points generated from the novel strategy earlier and the other two CAD model-based intelligent sampling strategies, named triangle- and rectangle-patch adaptive subdivision sampling, respectively [30].

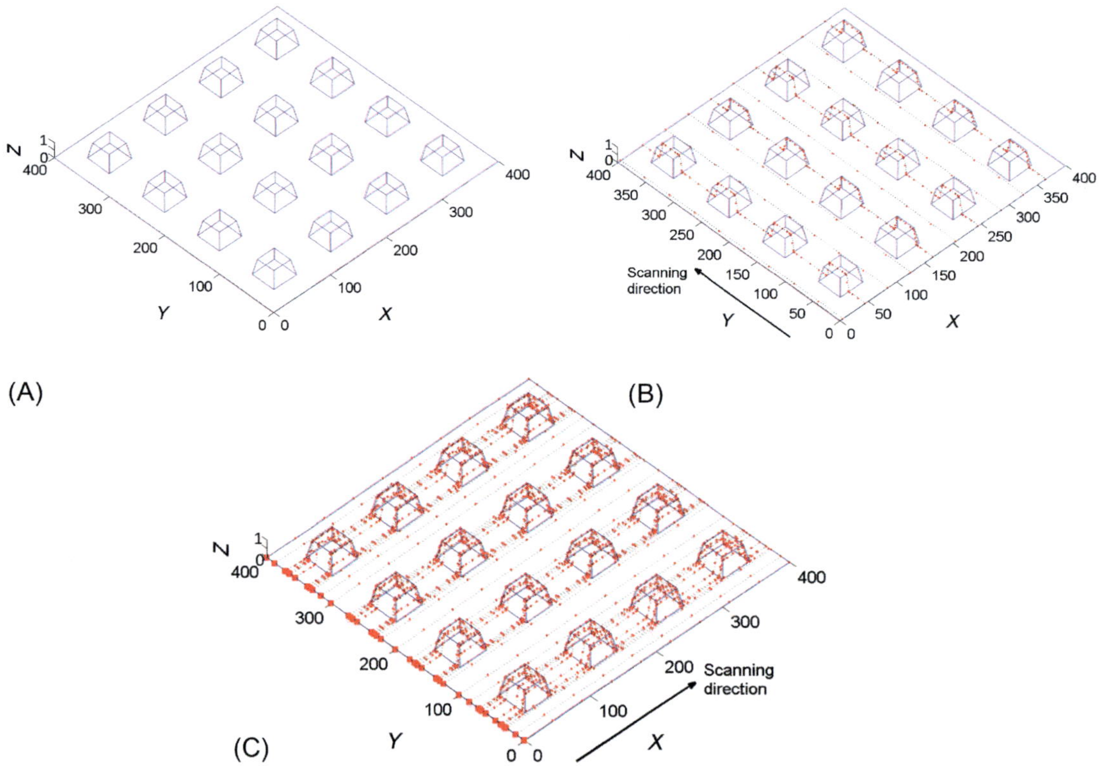

Fig. 3.9 Main processes of the sequential-profiling adaptive sampling. (A) A structured surface to be measured. (B) Ten profile adaptive compression sampling *(dashed lines)* and the selected key sample positions for *y*-axis *(red dots and squares)*. (C) Profile adaptive compression sampling on main (*x*-) axis at each selected key (*y*-) position.

3.5.2 Performance validation

3.5.2.1 Experimental settings

Free-form surfaces with structured features usually contain three basic types such as linear patterns, tessellations, and rotationally symmetric patterns [1]. Three representative high-precision structured surface specimens were used to validate the performance of different sampling methods. High-precision measurements of the three surfaces are presented in Table 3.2a–c, which include a five-parallel-grooves calibration artefact, a nine-pits-crossed-grating calibration artefact, and a Fresnel lens central patch. These high-density sampled results are used as the references for comparison. Evaluation of the parameter errors from the standard results can then be carried out. Specifically the detailed

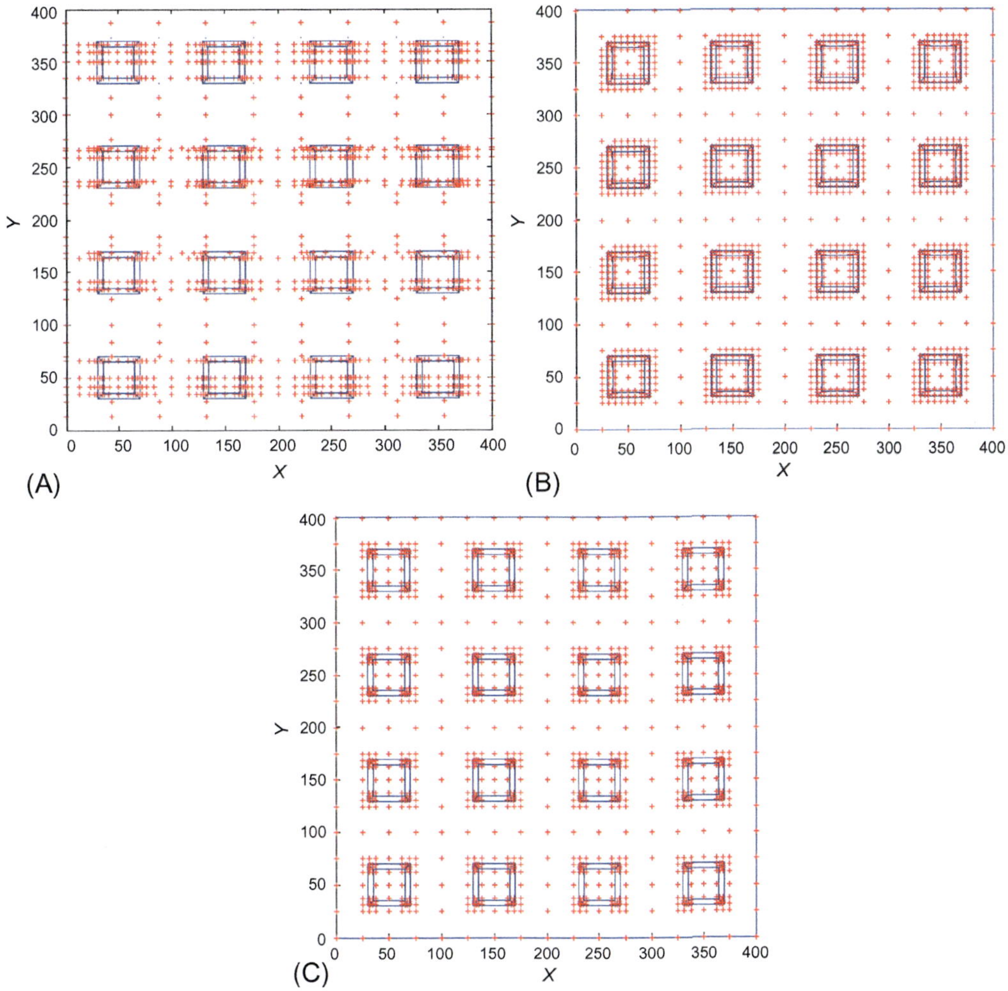

Fig. 3.10 Sampling patterns produced by the three adaptive sampling methods (≈1500 sample points). (A) Sequential profiling adaptive sampling. (B) Triangle patch adaptive subdivision sampling. (C) Rectangle patch adaptive subdivision sampling.

evaluation parameters for each surface sample are given in Table 3.2 with consideration of their main functions. Both the RMS height residuals and feature-based parameters are evaluated.

Seven sampling methods and six different sample sizes were used (the original and tested sample size can be seen in Table 3.1 on each of the specimens). The seven tested sampling methods include uniform sampling, jittered uniform sampling, Hammersley

Table 3.2 The three typical structured surface specimens and the experimental settings.

	Specimen 1	Specimen 2	Specimen 3
Specimen names	Five-parallel-grooves calibration artefact	Nine-pits-crossed-grating calibration artefact	Fresnel lens central patch
Rendering of the high-density sampled results	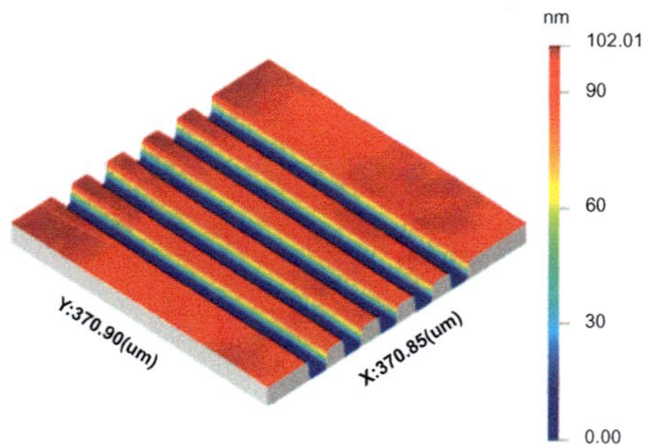 (a) Five-parallel-grooves (linear pattern) (1024 × 1024)	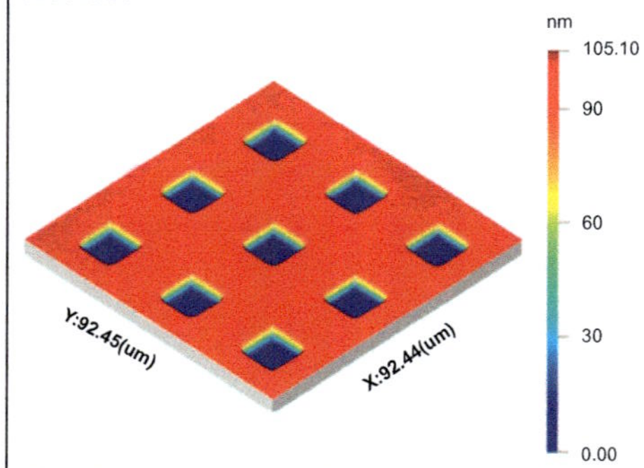 (b) Nine-pits-crossed-grating (tessellation) (256 × 256)	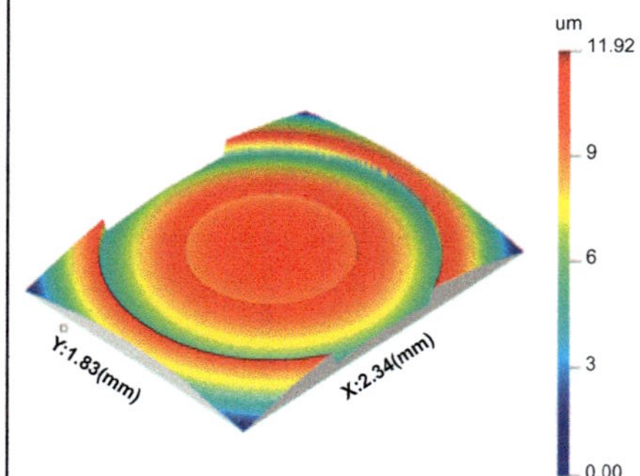 (c) Fresnel lens (rotational symmetric) (358 × 240)
Evaluation parameters	**1.** The RMS height deviation **2.** The mean groove width **3.** The step height	**1.** The RMS height deviation **2.** The mean pitch distance **3.** The step height	**1.** The RMS height deviation **2.** The radius the central lens edge **3.** The roundness of the central lens edge
Tested sample sizes	Six sizes: 2.5k, 10k, 40k, 90k, 160k, and 250k[a]	Six sizes: 1.2k, 2.5k, 5k, 10k, 15.6k, and 22.5k[a]	Five sizes: 1.2k, 2.5k, 5k, 10k, 22.5k, and 40k[a]

[a]The tested sample sizes are selected based on the following criteria: (1) The tested samples sizes are representatively selected, which indicates that they might be normally used in practical measurements, (2) the tested sample cannot be too small in which case giant reconstruction distortion occurs, and (3) the tested sample size cannot be too large in which case reconstruction error has minor fluctuations and the evaluation process may be time consuming.

pattern sampling, Halton pattern sampling, rectangular subdivision adaptive sampling, triangle subdivision adaptive sampling, and the sequential profiling adaptive sampling proposed earlier. The numerical sampling tests are carried out, and the sample results are stored. Potential reconstruction methods are then employed to reconstruct the continuous surface with the sample amount as dense as the originals. The best reconstruction results with the lowest RMS height residuals are selected. Aiming for the best reconstructed results, the RMS height residuals are recorded, and the feature-related parameters can be extracted.

3.5.2.2 Results and discussion

Fig. 3.11 shows the different evaluation indicators for the seven selected sampling methods, which were tested on three structured surfaces: a linear pattern, a tessellation pattern, and a rotational symmetric pattern (see [52] for a graphical view). The three graphs in the first row of Fig. 3.11 show the RMS height deviations of the sampling-reconstructed surface from the standard measurement results. The second row shows the evaluation errors of the key feature attributes for the three specimens, including, respectively, the groove width for linear pattern, the pitch distance for tessellation, and the radius for the rotational symmetric pattern. The third row presents the evaluation errors of step heights for the linear and the tessellation patterned specimen and the roundness of the boundary for the central lens of Specimen 3. Most of the evaluation indicators decrease in the form of a power function versus the number of sample points. Therefore the evaluation data were fitted using power functions, and all the graphs in Fig. 3.11 were shown in log-log forms. A few exceptions to decreasing performance occurred such as in Fig. 3.11h where the triangle and rectangle subdivision adaptive sampling methods performed worse when sample points increased. The exceptions may originate from limited data points for fitting, for example, due to the fact that the circle boundary recognition fails when there are only three valid indicator points in the test of rectangle subdivision adaptive sampling method.

It can be seen from Fig. 3.11 that the low-discrepancy pattern sampling methods have a similar performance to uniform or jittered uniform sampling. For the measurement of structured surfaces, the advantages of low-discrepancy pattern sampling are not apparent, and sometimes, these methods may not perform better than uniform methods. In some situations, uniform sampling may be a better solution compared with other fixed patterns.

The advantages of approximation optimized sampling methods can be easily appreciated. These methods allocate the sampling effort according to their earlier sample results or given models. In other words, they can adapt the sampling effort to key positions, which have higher impact factors on enhancing the reconstruction accuracy than others. Although adaptive sampling methods have no clear advantages on measuring the pitch distance of crossed-gratings (see Fig. 3.11E), they have been

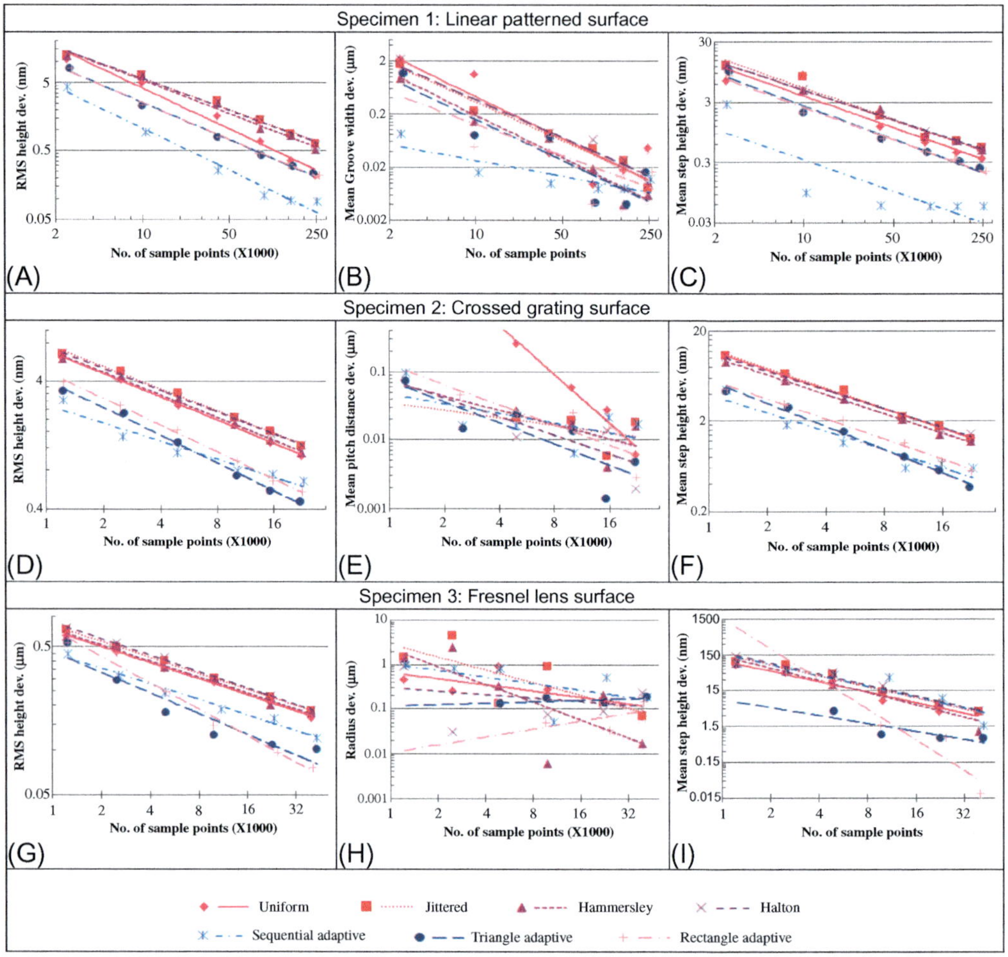

Fig. 3.11 Deviations of the evaluation parameters from the standard result for different specimens (log-log plots): linear patterned surface (A–C), crossed grating surface (D–F), and Fresnel lens surface (G–J).

shown to be effective in the measurement of other structured surfaces and with other parameters.

There are challenges, however, in applying adaptive sampling to practical measurement. Sequential profiling adaptive sampling may suffer from the mechanical limitations in stability (e.g., thermal drift) and accuracy in the y direction scanning. Most of the other approximation optimized sampling methods are difficult to implement within the operation envelope of scanning instruments, regarding the complex scan route designs and redundant scan durations. In terms of interferometers, many of the reviewed sampling methods may be promising, with the assistance of a high-resolution CCD and pixel stratification, or lens autoswitch systems. Considering the unavoidable positioning errors in the installation of the specimens and optical resolution constraints, specialized research work may be required in intelligent sampling for interferometers. In addition, more theoretical work is necessary to further the research on intelligent sampling. For example, nonregular sample data storage solutions need to be reconsidered contrary to the current grid data storage. The reverse problem of sampling and reconstruction needs to be fully investigated for the applications of geometric measurement. Also, determination of the sample size for a sampling strategy is a complex problem that requires further attention.

3.6 Triangular mesh sampling

In this last section the sampling of the free-form surface available in the format of a triangular mesh is discussed. The representation of a complex geometry, after the acquisition using a structured light scanner or a compute tomography (CT) device, is usually described by a polygonal mesh. Performing sampling on a polygonal mesh means reducing the number of vertices and faces preserving the topology and the features of interest. In the seminal paper of Garland and Heckbert [53], the authors proposed a quadratic metric, described later, to preserve the area of the original mesh during the simplification stage. The algorithm was then optimized by Lindstrom and Turk [54]; the original algorithm contracts arbitrary pairs of vertices; this can lead to a nonmanifold mesh.

The simplification method is divided into two stages:

1. Collection stage: the cost of removing each edge is computed.
2. Collapsing stage: the edge with lowest cost is processed and the new vertex position is computed.

Fig. 3.12 shows an example of a mesh before and after the simplification of the edge *e*. For each edge the removal cost is computed by a quadratic function that measures the distance between the mesh before and after the collapsing stage. The objective function, for each edge, is computed as

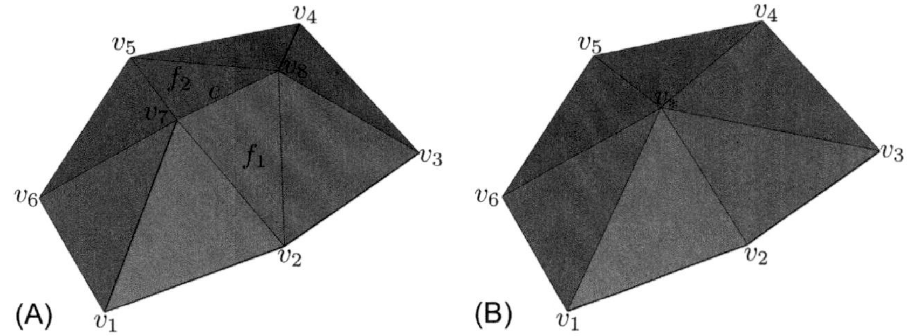

Fig. 3.12 Simplification step: before (A) and after (B) the collapsing stage [55]. *(Credit: https://www.sciencedirect.com/science/article/pii/S0263224118305773.)*

$$f(e, \boldsymbol{v}) = \frac{1}{2}\boldsymbol{v}^T \boldsymbol{H}\boldsymbol{v} + \boldsymbol{c}^T \boldsymbol{v} + \frac{1}{2}k$$

where e is the edge to be removed, $\boldsymbol{v}$ is the position of the new vertex, $\boldsymbol{H}$ is the Hessian matrix, $\boldsymbol{Hv}+\boldsymbol{c}$ is the gradient of $f(e, \boldsymbol{v})$, and k is a constant. During the collapsing stage the edge e and the two adjacent faces f_1 and f_2 are removed; a new vertex $\boldsymbol{v}^*$ is computed to minimize the objective function.

In the original paper of Garland and Heckbert [53], the authors proposed to use the following cost function for each removed vertex:

$$f(\boldsymbol{v}_i) = \sum_{\boldsymbol{p} \in planes(\boldsymbol{v}_i)} \left(\boldsymbol{p}^T \boldsymbol{v}_{i,h}^*\right)^2$$

where $\boldsymbol{p} = (a, b, c, d)^T$ represents the vector with the coefficients of the plane $a\,x + b\,y + c\,z = 0$ and $planes(\boldsymbol{v})$ represent all the planes of the faces around the vertex $\boldsymbol{v}$. The vertex that achieve the minimum value (lower removal cost) is processed first.

In Lindstrom and Turk [54] the objective function is constructed to preserve the volume and the boundaries of the mesh too; it is described by

$$f(e, \boldsymbol{v}) = \lambda f_V(e, \boldsymbol{v}) + (1 - \lambda)\, L^2(e) f_B(e, \boldsymbol{v})$$

where $f_V(e, \boldsymbol{v})$ and $f_B(e, \boldsymbol{v})$ are the objective functions preserving the volume and the boundaries, respectively; $L(e)$ is the length of the edge; and λ is a weight, usually set to 0.5, measuring the relative importance between volume, and the presentation of the volume and the boundaries.

The volume preserving objective function is computed as

$$f_V(e, \boldsymbol{v}) = \frac{1}{18}\left[\frac{1}{2}\boldsymbol{v}^T\left(\sum_i \boldsymbol{n}_i \boldsymbol{n}_i^T\right)\boldsymbol{v} - \sum_{i \in t(e)} \det\left(\left[\boldsymbol{v}_1^i, \boldsymbol{v}_2^i, \boldsymbol{v}_3^i\right]\right)\boldsymbol{n}_i^T \boldsymbol{v} + \frac{1}{2}\sum_{i \in t(e)} \det\left(\left[\boldsymbol{v}_1^i, \boldsymbol{v}_2^i, \boldsymbol{v}_3^i\right]\right)^2\right]$$

where $t(e)$ represents the indices of the modified triangles, $\boldsymbol{n}_i$ is the outward normal of the ith triangle, and $det([\boldsymbol{v}_1^i, \boldsymbol{v}_2^i, \boldsymbol{v}_3^i])$ is the determinant of the matrix $[\boldsymbol{v}_1^i, \boldsymbol{v}_2^i, \boldsymbol{v}_3^i]$.

The boundary preserving objective function is defined as

$$f_B(e, \boldsymbol{v}) = \frac{1}{2}\left[\frac{1}{2}\boldsymbol{v}^T\left(\sum_{i \in t(e)} \boldsymbol{E}_{1i}\boldsymbol{E}_{1i}^T\right)\boldsymbol{v} + \sum_{i \in t(e)} (\boldsymbol{e}_{1i} \times \boldsymbol{e}_{2i})^T \boldsymbol{v} + \frac{1}{2}\sum_{i \in t(e)} \boldsymbol{e}_{2i}^T \boldsymbol{e}_{2i}\right]$$

where $\boldsymbol{e}_{1i} = \boldsymbol{v}_2^{e_i} - \boldsymbol{v}_1^{e_i}$, $\boldsymbol{e}_{2i} = \boldsymbol{v}_2^{e_i} \times \boldsymbol{v}_1^{e_i}$ and

$$\boldsymbol{E}_{1i} = \begin{bmatrix} 0 & -e_{1i,z} & e_{1i,y} \\ e_{1i,z} & 0 & -e_{1i,x} \\ -e_{1i,y} & e_{1i,x} & 0 \end{bmatrix}$$

An example of a mesh decimation is shown in Fig. 3.13. It represents a portion of an additively manufactured lattice mesh measured with a CT device, and the conversion

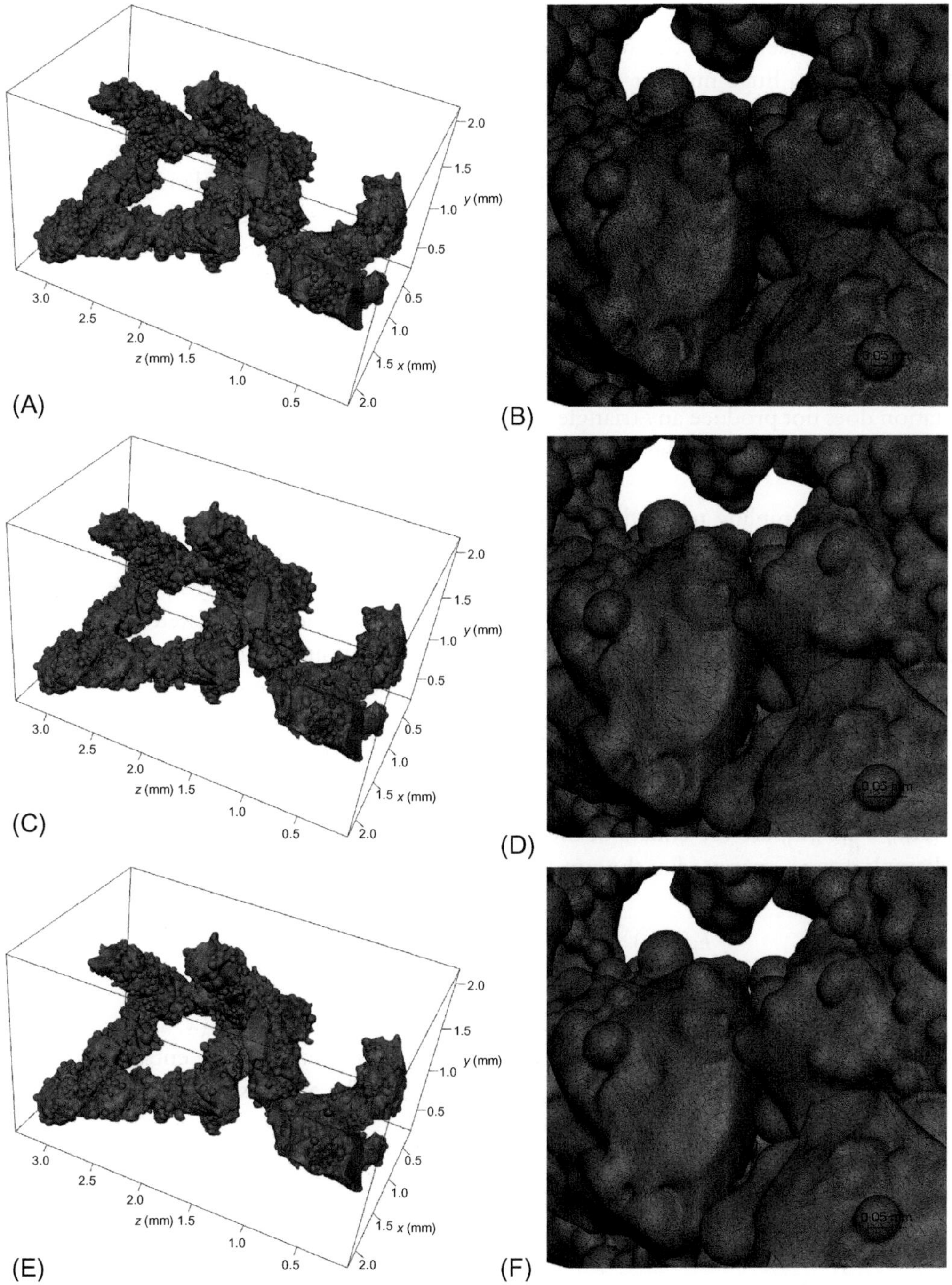

Fig. 3.13 Example of mesh decimation. Left column whole mesh, right column decimation. First row (A, B): original mesh, second row (C, D): decimation using CGAL library, third row (E, F): decimation using VCG.

from volume to mesh was performed using the marching cube algorithm [56]. As it is possible to observe, in the magnified figure, the marching cube algorithm produces meshes with a huge number of small triangles, 4,814,392 in total. An implementation of the algorithm proposed by Lindstrom and Turk [54] can be found in CGAL [57]; the resulting mesh after a reduction of 90% of its triangles is shown in the second row of Fig. 3.13. An implementation of the algorithm proposed by Garland and Heckbert [53] is available in the VCG library [58]. The authors apply some local optimization operations such as edge flip, and the operation is marked as unfeasible if it leads to a face flip; the result of the decimation is shown in Fig. 3.13 in the third row. Observing the full mesh, it is not qualitatively possible to see any difference between the original and the decimated meshes, but if the magnifications are analyzed, it is should be observed that the number of triangles is lower by an significant amount and that the VCG implementation does not produce any triangle flips, while there are some triangles that change orientation using CGAL. The quality of the decimated mesh is higher if the VCG library is used.

The last sampling method presented regards the sampling based on spectral analysis [59]. The method described by the authors is based on the spectral analysis on a mesh. Starting from a point cloud, the point that, after its removal, modify less the spectra of the heat operator kernel is removed; the operation can then be iterated until the final number of points is reached. The importance of each point is defined by starting from a discretization of the approximation of the heat operator, given by

$$(\boldsymbol{H}_t\boldsymbol{f})_i = \sum_j h_t(\boldsymbol{x}_i, \boldsymbol{x}_j)\boldsymbol{f}(\boldsymbol{x}_j)$$

where $h_t(\boldsymbol{x}, \boldsymbol{y})$ is the head kernel, $\boldsymbol{f}(\boldsymbol{x}_i)$ is a function defined on the surface, and $(\boldsymbol{H}_t)_{ij} = h_t(\boldsymbol{x}_i, \boldsymbol{x}_j)$ are the entries of the heat kernel matrix. It is possible to show that heat kernels can be expressed as the products of two functions in possibly infinite space:

$$h_t(\boldsymbol{x}, \boldsymbol{y}) = \sum_{i=0}^{\infty} e^{-\lambda_i t} u_i(\boldsymbol{x})\, u_i(\boldsymbol{y}) = \boldsymbol{\varphi}_t^T(\boldsymbol{x})\, \boldsymbol{\varphi}_t(\boldsymbol{y})$$

where $\boldsymbol{\varphi}(\boldsymbol{x})$ is a vector which ith component is given by $\sqrt{e^{-\lambda_i t}}\, u_i(\boldsymbol{x})$. $(\lambda_i, u_i(\boldsymbol{x}))$ is the eigenvalue-eigenfunction pair of the Laplace-Beltrami operator. The entries of the heat matrix can then be computed as $(\boldsymbol{H})_{ij} = \boldsymbol{\varphi}^T(\boldsymbol{x}_i)\, \boldsymbol{\varphi}(\boldsymbol{x}_j)$. The authors demonstrated that the importance of each point depends on the quantity

$$s(\boldsymbol{x}) = 1 - \frac{\boldsymbol{h}^T \boldsymbol{H}^{-1} \boldsymbol{h}}{h(\boldsymbol{x}, \boldsymbol{x})}$$

where $(\boldsymbol{h})_i = h(\boldsymbol{x}, \boldsymbol{x}_i)$. The correspondence between the spectrum and $s(\boldsymbol{x})$ is shown in Fig. 3.14: on the left the values of the eigenvalues for the six black dots on the right figure are plotted. On the right, it is also shown the value of $s(\boldsymbol{x})$ (low values in blue and high in red). As it is possible to observe if the triangle point is added (high $s(\boldsymbol{x})$), the values of the

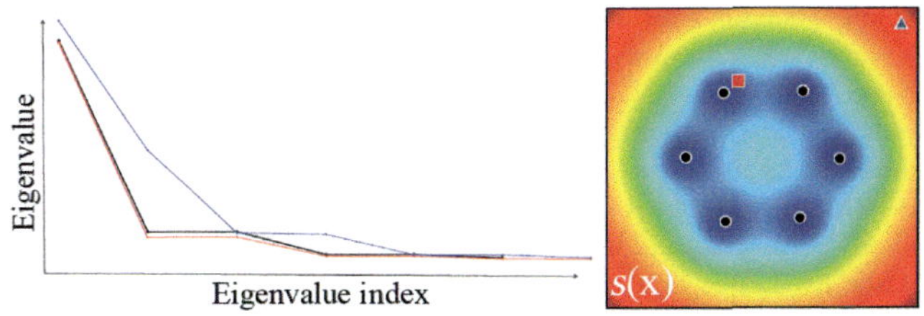

Fig. 3.14 Effect on the spectrum of the heat kernel when a point is added [59]. *(Credit: https://dl.acm.org/citation.cfm?id=1866190.)*

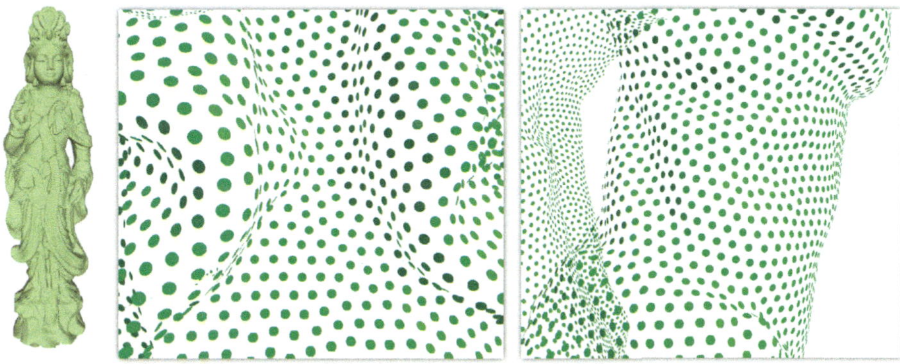

Fig. 3.15 Example of uniform resampling [59]. *(Credit: https://dl.acm.org/citation.cfm?id=1866190.)*

eigenvalues change a lot, while if the square point is added, the spectrum is similar to the initial one. An example of the uniform resampling using the sampling based on the spectral analysis is shown in Fig. 3.15.

3.7 Summary

In this chapter an overview of the sampling strategies for free-form surfaces has been presented. The sampling strategies were organized by the surface representation: lattice data, parametric function, and triangular mesh. The sampling strategies of known surfaces are first analyzed, which refer to nominal designed surface. Adaptive sampling of height map surfaces is then investigated, and finally the mesh decimation algorithms based on mesh and spectral methods were reviewed.

References

[1] Jiang X, Scott PJ, Whitehouse DJ, Blunt L, editors. Paradigm shifts in surface metrology. Part II. The current shift. Proceedings of the Royal Society of London A: Mathematical, Physical and Engineering Sciences; 2007.

[2] Jiang XJ, Whitehouse DJ. Technological shifts in surface metrology. CIRP Ann 2012;61(2):815–36.

[3] Unser M. Sampling-50 years after Shannon. Proc IEEE 2000;88(4):569–87.
[4] Zhao F, Xu X, Xie SQ. Computer-aided inspection planning—the state of the art. Comput Ind 2009;60(7):453–66.
[5] Elkott DF, Elmaraghy HA, Elmaraghy WH. Automatic sampling for CMM inspection planning of free-form surfaces. Int J Prod Res 2002;40(11):2653–76.
[6] Wang J, Jiang X, Blunt LA, Leach RK, Scott PJ. Intelligent sampling for the measurement of structured surfaces. Meas Sci Technol 2012;23(8)085006.
[7] Ainsworth I, Ristic M, Brujic D. CAD-based measurement path planning for free-form shapes using contact probes. Int J Adv Manuf Technol 2000;16(1):23–31.
[8] Cho MW, Seo TI. Inspection planning strategy for the on-machine measurement process based on CAD/CAM/CAI integration. Int J Adv Manuf Technol 2002;19(8):607–17.
[9] Cho MW, Kim K. New inspection planning strategy for sculptured surfaces using coordinate measuring machine. Int J Prod Res 1995;33(2):427–44.
[10] Fisker R, Clausen T, Deichmann N, Ojelund H. Adaptive 3D scanning. Google Patents; 2014.
[11] Shannon CE. Communication in the presence of noise. Proc IRE 1949;37(1):10–21.
[12] Li Y, Gu P. Free-form surface inspection techniques state of the art review. Comput Aided Des 2004;36 (13):1395–417.
[13] Savio E, De Chiffre L, Schmitt R. Metrology of freeform shaped parts. CIRP Ann 2007;56(2):810–35.
[14] Pharr M, Jakob W, Humphreys G. Physically based rendering: from theory to implementation. Burlington: Morgan Kaufmann Publishers Inc.; 2016. 1266 p.
[15] Thompson SK, Seber GAF. Adaptive sampling. Wiley; 1996.
[16] Wang J, Leach RK, Jiang X. Advances in sampling techniques for surface topography measurement—a review. National Physical Laboratory; 2014.
[17] Woo TC, Liang R. Dimensional measurement of surfaces and their sampling. Comput Aided Design 1993;25(40):233–9.
[18] Woo TC, Liang R, Hsieh CC, Lee NK. Efficient sampling for surface measurements. J Manuf Syst 1995;14(5):345–54.
[19] Lee G, Mou J, Shen Y. Sampling strategy design for dimensional measurement of geometric features using coordinate measuring machine. Int J Mach Tools Manuf 1997;37(7):917–34.
[20] Capello E, Semeraro Q. The effect of sampling in circular substitute geometries evaluation. Int J Mach Tools Manuf 1999;39(1):55–85.
[21] Kim W-S, Raman S. On the selection of flatness measurement points in coordinate measuring machine inspection. Int J Mach Tools Manuf 2000;40(3):427–43.
[22] Fang K-T, Wang S-G. A stratified sampling model in spherical feature inspection using coordinate measuring machines. Stat Probab Lett 2001;51(1):25–34.
[23] Cho M, Lee H, Yoon G, et al. A computer-aided inspection planning system for on-machine measurement—part II: local inspection planning. KSME Int J 2004;18(8):1358–67.
[24] Cho M, Lee H, Yoon G, et al. A feature-based inspection planning system for coordinate measuring machines. Int J Adv Manuf Technol 2005;36(9-10):1078–87.
[25] Barari A, ElMaraghy HA, Knopf GK. Search-guided sampling to reduce uncertainty of minimum deviation zone estimation. J Comput Inf Sci Eng 2007;7(4):360–71.
[26] Barari A, ElMaraghy HA, Knopf GK. Evaluation of geometric deviations in sculptured surfaces using probability density estimation. In: Models for computer aided tolerancing in design and manufacturing, Berlin: Springer; 2007. p. 135–46.
[27] Barini EM, Tosello G, De Chiffre L. Uncertainty analysis of point-by-point sampling complex surfaces using touch probe CMMs: DOE for complex surfaces verification with CMM. Precis Eng 2010; 34(1):16–21.
[28] Pahk HJ, Jung MY, Hwang SW, Kim YH, Hong YS, Kim SG. Integrated precision inspection system for manufacturing of moulds having CAD defined features. Int J Adv Manuf Technol 1995; 10(3):198–207.
[29] ElKott DF, Veldhuis SC. Isoparametric line sampling for the inspection planning of sculptured surfaces. Comput Aided Des 2005;37(2):189–200.
[30] Shih CS, Gerhardt LA, Chu WC-C, Lin C, Chang C-H, Wan C-H, et al, editors. Non-uniform surface sampling techniques for three-dimensional object inspection. SPIE; 2008.

[31] Wang J, Jiang X, Blunt LA, Leach RK, Scott PJ. Efficiency of adaptive sampling in surface texture measurement for structured surfaces. J Phys Conf Ser 2011;311(1):012017.
[32] Pagani L, Scott PJ. Curvature based sampling of curves and surfaces. Comput Aided Geom Design 2018;59:32–48.
[33] Edgeworth R, Wilhelm RG. Adaptive sampling for coordinate metrology. Precis Eng 1999; 23(3):144–54.
[34] Li YF, Liu ZG. Method for determining the probing points for efficient measurement and reconstruction of freeform surfaces. Meas Sci Technol 2003;14(8):1280.
[35] Huang Y, Qian X. A dynamic sensing-and-modeling approach to three-dimensional point- and area-sensor integration. J Manuf Sci Eng 2006;129(3):623–35.
[36] Menq C, Yau H, Lai G. Automated precision measurement of surface profile in CAD-directed inspection. IEEE Trans Robot Autom 1992;8(2):268–78.
[37] Yau H-T, Menq C-H. An automated dimensional inspection environment for manufactured parts using coordinate measuring machines. Int J Prod Res 1992;30(7):1517–36.
[38] Lin T-Y. Characterisation, sampling and measurement variation of surface topography: a viewpoint from standardisation. University of Birmingham; 1993.
[39] Lin TY, Blunt L, Stout KJ. Determination of proper frequency bandwidth for 3D topography measurement using spectral analysis. Part I: isotropic surfaces. Wear 1993;166(2):221–32.
[40] Zhang YF, Nee AYC, Fuh JYH, Neo KS, Loy HK. A neural network approach to determining optimal inspection sampling size for CMM. Comput Integr Manuf Syst 1996;9(3):161–9.
[41] Jiang BC, Chiu S-D. Form tolerance-based measurement points determination with CMM. J Intell Manuf 2002;13(2):101–8.
[42] Barker R, Cox M, Forbes A, Harris P. Best practice guide no. 4 software support for metrology: discrete modelling and experimental data analysis. Teddington, UK: National Physical Laboratory; 2004.
[43] Piegl L, Tiller W. The NURBS Book. Springer-Verlag New York, Inc.; 1997.
[44] Delaunay B. Sur la sphere vide. Izvestia Akademia Nauk SSSR, VII Seria, Otdelenie Matematicheskii i Estestvennyka Nauk 1934;7:793–800.
[45] Cazals F, Giesen J. Delaunay triangulation based surface reconstruction: ideas and algorithms. INRIA; 2004.
[46] Sandwell DT. Biharmonic spline interpolation of GEOS-3 and SEASAT altimeter data. Geophys Res Lett 1987;14(2):139–42.
[47] Hernández-Mederos V, Estrada-Sarlabous J. Sampling points on regular parametric curves with control of their distribution. Comput Aided Geom Design 2003;20(6):363–82.
[48] do Carmo MP. Differential geometry of curves and surfaces. Prentice-Hall; 1976.
[49] Santner TJ, Williams BJ, Notz WI. The Design and Analysis of Computer Experiments. New York: Springer-Verlag; 2003.
[50] Akima H. A new method of interpolation and smooth curve fitting based on local procedures. J ACM 1970;17(4):589–602.
[51] Wong T-T, Luk W-S, Heng P-A. Sampling with Hammersley and Halton points. J Graph Tools 1997;2(2):9–24.
[52] Welch G, Bishop G. An introduction to the Kalman filter. J Proc SIGGRAPH 2001;8 (27599–3175):59.
[53] Garland M, Heckbert PS, editors. Surface simplification using quadric error metrics. Proceedings of the 24th annual conference on computer graphics and interactive techniques. New York, NY: ACM Press/Addison-Wesley Publishing Co; 1997.
[54] Lindstrom P, Turk G. Fast and memory efficient polygonal simplification. IEEE Vis 1998;279–86.
[55] Pagani L, Jiang X, Scott PJ. Investigation on the effect of sampling on areal texture parameters. Measurement 2018;128:306–13.
[56] Lorensen WE, Cline HE. Marching cubes: a high resolution 3D surface construction algorithm. ACM Comput Graph 1987;21(4):163–9.
[57] Cacciola F. Triangulated surface mesh simplification. CGAL user and reference manual. 4.9 ed. CGAL Editorial Board; 2016.
[58] Visual Computing Lab. Visualization and computer graphics library. Available from: https://github.com/cnr-isti-vclab/vcglib/; 2016.
[59] Öztireli AC, Alexa M, Gross M. Spectral sampling of manifolds. ACM Trans Graph 2010;29(6):168.

CHAPTER 4

Geometrical fitting of free-form surfaces

4.1 Introduction

A manufactured surface must be measured to guarantee its quality. The measured data of discrete points must be converted into a mathematical representation and imported back into the optical design to examine for the performance specifications. Geometrical fitting refers to the operation of fitting geometrical functions specifying the shapes of the non-ideal features to the sampled discrete point sets in accordance with specified criteria. In ISO 17450-1, it is also termed "association." The criteria of geometrical fitting give an objective for a characteristic and can set constraints. The constraints fix the values of the characteristics or set limits to the characteristics [1]. The purpose of geometrical fitting is to work out the underlying intrinsic characteristics such as the radius of a sphere or situation characteristics such as the center of a sphere.

4.2 Geometrical representations

Free-form surfaces are widely used in optical design to compensate for imaging aberrations. In optical design software like ZEMAX and CODE V, the standard representations are recommended [2]:

$$z(x,y)=\frac{\dfrac{x^2}{R_x}+\dfrac{y^2}{R_y}}{1+\sqrt{1-(1+k_x)\left(\dfrac{x}{R_x}\right)^2-(1+k_y)\left(\dfrac{y}{R_y}\right)^2}}+A_4x^4+B_4y^4+A_6x^6+B_6y^6+\cdots \quad (4.1)$$

It is called the biconic function. Here the power series are used to correct high-order aberrations. If the order of polynomials is large enough, they form a complete set of optical surface approximations of the required accuracies.

Unfortunately the power series are numerically unstable, because the surface approximations are achieved by heavy cancellation of the terms. The resulting least squares approximation and the Gram matrix will become heavily ill conditioned. Some researchers have addressed this issue by applying normalization of the basis, but this

* For more examples and practical implementation, check the website: www.digitalsurf.com/freeform.

Advanced Metrology
https://doi.org/10.1016/B978-0-12-821815-0.00004-6

method cannot solve the cancellation problem; hence the practical effectiveness is limited. As a consequence the Zernike polynomials can in turn be utilized for the sag correction [3]:

$$z'(\rho,\varphi) = \sum_{n=0}^{\infty}\sum_{m=1}^{n} c_n^m Z_n^m(\rho,\varphi) \tag{4.2}$$
$$= \sum\nolimits_{n=0}^{\infty}\sum\nolimits_{m=1}^{n} \alpha_{nm} R_n^m(\rho)\cos(m\varphi) + \sum\nolimits_{n=0}^{\infty}\sum\nolimits_{m=1}^{n} \beta_{nm} R_n^m(\rho)\sin(m\varphi)$$

with $R_n^m(\rho) = \sum\nolimits_{s=0}^{\frac{n-m}{2}} (-1)^s \dfrac{(n-s)!}{s!\left(\dfrac{n+m}{2}-s\right)!\left(\dfrac{n-m}{2}-s\right)!}\rho^{n-2s}$.

Here, ρ is the normalized radial coordinate, and the φ is the corresponding angular coordinate. The radial polynomial defined on a unit disk always has the value 1 at the pupil edge.

The Zernike polynomials are sag orthogonal to each other on a unit disk; thus they are numerically stable. Theoretically, they can represent arbitrary smooth surfaces without constraints to symmetry. As a result, Zernike polynomials are increasingly used in optical design and surface assessment. More importantly, it is interesting that the low-order Zernike terms are linked to the principal Seidel aberrations, as listed below (Table 4.1).

The Zernike polynomials are sag orthogonal. But in fact the imaging functionalities of optical components are more relevant to the surface slopes, instead of the sag heights, because the ray reflection or refraction at the surfaces depends directly on the surface normals. Moreover, from the viewpoint of fabrication, the tools are located normal to workpiece surface in precision grinding and polishing; thus the positioning adjustment and material removal are also along the surface normal directions. Henceforth the form

Table 4.1 Zernike polynomials and wave aberrations.

Zernike polynomials	Wave aberrations
$Z_1=1$	Piston
$Z_2=r\cos\varphi$	Tilt y
$Z_3=r\sin\varphi$	Tilt x
$Z_4=2r^2\text{-}1$	Defocus + piston
$Z_5=r^2\cos 2\varphi$	Astigmatism y
$Z_6=r^2\sin 2\varphi$	Astigmatism x
$Z_7=(3r^3\text{-}2r)\cos\varphi$	Coma y
$Z_8=(3r^3\text{-}2r)\sin\varphi$	Coma x
$Z_9=6r^4\text{-}6r^2+1$	Third-order spherical + defocus + piston
$Z_{16}=20r^6\text{-}30r^4+12r^2+1$	Fifth-order spherical + third-order spherical + defocus + piston
$Z_{25}=70r^8-140r^6+90r^4-20r^2+1$	Seventh-order spherical + fifth-order spherical + third-order spherical + defocus + piston

deviation along the normal vectors is of more concern than the deviation of sag height. As a consequence a slope-orthogonal representation, the Zernike difference polynomials, is developed [4] by a proper combination of Zernike polynomials.

$$D_n^m = Z_n^m - D_{n-2}^m$$

The Zernike polynomials with $n < 2$ or $n < m$ are assumed to be zero. The terms of the Zernike differences directly correspond to the transverse aberrations as a derivative of the wave aberration and therefore show good performance for the optimization of spot criteria. The sum of the squared coefficients is equal to the RMS value of the slope deviation. Thus this representation gives an interesting opportunity in the design to constrain the slope of a surface, which usually guarantees a smaller sensitivity against tolerances.

G Forbes proposed another slope-orthogonal representation, called the Q polynomials [5].

$$z(x,y) = \frac{r^2/R}{1+\sqrt{1-(r/R)^2}} \frac{1}{1+\sqrt{1-(r/R)^2}} \left[\frac{r^2}{a^2}\left(1-\frac{r^2}{a^2}\right) \sum_{n=0} a_n^0 Q_n^0 \left(\frac{r^2}{a^2}\right) + \sum_{m=1} \left(\frac{r}{a}\right)^m \sum_{n=0} \left[a_n^m \cos(m\theta) + b_n^m \sin(m\theta) \right] Q_n^m \left(\frac{r^2}{a^2}\right) \right] \tag{4.3}$$

In the equation, a is the maximum radial coordinate used for normalization.

This representation is more meaningful and advantageous in the tolerance design of complex optical components. The basic shape is spherical, instead of conicoidal; hence, it applies to mild aspheres only. From the viewpoint of manufacturing, any deviation from a circular symmetric shape affects the tool feed. If this deviation amount for the sag is directly expressed by the equation, it is an attractive advantage. On the other hand, it makes sense to keep the expansion contribution small to separate the first-order functionality and higher-order correction requirements in the optical design. Therefore a biconic basic shape is more beneficial from the viewpoint of metrology [6].

4.3 Algebraic fitting and geometrical fitting

In most of the commercial metrological software, only the deviation in the z direction is considered, which is called the *algebraic fitting* [7]. This approach is extensively applied because of its ease of implementation. However, its definition of error distance does not coincide with measurement guidelines. The estimated fitting parameters will be biased, especially in the case where there are errors in the explanatory variables [8]. Consequently, researchers have developed the *orthogonal distance fitting* (also termed *geometric fitting*) method (Fig. 4.1). This technique intends to minimize the sum of the squared orthogonal distances from the measurement points to the nominal surface.

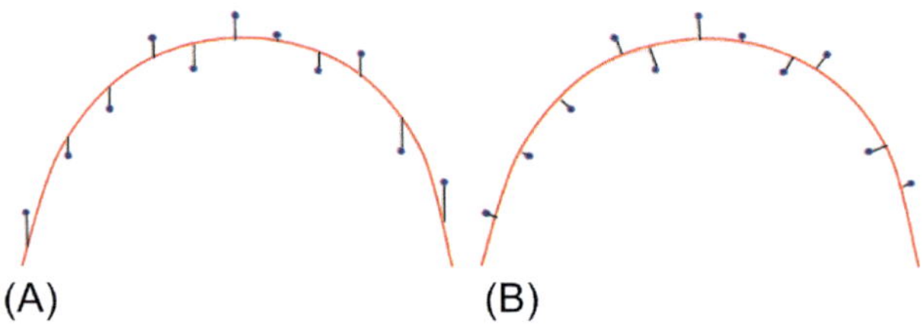

Fig. 4.1 Comparison of algebraic and geometric fitting: (A) Algebraic fitting, (B) orthogonal distance fitting.

It can effectively overcome the bias problem of the algebraic fitting. Most researchers paid attention to the fitting of simple geometries like quadric surfaces, whose orthogonal distances can be directly calculated via closed-form methods. But the orthogonal distances are not so straightforward to find for free-form surfaces, and the computational cost may be dramatically increased if the orthogonal projection points with recursive techniques are calculated.

4.3.1 Geometric fitting for explicit functions

Suppose the explicit function is given as $z=f(x, y; \mathbf{a})$. The intrinsic characteristics $\mathbf{a}$, namely, the shape parameters, an optimal rotation matrix $\mathbf{R}$, and a translation vector $\mathbf{t}$, need to be optimized to minimize the sum of squared orthogonal distances from all the measurement points to the nominal function:

$$\min_{\mathbf{a},\mathbf{R},\mathbf{t}} \sum_{i=1}^{N} \|\mathbf{R}\mathbf{p}_i + \mathbf{t} - \mathbf{q}_i\|^2 = \min_{\mathbf{a},\mathbf{R},\mathbf{t}} \sum_{i=1}^{N} \|\mathbf{p}'_i - \mathbf{q}_i\|^2 \tag{4.4}$$

Here, $\mathbf{p}'_i = \mathbf{R}\mathbf{p}_i + \mathbf{t} = [x'_i, y'_i, z'_i]^T$ is an arbitrary measurement point after motion, and $\mathbf{q}_i$ is the closest point on the template corresponding to $\mathbf{p}'_i$.

The coordinates of $\mathbf{q}_i$ are represented as

$$\mathbf{q}_i = [x'_i + \xi_i, y'_i + \zeta_i, f(x'_i + \xi_i, y'_i + \zeta_i; \mathbf{a})]^T$$

and the weighting technique is incorporated. Then the error metric in Eq. (4.4) can be rewritten as

$$E = \sum_{i=1}^{N} w_i^2 [z'_i - f(x'_i + \xi_i, y'_i + \zeta_i; \mathbf{a})]^2 + \sum_{i=1}^{N} u_i^2 \xi_i^2 + \sum_{i=1}^{N} v_i^2 \zeta_i^2 = \mathbf{G}^T\mathbf{G} + \mathbf{H}^T\mathbf{H} \tag{4.5}$$

with

$$\mathbf{G} \in \mathfrak{R}^{N\times 1} : g_i = w_i [z'_i - f(x'_i + \xi_i, y'_i + \zeta_i; \mathbf{a})], \ i = 1, \ldots, N$$

and

$$\mathbf{H}\in\mathfrak{R}^{2N\times 1}:h_j=\begin{cases}-u_{(j+1)/2}\xi_{(j+1)/2}, & j=1,3,\ldots,2N-1\\ -v_{j/2}\zeta_{j/2}, & j=2,4,\ldots,2N\end{cases}$$

If regarding $\{\xi_i\}$ and $\{\zeta_i\}$ as unknown variables, we can solve them simultaneously with the intrinsic characteristics $\mathbf{a}\in\mathfrak{R}^{p\times 1}$ and the six motion parameter θ_x, θ_y, θ_z, t_x, t_y, and t_z.

Boggs et al. [9] proposed a method to solve this nonlinear least squares problem based on the Levenberg-Marquardt algorithm. Denoting the shape and motion parameters with a vector $\mathbf{b}\in\mathfrak{R}^{(p+6)\times 1}$ and denoting $\{\xi_i\}$ and $\{\zeta_i\}$ with a vector $\boldsymbol{\beta}\in\mathfrak{R}^{2N\times 1}$ mean Eq. (4.5) can be solved iteratively by

$$\begin{bmatrix}\mathbf{J} & \mathbf{V}\\ 0 & \mathbf{D}\\ \lambda^{1/2}\mathbf{S} & 0\\ 0 & \lambda^{1/2}\mathbf{T}\end{bmatrix}\begin{bmatrix}\delta\mathbf{b}\\ \delta\boldsymbol{\beta}\end{bmatrix}=-\begin{bmatrix}\mathbf{G}\\ \mathbf{H}\\ 0\\ 0\end{bmatrix} \tag{4.6}$$

with

$$\mathbf{J}\in\mathfrak{R}^{N\times(p+6)}:J_{ij}=\partial g_i/\partial b_j, i=1,\ldots,N,\ j=1,\ldots,p+6$$

$$\mathbf{V}\in\mathfrak{R}^{N\times 2N}:V_{ij}=\partial g_i/\partial\beta_j=-w_i\partial f_i/\partial\beta_j,\quad i=1,\ldots,N, j=1,\ldots,2N$$

and

$$\mathbf{D}\in\mathfrak{R}^{2N\times 2N}:\mathbf{D}=-diag\{u_1, v_1, u_2, v_2, \cdots, u_N, v_N\}$$

Here, $\lambda>0$ is a damping factor, and $\mathbf{S}\in\mathfrak{R}^{(p+6)\times(p+6)}$ and $\mathbf{T}\in\mathfrak{R}^{2N\times 2N}$ are two scaling matrices. Employing damping terms can greatly improve the numerical stability of the system and guarantee the convergence of its solution. The damping factor λ needs to be selected properly according to the specific problem. If the value of λ is too large, the step length of the updates $\delta\mathbf{b}$ and $\delta\boldsymbol{\beta}$ will be very small, consequently leading to a rather slow convergence speed; on the other hand, a λ too small cannot significantly affect the system's numerical property. The normal function of Eq. (4.6) is

$$\begin{bmatrix}\mathbf{J}^T\mathbf{J}+\lambda\mathbf{S}^2 & \mathbf{J}^T\mathbf{V}\\ \mathbf{V}^T\mathbf{J} & \mathbf{V}^T\mathbf{V}+\mathbf{D}^2+\lambda\mathbf{T}^2\end{bmatrix}\begin{bmatrix}\delta\mathbf{b}\\ \delta\boldsymbol{\beta}\end{bmatrix}=-\begin{bmatrix}\mathbf{J}^T\mathbf{G}\\ \mathbf{V}^T\mathbf{G}+\mathbf{DH}\end{bmatrix},$$

so

$$\delta\mathbf{b}=-\left(\mathbf{J}^T\mathbf{MJ}+\lambda\mathbf{S}^2\right)^{-1}\mathbf{J}^T\mathbf{M}\left(\mathbf{G}-\mathbf{VE}^{-1}\mathbf{DH}\right) \tag{4.7}$$

and

$$\delta\boldsymbol{\beta}=-\mathbf{E}^{-1}\left[\mathbf{V}^T\mathbf{M}\left(\mathbf{G}+\mathbf{J}\delta\mathbf{b}-\mathbf{VE}^{-1}\mathbf{DH}\right)+\mathbf{DH}\right] \tag{4.8}$$

In the equation, $\mathbf{M}=\mathbf{I}-\mathbf{V}(\mathbf{V}^T\mathbf{V}+\mathbf{E}^2)^{-1}\mathbf{V}^T$. In fact, it is a diagonal matrix;

$$\mathbf{M}=diag\left\{1\Big/\left[1+\left(w_i\frac{\partial f_i}{\partial \xi_i}\right)^2\Big/E_{2i-1}+\left(w_i\frac{\partial f_i}{\partial \zeta_i}\right)^2\Big/E_{2i}\right]\right\},\ i=1,\ldots,N$$

and $\mathbf{E}=\mathbf{D}^2+\lambda\mathbf{T}^2$ is diagonal as well.

4.3.2 Geometric fitting for implicit/parametric functions

The previous algorithm works well for the orthogonal distance fitting (ODF) of explicit surfaces. However, explicit functions are not always available for free-form surfaces. Parametric representations are more common, for instance, NURBS surfaces:

$$\begin{cases} X=X(u,v) \\ Y=Y(u,v) \\ Z=Z(u,v) \end{cases}$$

Here, u and v are called the location parameters. In most situations the Cartesian coordinates x, y, and z are nonlinear with respect to the location parameters u and v. Additionally the number of measurement points in practice is probably very large, sometimes over a million points, which makes the size of the observation matrix increase dramatically. Then the computational cost and memory usage will be rather tedious; therefore it is practical to use a nested iteration scheme—to solve the foot-point parameters alternately with the motion parameters [10]:

$$\min_{\mathbf{m}}\sum_{i=1}^{N}\min_{u_i,v_i}\|\mathbf{p}_i-\mathbf{q}_i\|^2 \tag{4.9}$$

That means firstly locating the closest projection point corresponding to each measurement point in the inner iteration and then working out the optimal motion parameters and intrinsic characteristics at the outer iteration.

Some closed-form techniques have been developed to calculate the orthogonal projection points for simple geometries [8], whereas for general-shaped parametric functions, the iterative Newton-Raphson algorithm can be adopted. For example, the two-stage approach can be employed to solve the point projection problem of NURBS surfaces.

After all the projection points $\{\mathbf{q}_i\}$ are obtained, the motion and shape parameters will be updated subsequently. We define $\mathbf{g}\in\mathbb{R}^{3N\times 1}$,

$$g_k=\begin{cases} x_i'-X_i & k=3i-2 \\ y_i'-Y_i & k=3i-1, \quad i=1,\ldots,N \\ z_i'-Z_i & k=3i \end{cases}$$

where $[X_i, Y_i, Z_i]^T=\mathbf{q}_i$ is the projection point associated with $\mathbf{p}_i$.

Hence Eq. (4.9) can be rewritten as

$$\min_{\mathbf{m}} \sum_{k=1}^{3N} g_k^2 = \min_{\mathbf{m}} \mathbf{g}^T \mathbf{g} \tag{4.10}$$

It can be solved with the Levenberg-Marquardt algorithm:

$$\left[\left(\frac{\partial \mathbf{g}}{\partial \mathbf{m}} \right)^T \left(\frac{\partial \mathbf{g}}{\partial \mathbf{m}} \right) + \lambda \mathbf{I} \right] \delta \mathbf{m} = -\left(\frac{\partial \mathbf{g}}{\partial \mathbf{m}} \right)^T \mathbf{g} \tag{4.11}$$

The three rows of the Jacobian matrix $\mathbf{J} = \partial \mathbf{g} / \partial \mathbf{m} \in \mathbb{R}^{3N \times (p+6)}$ associated with the point $\mathbf{p}_i$ are

$$\begin{bmatrix} J_{3i-2} \\ J_{3i-1} \\ J_{3i} \end{bmatrix} = \frac{\partial \mathbf{p}_i'}{\partial \mathbf{m}} - \frac{\partial \mathbf{q}_i}{\partial \mathbf{u}_i} \frac{\partial \mathbf{u}_i}{\partial \mathbf{m}} \tag{4.12}$$

Here, $\mathbf{u}_i = [u_i, v_i]^T$ is the foot-point parameters of $\mathbf{q}_i$. When the measurement point $\mathbf{p}_i$ moves, its closest point $\mathbf{q}_i$ moves as well, so that the corresponding foot-point parameters $\mathbf{u}_i$ shall be updated simultaneously. That means the foot-point parameters of $\{\mathbf{q}_i\}$ are relevant with the motion parameters $\mathbf{m}$ in each iteration. Here the parameter dependency $\partial \mathbf{u}_i / \partial \mathbf{m}$ will be derived as follows. Each pair of points is nearest to each other, and the following relation always holds true:

$$\frac{\partial}{\partial \mathbf{u}_i} (\mathbf{p}_i - \mathbf{q}_i)^T (\mathbf{p}_i - \mathbf{q}_i) = -2 \left(\frac{\partial \mathbf{q}_i}{\partial \mathbf{u}_i} \right)^T (\mathbf{p}_i - \mathbf{q}_i) = \mathbf{0}.$$

For the sake of clarity, the subscript i is omitted, and the partial derivatives $\partial \mathbf{q} / \partial \mathbf{u}$ are written as $\mathbf{q}_{\mathbf{u}}$, so that

$$\frac{\partial}{\partial \mathbf{m}} \left(-\mathbf{q}_{\mathbf{u}}^T (\mathbf{p} - \mathbf{q}) \right) = \mathbf{q}_{\mathbf{u}}^T \mathbf{q}_{\mathbf{u}} \mathbf{u}_{\mathbf{m}} - \begin{bmatrix} \mathbf{q}_{uu}^T (\mathbf{p} - \mathbf{q}) & \mathbf{q}_{uv}^T (\mathbf{p} - \mathbf{q}) \\ \mathbf{q}_{uv}^T (\mathbf{p} - \mathbf{q}) & \mathbf{q}_{vv}^T (\mathbf{p} - \mathbf{q}) \end{bmatrix} \mathbf{u}_{\mathbf{m}} - \mathbf{q}_{\mathbf{u}}^T \mathbf{p}_{\mathbf{m}} = 0.$$

We obtain

$$\mathbf{u}_{\mathbf{m}} = \left\{ \mathbf{q}_{\mathbf{u}}^T \mathbf{q}_{\mathbf{u}} - \begin{bmatrix} \mathbf{q}_{uu}^T (\mathbf{p} - \mathbf{q}) & \mathbf{q}_{uv}^T (\mathbf{p} - \mathbf{q}) \\ \mathbf{q}_{uv}^T (\mathbf{p} - \mathbf{q}) & \mathbf{q}_{vv}^T (\mathbf{p} - \mathbf{q}) \end{bmatrix} \right\}^{-1} \mathbf{q}_{\mathbf{u}}^T \mathbf{p}_{\mathbf{m}} \tag{4.13}$$

Substituting Eq. (4.13) into (4.12), then the increment of the solution in Eq. (4.11) will be obtained.

A necessary condition for Eq. (4.11) to converge at a local minimum is that the observation matrix $\mathbf{A} = \mathbf{J}^T \mathbf{J} + \lambda \mathbf{I}$ is positive definite, so that the damping factor λ can be selected according to the smallest singular value of the matrix $\mathbf{J}^T \mathbf{J}$.

For a uniform NURBS or B-spline surface, the explicit representations of the second-order derivatives in Eq. (4.13) can be obtained. They are even simpler to calculate than the first-order derivatives and will be reserved when the template is represented as a

NURBS or B-spline surface, thus leading to a damped Newton minimization. However, for most general-shaped parametric surfaces, the second-order derivatives are rather tedious to be derived from the complex surface functions. Therefore they will be neglected when the residuals are very small and the surface is very smooth, that is, $\begin{bmatrix} \mathbf{q}_{uu}^T(\mathbf{p}-\mathbf{q}) & \mathbf{q}_{uv}^T(\mathbf{p}-\mathbf{q}) \\ \mathbf{q}_{uv}^T(\mathbf{p}-\mathbf{q}) & \mathbf{q}_{vv}^T(\mathbf{p}-\mathbf{q}) \end{bmatrix}$ is much smaller than $\mathbf{q}_\mathbf{u}^T\mathbf{q}_\mathbf{u}$. To guarantee convergence in this case, the damping factor needs to be adapted carefully.

Due to the slow convergence rate and local minimum problem of ICP, the *derivative-based algorithms* are instead employed to fit smooth free-form surfaces. Here, derivative information is needed for calculating the increment of the solution; therefore a continuous representation should be provided for the nominal template. If the template is in a form of discrete points or a mesh, a reconstruct procedure will be undertaken.

4.3.2.1 Experimental validation

Fig. 4.2A shows a bearing surface of a total knee replacement bearing model represented using NURBS. A nearly spherical part on this joint replacement is measured using a Zeiss PRISMO CMM, as denoted with blue dots in Fig. 4.2B.

The LM algorithm is also applied for the final optimization of the knee joint. After seven iterations and running time of 7.495 s, the program converges. The final fitting result is shown in Fig. 4.3. The resulting Sa, Sq, and St of the residual map are 5.2 μm, 6.2 μm, and 34.3 μm, respectively.

4.4 Robust estimators

4.4.1 Statistical fundaments

Before introducing robust regression techniques, two critically important terminologies are defined:

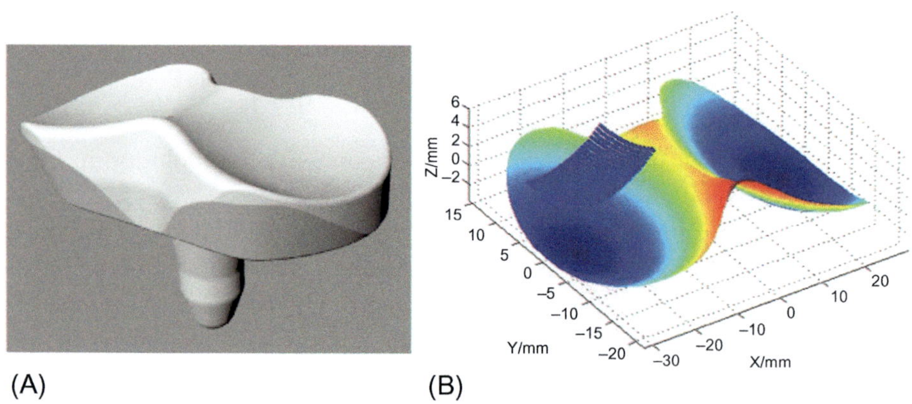

Fig. 4.2 Matching knee joint replacement: (A) Knee joint model, (B) initial position.

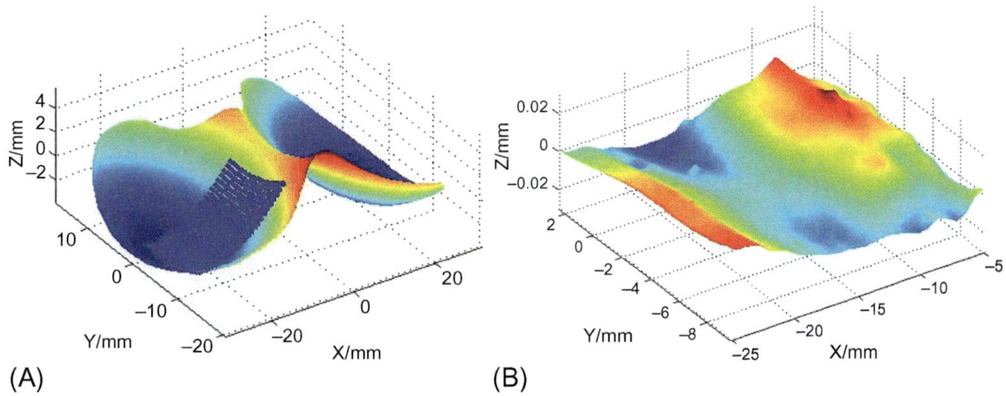

Fig. 4.3 Final fitting result of knee joint: (A) Final position, (B) final residual.

Robustness—a statistical procedure is described as robust if it is not very sensitive to departure from the assumptions on which it depends [11].

Breakdown point—the breakdown point of an estimator is the smallest fraction or percentage of discrepant data (i.e., outliers or data grouped at the extreme end of the tail of the distribution) that the estimator can tolerate without producing an arbitrary result [12]. It is a common measure of the robustness of an estimator.

In practice, most solutions are based on the least squares method as follows:

$$E = \sum_{i=1}^{N} \rho(r_i) = \sum_{i=1}^{N} r_i^2 \tag{4.14}$$

In the equation, ρ is called the *cost function* (also called the *loss function*), and r_i is the residual error associated with the datum $\mathbf{x}_i = [x_i, y_i]^T$.

The least squares method is applied very extensively because of its ease of implementation. More importantly, it is unbiased when the measurement error obeys a Gaussian distribution (Normal distribution), as asserted by the Gauss-Markov theorem [13].

However, the assumption of normality does not always hold true. In practice, deviations such as gross errors (outliers), rounding and grouping errors, and departure from an assumed sample distribution will take place because of defects, improper operation, external influence, and some functional properties, for example, lubrication and friction.

Gross errors are data severely deviating from the pattern set by the majority of the data. They are the most dangerous type of errors, and one single outlier could make the least squares fitting result break down. It is worth noting that outliers do not necessarily mean bad data or wrong data. Outliers should never be blindly discarded. They are usually analyzed separately with the clean data. Rounding and grouping errors result from the inherent inaccuracy in the collection and recording of data that is usually rounded, grouped, or even coarsely classified.

The departure from an assumed model means that real data probably deviated from the often-assumed normal distribution. The actual distribution may be skewed or with a long tail. Estimators are required to be consistent with the error distributions.

4.4.2 Commonly used error metrics

4.4.2.1 M-Estimators

M-estimators are generalizations of the usual maximum likelihood estimates. M-estimators will be very robust when formulated properly and more efficient than other robust regression methods for large samples. The cost function must be strictly convex to ensure the uniqueness of the solution.

Huber [14] proposed a robust estimator, now called the *Huber estimator*:

$$\rho(r;\theta)=\begin{cases} r^2/2 & |r|\le c \\ c|r|-c^2/2 & |r|>c \end{cases} \tag{4.15}$$

The performance of this estimator relies on the value of the threshold c. When $c\to\infty$, it reduces to the least squares estimator; as $c\to 0$ a l_1 norm estimator is obtained. In practice, $c=2$ MAD is recommended, where MAD is the median absolute deviation.

Mosteller and Tukey [15] proposed the *biweight estimator* (also called Tukey's bisquare estimator):

$$\rho(r;\theta)=\begin{cases} \dfrac{c^2}{6}\left\{1-\left[1-\left(\dfrac{r}{c}\right)^2\right]^3\right\} & |r|\le c \\ c^2/6 & |r|>c \end{cases} \tag{4.16}$$

$c=7$ MAD is recommended. The biweight and the Huber estimators behave similarly for most of the distribution, except in the very center and at the extreme tails of the distribution. For larger errors the bisquare estimator tapers off.

The *fair estimator* is defined as follows [11]:

$$\rho(r;\theta)=c^2\left[\frac{|r|}{c}-\log\left(1+\frac{|r|}{c}\right)\right] \tag{4.17}$$

with $c=2$ MAD. It is a third-order differentiable function everywhere. Its performance is between the least squares and the least absolute value regression.

Compared with the least squares method, these estimators pay less attention to the gross error. In fact, they can be regarded as the iterative reweighted least squares: $\rho(r;\theta)=wr^2$.

The weighting parameters are assigned inversely proportional to the residual errors; that is, smaller weighting parameters are assigned onto larger residuals and vice versa.

Additionally, many new estimators have been developed based on the M-estimators, for example, GM-estimators and MM-estimators [12].

4.4.2.2 L-Estimators

L-estimators are linear combinations of order statistics and firstly proposed by Lloyds [11]. The *k*th order statistic of a statistical sample is equal to its *k*th smallest value.

The first L-estimator is the least absolute values (LAV), also known as l_1 norm, which intends to minimize the sum of the absolute deviations. This will be introduced in the l_p norm estimators later.

Another famous L-estimator is the least median of squares (LMS) proposed by Rousseeuw [16]:

$\theta = \arg\min Median(r_i^2)$.

It is resistant to outliers, and the resulting *breakdown point* may be as high as 0.5, which is the highest possible value of all regression techniques. However, it is very complex to solve and two times slower than the ordinary least squares.

Rousseeuw developed a least trimmed squares (LTS) regression method [16], which minimizes the sum of the trimmed squared residuals:

$$E = \sum_{i=1}^{q} r_i^2$$

where $q = N(1-\alpha)+1$ is the number of data points included in the error metric and α is the proportion of trimming. The previous case is sometimes called the α-least trimmed squares. Its breakdown point is α, and it is more than 10 times slower than ordinary least squares. It is so slow that LTS is not commonly applied in practice.

4.4.2.3 R-Estimators

In *R-estimators* the residuals are weighted based on their ranks:

$$\theta = \arg\min \sum_{i=1}^{N} a_N(R_i) r_i \tag{4.18}$$

where R_i is the rank of the *i*th residual in $\{r_1, r_2, \ldots, r_N\}$ and a_N is a nondecreasing score function satisfying $\sum_{i=1}^{N} a_N(i) = 0$. Many forms have been proposed for the score function, such as the following [12]:

- Wilcoxon score $a_N(i) = i - \frac{N+1}{2}$
- Median score $a_N(i) = \sin\left(i - \frac{N+1}{2}\right)$
- Van der Waerden score $a_N(i) = \Phi^{-1}\left(\frac{i}{n+1}\right)$, where Φ is the normal probability density function.

An advantage of R-estimators over others is that they are scale invariant. Nevertheless the choice of the optimal score function is not clear. Additionally, they are not easy to solve, and their breakdown points never achieve more than 0.20.

4.4.2.4 l_p Norm estimators

The ordinary least squares regression can be extended into the l_p norm [17]:

$$E_p = \frac{1}{N}\left(\sum_{i=1}^{N}|r_i|^p\right)^{1/p}, \quad p > 0 \tag{4.19}$$

The cost function $\rho_p = |r|^p$ and the corresponding probability density function (PDF) $\mathrm{PDF}_p = \frac{\exp\{-|r|^p\}}{A_p}$ for different p values are depicted in Fig. 4.4.

It can be seen that the l_p norm with a smaller p value assigns a smaller cost function for the wild points, therefore, and is more robust against outliers. It is good at dealing with long-tailed errors. If the errors are uniformly distributed or with a sharply truncated PDF, $p = \infty$ will be a good choice.

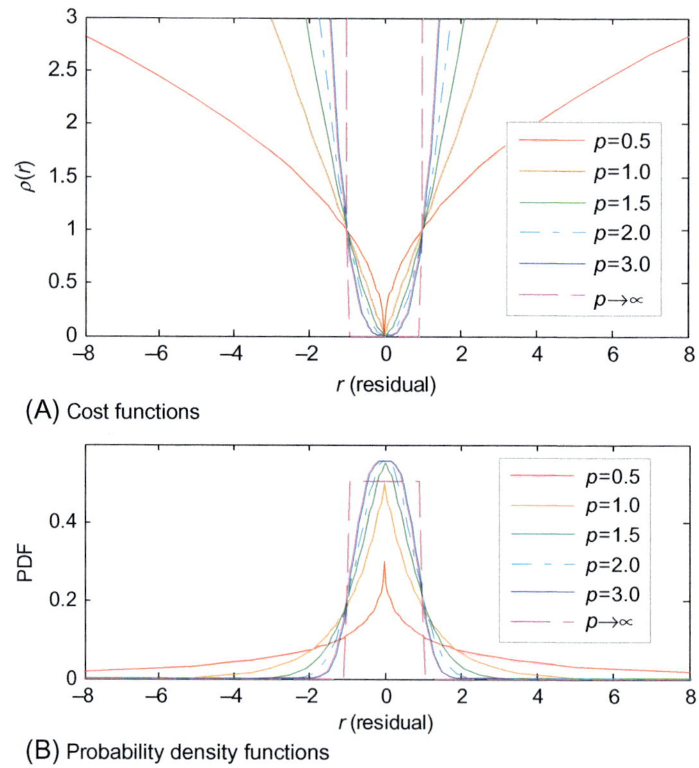

Fig. 4.4 Comparison of different p values: (A) Cost functions, (B) probability density functions.

When $p < 1$, the objective function is concave, thus, not of our interest. If the errors are skewly distributed, it may be useful.

If $p = 1$, it is the well-known least absolute values (also called minsummod) mentioned in the preceding text in L-estimators. The object function is not strictly convex; thus the solution cannot be guaranteed to be unique. Furthermore, it is not differentiable at $r = 0$, so the l_1 norm cannot be directly solved based on the derivatives. A common approach is to transform it into a minimization problem with inequality constraints and solve it with linear programming techniques.

However, linear programming methods are not straightforward to use. To improve the calculation efficiency, some approximates of the l_1 norm have been proposed, such as

$$\rho(r) = \sqrt{r^2 + \varepsilon^2} \text{ and } \rho(r) = \begin{cases} r^2/2 & |r| < \varepsilon \\ \varepsilon(|r| - \varepsilon/2) & \text{otherwise} \end{cases}.$$

When the threshold $\varepsilon \to 0$, the approximations approach the l_1 norm.

The l_p norm with $1 < p < 2$ goes to the category of M-estimators. They are smooth and differentiable with respect to the residual r. Most of the authors handled this problem using Newton or quasi-Newton algorithms. Cooper and Mason transformed it into a reweighed least squares problem [18].

At the kth iteration the weighting is assigned as

$$w_i^{(k)} = \left(\frac{w_i^{(k-1)}}{\left| r_i^{(k-1)} \right|} \right)^{\frac{2-p}{3-p}} \tag{4.20}$$

In the equation, $w_i^{(k-1)}$ and $r_i^{(k-1)}$ are the weighting and residual error at the previous iteration. When k becomes larger, $\frac{|r_i^{(k)}|}{|r_i^{(k-1)}|} \to 1$ and $\frac{w_i^{(k)}}{w_i^{(k-1)}} \to 1$, so $w_i^{(k)} \to |r_i^{(k)}|^{p-2}$, that is, $w_i^{(k)}(r_i^{(k)})^2 \to |r_i^{(k)}|^p$. To avoid infinite or very large weights caused from Eq. (4.20), an upper bound is set for them. After all the weightings are worked out, they are then normalized. At the first iteration the weightings are initialized as $w_i^{(0)} = 1/N$; here, N is the number of data points.

$p = 2$ is the ordinary least squares problem. More details will not be presented here.

As summary the relative performance of different robust estimators is listed in Table 4.2, and [12].

4.4.3 Surrogate functions

The l_1 norm is robust because it pays less attention to the wild points and concentrates on the vast majority of the data points. However, it has discontinuous derivatives and is consequently difficult to solve. Hunter and Lange [19] proposed an algorithm based on the majorize-minimize theory. A continuous surrogate function is adopted to approximate the initial l_1 norm objective function.

Table 4.2 Comparison of various estimators.

Estimator	Breakdown point	Bounded influence	Asymptotic efficiency
Ordinary least Squares	0	No	100
Least absolute value	0	Yes	64
Least median squares	0.5	Yes	37
Least trimmed squares	α	Yes	8
Least trimmed median	0.5	Yes	66
Bounded R-estimator	<0.2	Yes	90
M-estimator (Huber, Biweight)	0	No	95
GM-estimator (Mallows, Schweppe)	$1/(p+1)$	Yes	95
GM-estimator (Schweppe one-step estimator)	0.5	Yes	95
S-estimator	0.5	Yes	33
GS-estimator	0.5	Yes	67
MM-estimator	0.5	Yes	75
Generalized estimator	0.5	Yes	95

For the l_1 norm fitting problem,

$$E = \sum_{i=1}^{N} \|\mathbf{R}\mathbf{p}_i + \mathbf{t} - \mathbf{q}_i\| = \sum_{i=1}^{N} r_i \tag{4.21}$$

A small perturbation $\varepsilon > 0$ is introduced:

$$E^{\varepsilon} = \sum_{i=1}^{N} \rho^{\varepsilon}(r_i) = \sum_{i=1}^{N} [r_i - 2\varepsilon \ln(\varepsilon + |r_i|)] \tag{4.22}$$

It turns out that, for a given residual $r^{(k)}$ at the kth iteration, $\rho^{\varepsilon}(r)$ is majorized by $\psi^{\varepsilon}\left(r|\, r^{(k)}\right) = \frac{r^2}{\varepsilon + |r^{(k)}|} + c$, that is, $\begin{cases} \psi^{\varepsilon}\left(r^{(k)}|\, r^{(k)}\right) = \rho^{\varepsilon}\left(r^{(k)}\right) \\ \psi^{\varepsilon}\left(r|\, r^{(k)}\right) \geq \rho^{\varepsilon}(r) \text{ for all } r \end{cases}$; thus the new objective function will be

$$E^{\varepsilon} = \sum_{i=1}^{N} \psi^{\varepsilon}(r_i) = \sum_{i=1}^{N} \left[\frac{r_i^2}{\varepsilon + |r_i^{(k)}|} + c_i \right] \tag{4.23}$$

It can be proved that $|E^{\varepsilon} - E| \leq -2\varepsilon N \ln \varepsilon$; hence the constant ε can be determined according to the overall error threshold $\tau = -2\varepsilon N \ln \varepsilon$.

Obviously the constants $\{c_i\}$ do not affect the solution. Therefore the l_1 norm optimization becomes a reweighted least squares problem with $w_i = 1/(\varepsilon + |r_i^{(k)}|)$. Now, recall the weights of the ODF algorithm. We set them as $\begin{cases} u_i = aw_i \\ v_i = bw_i \end{cases}, i = 1, 2, \ldots, N$. Here, a, $b \geq 0$ is utilized to measure the relative influence of the lateral and vertical deviations onto the error metric. In practice, if there is noise in all the x, y, and z coordinates, we set $a, b > 0$. On the contrary, if only z coordinates contain errors, we set $a = b = 0$; that is, algebraic fitting will be adopted.

4.4.3.1 Experimental validation

The Matlab built-in function *peaks* is adopted as a nominal surface:

$z = 4peaks\left(\frac{x}{20}, \frac{y}{20}\right)$.

It is represented using a bicubic NURBS surface with 18×18 control points, as shown in Fig. 4.5.

Gaussian noise of $N(0, (0.6\ \mu m)^2)$ and outliers are added onto the data points (Fig. 4.6). The Monte Carlo simulation is employed, and the fitting procedure is run, 15 iterations 300 times.

The dependency between the foot-point parameters and motion parameters is derived from the closest point constraints. Three algorithms are compared here: the robust orthogonal distance fitting, the robust singular value decomposition, and the l_2 norm orthogonal distance fitting. The corresponding fitting bias and uncertainty the rotation angles and translation components are listed in Table 4.3. It is obvious that the SVD method obtains the worst result. At each iterate, it endeavors to minimize the distances between the corresponding point pairs. However, the projection point is already the closest one on the

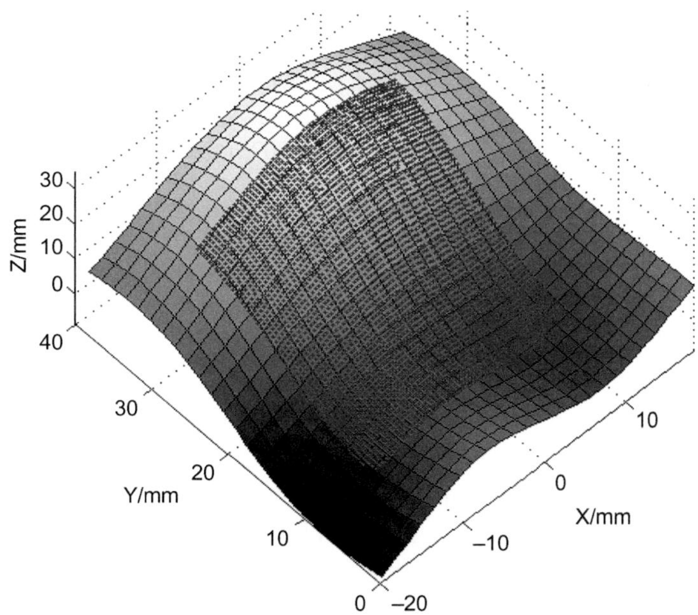

Fig. 4.5 Nominal surface and sampled data.

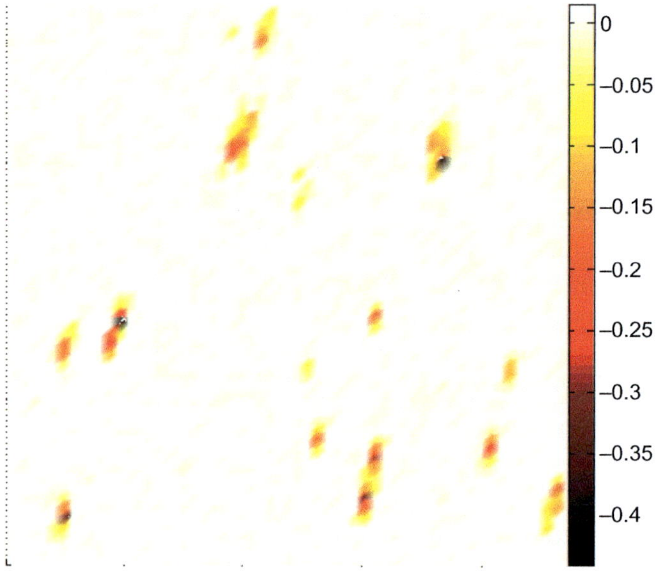

Fig. 4.6 Defects and noise.

Table 4.3 Comparison of three fitting methods.

Method		Robust ODF	Robust SVD	l_2 Norm ODF
Bias	ϑ_x	3.068×10^{-5} degree	0.2562°	-2.5834×10^{-3} degree
	ϑ_y	-2.802×10^{-6} degree	0.5211°	-5.6011×10^{-3} degree
	ϑ_z	5.259×10^{-5} degree	2.2824°	-4.2877×10^{-3} degree
	t_x	−0.04909 μm	0.8329 mm	−1.8408 μm
	t_y	−0.01835 μm	−0.2093 mm	−2.1993 μm
	t_z	0.10678 μm	0.5293 mm	2.3557 μm
Uncertainty	ϑ_x	1.371×10^{-4} degree	6.1952×10^{-3} degree	2.496×10^{-4} degree
(σ)	ϑ_y	2.687×10^{-4} degree	8.0340×10^{-3} degree	4.921×10^{-4} degree
	ϑ_z	3.400×10^{-4} degree	9.2670×10^{-3} degree	6.625×10^{-4} degree
	t_x	0.0957 μm	1.4528 μm	0.1843 μm
	t_y	0.0617 μm	0.7378 μm	0.0982 μm
	t_z	0.0559 μm	1.6689 μm	0.0777 μm
Running time		47.2057 s	42.9947 s	45.7937 s

template associated with each measurement point. Thus this algorithm will be trapped at a local minimum and lead to an incorrect result. Therefore it is not proper to directly neglect the dependency between the projection points and the transformation parameters. The ordinary least squares technique is also biased, especially for the rotation angles. Adopting the robust estimator the influence of the defects can be greatly reduced, and the fitting accuracy of the motion parameters may be two orders higher. It can be seen that the uncertainty is roughly in the same order for the three algorithms, since it is mainly determined by the amplitude of the introduced random noise.

4.5 Minimum zone fitting

4.5.1 Definition of tolerance zone

Currently the peak-to-valley (PV) value is the most widely adopted parameter for the assessment of form accuracies of optical components, although the root-mean-square (RMS) parameter of the form deviation is more closely related to their optical functionalities. The form errors are evaluated by fitting a nominal function to the measured data set. In coordinate metrology, it is the required form error parameters that determine the objective functions of the fitting program. The Chebyshev fitting should be implemented to calculate the PV form error according to the numerical optimization theory. However, the Chebyshev fitting is not continuously differentiable, thus, very difficult to be solved. At present, most researchers and commercial software in precision engineering implement least squares fitting due to its simplicity, and the difference between the maximal and minimal residuals is taken as the PV parameter. But the solution is not consistent with the definition of form errors in ISO standards [20], and the PV values will be seriously overestimated, which will lead to unnecessary rejection of acceptable parts.

For simple geometries, like spheres, cylinders, and cones, the tolerance zone is defined as the radial separation between two concentric elements. If considering symmetric tolerance only, it is equivalent to minimizing the maximal orthogonal distance from the data points to the fitted surface:

$$\min_{\mathbf{x}} \max_{1 \le j \le N} \left\| \mathbf{p}_j - \mathbf{q}_j \right\| \tag{4.24}$$

where $\mathbf{q}_j$ is the projection point of a data point $\mathbf{p}_j$ onto the surface. $x \in \mathrm{R}^n$ are the unknown motion parameters $\boldsymbol{m}$ (rotation angles and translation vector) and shape parameters $\boldsymbol{s}$. For the sake of computational simplicity, the transformation is always performed on the data points, and the surface stays at the standard position.

For complex surfaces the shapes of the two banding surfaces do not keep similar; therefore the geometric meaning of the tolerance zone is generalized as the covered space by a moving sphere whose center travels on the nominal surface. Therefore the minimum zone fitting is equivalent to minimizing the radius/diameter of the sphere, and the resultant PV parameter is the sphere diameter d, as shown in Fig. 4.7.

Various methods have been proposed to calculate the MZ form errors of simple geometries, for example, computational geometry methods [21], support vector machines [22], and simplex methods [23]. But these techniques have serious limitations in restricted applicability. They usually need to identify different cases, and then appropriate manipulation will be implemented according to the specific shape or point distribution. This means they are very inconvenient to be applied in practice.

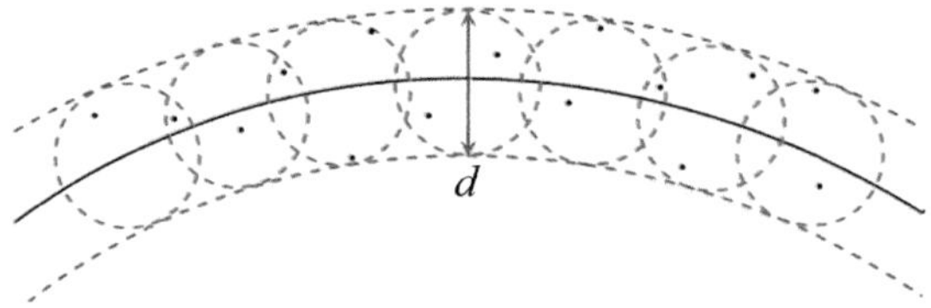

Fig. 4.7 Minimum zone fitting of complex surfaces.

From the point of view of numerical optimization, the MZ fitting is a discrete minimax problem. It can be approximated as a differentiable p-norm fitting problem. The solution approaches the Chebyshev fitting when $p \rightarrow \infty$. Subsequently the problem can be solved using the reweighted least squares technique [24]. At the kth iteration the weighting is calculated as

$$w_i^{(k)} = \left(w_i^{(k-1)} \left| r_i^{(k-1)} \right| \right)^{\frac{p-1}{p-2}} \tag{4.25}$$

Analogues to the circumstance of $1 < p < 2$, $w_i^{(k)} (r_i^{(k)})^2 \rightarrow |r_i^{(k)}|^p$ when k becomes larger.

Some researchers turned it into a sequential linear programming [25] or sequential quadratic programming [26] problem, and the Gauss-Newton or quasi-Newton algorithm can be employed iteratively. These algorithms suffer from slow convergence when the solution approaches the region of the optimum point.

4.5.2 Interior point method

The interior point method converts the non-smooth minimax problem into a constrained differentiable optimization problem, and a Newtonian approach is employed to iteratively solve the Lagrange function of the constrained minimization problem [27].

For an optimization problem with inequality constraints,

$$\min_{\boldsymbol{x} \in R^n} f(\boldsymbol{x}) \text{ s.t.} d_j(\boldsymbol{x}) \geq 0, j = 1, \ldots, N \tag{4.26}$$

where $f: R^n \rightarrow R$ and $\{d_j : R^n \rightarrow R\}$ are twice continuously differentiable functions. A primal-dual feasible interior point (IP) method can be applied to find the optimal solution x^*. This method needs an initial feasible solution $\boldsymbol{x}_0$ and decreases the objective function $f(\boldsymbol{x})$ at each iterate. A correct solution is thought to be found when the Karush-Kuhn-Tucker (KKT) conditions are satisfied:

$$\begin{gathered} \boldsymbol{g}(\boldsymbol{x}) - \boldsymbol{B}(\boldsymbol{x})^T = 0 \\ \boldsymbol{d}(\boldsymbol{x}) \geq 0 \\ \boldsymbol{D}(\boldsymbol{x})\boldsymbol{\omega} = 0 \\ \boldsymbol{\omega} \geq 0 \end{gathered} \tag{4.27}$$

Here, $g(x) \triangleq \partial f(x)/\partial x$, $B(x) \triangleq \partial d(x)/\partial x$, and $D(x) \triangleq \mathrm{diag}(d_j(x))$. The solution x is called stationary if there exists $\omega \in R^N$ such that the KKT conditions hold.

The procedures of iteratively updating the solution x and vector ω are shown as follows:

1: Initialize parameters $\xi \epsilon (0,1/2)$, $\eta \epsilon (0,1)$, $\nu > 2$, $\theta \epsilon (0,1)$, $\tau \epsilon (2,3)$, and $\kappa \epsilon (0,1)$ and thresholds $\omega_{\min}$, $\omega_{\max}$, and ε. Obtain an initial solution $\mathbf{x}_0$ and then solve $\boldsymbol{\omega}_0$ from $\min_{\omega'_0} \|g(x_0 - B(x_0))\omega'_0\|$ and $\omega_0 = \max(0.1, \omega'_0)$.

2: Computation of arc search:

(1) Compute $(\Delta \mathbf{x}, \Delta \boldsymbol{\omega})$ by solving the Lagrangian function $L(\mathbf{x}, \boldsymbol{\omega}, \mathbf{W}, \boldsymbol{\mu})$:

$$\begin{bmatrix} -W & B^T \\ \mathrm{diag}(\omega)B & \mathrm{diag}(d) \end{bmatrix} \begin{bmatrix} \Delta x \\ \Delta \omega \end{bmatrix} = \begin{bmatrix} g(x) - B^T \omega \\ \mu - \mathrm{diag}(d)\omega \end{bmatrix} \tag{4.28}$$

with $\boldsymbol{\mu} = 0$ and $W \approx \nabla^2 f(\mathbf{x}) - \sum_{j=1}^{N} \omega_j \nabla^2 d_j(\mathbf{x})$.

If $\|\Delta x\| \le \varepsilon$ and $\omega + \Delta \omega \ge 0$, then stop.

(2) Set $\varphi \triangleq \min(1, \max(0, -\omega - \Delta\boldsymbol{\omega} - 10^3 d(x)))$ and $\delta \triangleq \langle g(x), \Delta x \rangle + \sum_{j=1}^{N} \frac{\omega_j + \Delta \omega_j}{\omega_j} \varphi_j$.

Set $\mu = (1 - \psi)\varphi + \psi \rho \omega$ with $\rho \triangleq \|\Delta x\|^{\nu} + \|\varphi\|$ and

$$\psi \triangleq \begin{cases} 1 & \text{if } \sum_{j=1}^{N} \frac{\omega_j + \Delta\omega_j}{\omega_j} \left[\rho\omega_j - \varphi_j\right] \le 0 \\ \min\left\{1, \dfrac{(1-\theta)|\delta|}{\sum_{j=1}^{N} \frac{\omega_j + \Delta\omega_j}{\omega_j}\left[\rho\omega_j - \varphi_j\right]}\right\} & \text{otherwise} \end{cases}.$$

(3) Compute $(\Delta \mathbf{x}, \Delta \boldsymbol{\omega})$ by solving $L(\mathbf{x}, \boldsymbol{\omega}, \mathbf{W}, \boldsymbol{\mu})$.

(4) Set $I \triangleq \{j : d_j(x) \le \omega_j\}$.

(5) Set $\delta \mathbf{x}$ be the solution of

$$\min \langle \delta \mathbf{x}, \mathbf{W}\delta \mathbf{x} \rangle, \text{s.t.} d_j(\mathbf{x} + \Delta \mathbf{x}) + \langle \nabla d_j(\mathbf{x}), \delta \mathbf{x} \rangle = \max\left(\|\Delta \mathbf{x}\|^{\tau}, \max_{j \in I} \left|\rho_j\right|^{\kappa} \|\Delta \mathbf{x}\|^2\right), \forall j \in I$$

If this is infeasible or unbounded or $\|\delta x\| > \|\Delta x\|$, then set $\delta \mathbf{x} = 0$.

3: Arc search. Compute α, the first in the sequence $\{1, \eta, \eta^2, \ldots\}$ satisfying

$$f\left(\mathbf{x} + \alpha \Delta \mathbf{x} + \alpha^2 \delta \mathbf{x}\right) \le f(\mathbf{x}) + \xi \alpha \langle \mathbf{g}(\mathbf{x}), \alpha \Delta \mathbf{x} \rangle, \quad \forall d_j\left(\mathbf{x} + \alpha \Delta \mathbf{x} + \alpha^2 \delta \mathbf{x}\right) > 0, \ j = 1, \ldots, N$$

4: Update

$$\mathbf{x} \leftarrow \mathbf{x} + \alpha\Delta\mathbf{x} + \alpha^2\delta\mathbf{x} \text{ and } \boldsymbol{\omega} \leftarrow \min\left(\max\left(\max\left(\|\Delta\mathbf{x}\|^2, \boldsymbol{\omega} + \Delta\boldsymbol{\omega}\right), \omega_{\min}\right), \omega_{\max}\right)$$

Go to **Step 2**.

The global convergence condition for the sequence of solution $\{\mathbf{x}_k\}$ is given in any index K such that the sequence $\{\mathbf{x}_k | k \in K\}$ is bounded; there exists $\sigma_1, \sigma_2 > 0$ such that for all $k \epsilon K$, $k \geq 1$:

$$\left\langle \mathbf{u}, \left(\mathbf{W}_k + \sum_j \frac{\omega_k^j}{d_j(\mathbf{x}_k)} \nabla d_j(\mathbf{x}_k) \nabla d_j(\mathbf{x}_k)^T\right)\mathbf{u}\right\rangle \geq \sigma_1 \|\mathbf{u}\|^2, \forall \mathbf{u} \text{ and } \|\mathbf{W}_k\| \leq \sigma_2.$$

Therefore, at each iteration, it needs to check whether the matrix $\mathbf{W} + \mathbf{B}^T\text{diag}\{\omega_j/d_j\}\mathbf{B}$ is positive definite. If not, an appropriate identity matrix will be added onto it to guarantee convergence of the solution.

As the maximum inscribed element (MIE) and the minimum circumscribed element (MCE) fitting can be straightforwardly converted into a constrained optimization problem, this primal-dual interior point method can be applied to solve these problems as well.

4.5.2.1 Experimental validation

The effectiveness of this algorithm is demonstrated by several benchmark data sets [27]. Here, two spheres, two cylinders, and two cones are employed, respectively. In Table 4.4, it can be seen that the proposed algorithm can always obtain better or the same results with the techniques in literature.

In Table 4.5 the calculated results of MIE, MCE, and MZE for Cone 2 are all very close. The MZ form error is a little lower than the reference, which implies that the IP algorithm can find better results. Fig. 4.8A illustrates the fitted minimum zone cone of a data set. However, the minimum circumscribed cone (Fig. 4.8B) is not the one we want, and it is seriously tilted. This is because the data points at the upper right area are missing,

Table 4.4 Comparison on the evaluation results of spheres and cylinders.

Data set	MZ form error (μm)		MC radius (mm)		MC form error (μm)	MI radius (mm)		MI form error (μm)
	Literature	IP	Literature	IP		Literature	IP	
Sphere 1	8.327	8.327	1.0035	1.0035	8.346	0.9959	0.9960	8.365
Sphere 2	9.67	9.67	50.0379	50.0379	12.59	50.0302	50.0302	9.93
Cylinder 1	2.788	2.788	12.0002	12.0002	2.826	12.0016	12.0016	2.788
Cylinder 2	9.931	9.931	50.004	50.004	10.652	49.996	49.996	9.931

Table 4.5 Comparison on the evaluation results of cones.

		IP method		
Data set	**MZ in literature**	**MZ**	**MC**	**MI**
Cone 1	0.0032 in.	0.0032 in.	0.4292 in.	0.0038 in.
Cone 2	4.590 μm	4.589 μm	6.264 μm	5.387 μm

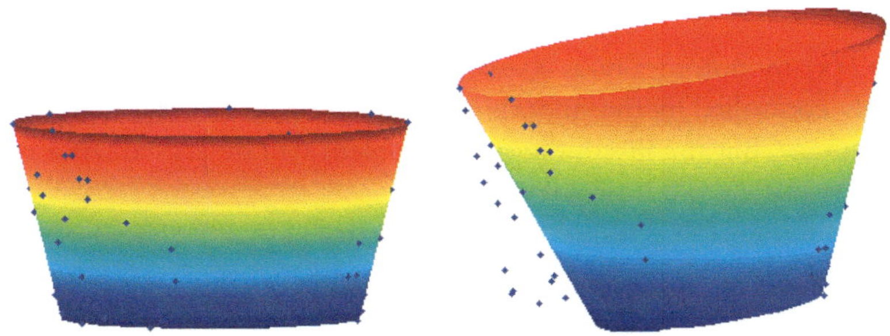

Fig. 4.8 Misalignment of maximum inscribed cone: (A) Minimum zone cone and (B) maximum inscribed cone.

and the resulted blank allows the truncated inscribed cone to expand outside, so that its volume is much bigger than the correct one. Consequently the axis of the cone is tilted, and the form error at the lower left region will become unacceptably large. That is to say the error metric of MI cones is not appropriate to be adopted when some data points are missing in a big region at the bottom or top of the cone.

It is also interesting to note that within a tolerance of 10^{-8}, there are 6, 7, and 8 data points getting their D values equal to the objective function values for the fitted spheres, cylinders, and cones, respectively, irrelevant with the fitting criteria. That means, for these simple geometries, the number of the data points determining the form error band is always 1 more than the number of variables in the optimization program. The optimization programs here normally converge within 40 iterates, and the time taken is less than 0.1 second. Thus this method is very computationally efficient and easy to implement.

4.5.3 Exponential penalty function

The original minimax optimization

$$\min_{\mathbf{x}} F(\mathbf{x}) \quad \text{with} \quad F(\mathbf{x}) = \max_{1 \leq j \leq N} d_j(\mathbf{x}) \tag{4.29}$$

can be converted into a continuously differentiable problem using the exponential penalty function (also termed the aggregate function) [28]:

$$\min_{\mathbf{x}} F_p(\mathbf{x}) \quad \text{with} \quad F_p(\mathbf{x}) = \frac{1}{p} \log \sum_{j=1}^{N} \exp\left\{ p d_j(\mathbf{x}) \right\} \tag{4.30}$$

From the numerical approximation theories, it is known that

$$F(\mathbf{x}) \leq F_p(\mathbf{x}) \leq F(\mathbf{x}) + \log N/p$$

$F_p(\mathbf{x}) \rightarrow F(\mathbf{x})$ monotonically as $p \rightarrow \infty$. Therefore the approximation error can be quantitatively controlled by the smoothing parameter p. A larger p is preferred in terms of fitting accuracy, but the ill-conditioning problem will be caused. As a consequence, proper trade-off needs to be carefully made.

Solve the differentiable optimization problem in Eq. (4.30) using the Newton's algorithm. Its gradient and Hessian matrices can be calculated as

$$\begin{cases} \nabla F_p(\mathbf{x}) = \sum_{j=1}^{N} \zeta_j \nabla d_j(\mathbf{x}) \\ \nabla^2 F_p(\mathbf{x}) = \sum_{j=1}^{N} \zeta_j \nabla^2 d_j(\mathbf{x}) + p\left(\sum_{j=1}^{N} \zeta_j \nabla d_j(\mathbf{x}) \nabla d_j(\mathbf{x})^T - \sum_{j=1}^{N} \zeta_j \nabla d_j(\mathbf{x}) \sum_{j=1}^{N} \zeta_j \nabla d_j(\mathbf{x})^T \right) \end{cases} \tag{4.31}$$

where $\zeta_j = \exp\left\{ p d_j(\mathbf{x}) \right\} / \sum_{j=1}^{N} \exp\left\{ p d_j(\mathbf{x}) \right\}$.

To improve the numerical stability and enlarge the convergence domain, the Hessian matrix is modified as

$$\mathbf{B}_p(\mathbf{x}) = \nabla^2 F_p(\mathbf{x}) + \mu \mathbf{I}$$

where $\mu = \max\{0, \delta - e(\mathbf{x})\}$ with $e(\mathbf{x})$ the smallest eigenvalue of $\nabla^2 F_p(\mathbf{x})$ and δ is a user-set parameter. In this way the Hessian matrix can always be guaranteed positive definite. As a result, divergence of the solutions can be effectively avoided.

Then the solution can be iteratively updated as

$$\mathbf{x} \leftarrow \mathbf{x} - \mathbf{B}_p^{-1} \nabla F_p(\mathbf{x})$$

During the optimization process, most of the computation effort is spent on calculating the gradient and Hessian matrices. To improve the efficiency, only a subset of data points is applied. It is obvious that, when p is sufficiently large and $d_j(\mathbf{x}) < F(\mathbf{x})$, the term $\exp\{pd_j(\mathbf{x})\} / \exp\{pF(\mathbf{x})\}$ is approximately equal to zero; thus the term $\exp\{pd_j(\mathbf{x})\}$ has little contribution to $F_p(\mathbf{x})$ and can be ignored. An ε-active set is defined as

$$\Omega_p := \left\{ j \mid F_p(x) - d_j(x) \leq \varepsilon, \quad 1 \leq j \leq N \right\}$$

Furthermore, the smoothing parameter p is adjusted adaptively during the optimization process. A simple and flexible adaptive active-set exponential penalty smoothing algorithm is presented as follows:

1. Initialize solution $\boldsymbol{x}^{(0)}$, configuration parameters α, β, $\kappa \in (0,1)$, $p_0 \geq 1$, $\varepsilon_0 > 0$, $\xi > 1$, $\varsigma > 1$, $i = j = 0$, and ε-active set $\boldsymbol{\Omega}_0 = \boldsymbol{\Omega}_{\varepsilon_0}(\mathbf{x}^{(0)})$.

//Here, α and β control the step length of the solution incremental, κ controls numerical stability of matrix inversion, ξ is used to update the smoothing parameter p, and ς is used to update the parameter ε. i counts the iterates of solution incremental, and j counts the iterates of adjustment of p and ε.

2. Find descent direction. Compute $\mathbf{B}_{p_i, \Omega_i}(\mathbf{x}^{(i)})$ and its Cholesky factor $\boldsymbol{R}$ such that $\mathbf{B}_{p_i, \Omega_i}(\mathbf{x}^{(i)}) = \mathbf{R}\mathbf{R}^T$ and the reciprocal condition number $c(\mathbf{R})$.

If $c(\mathbf{R}) \geq \kappa$, the inverse of the Hessian matrix is stable, then

$$\mathbf{h}^{(i)} = -\mathbf{B}_{p_i,\Omega_i}\left(\mathbf{x}^{(i)}\right)^{-1} \nabla F_{p_i,\Omega_i}\left(\mathbf{x}^{(i)}\right) \tag{4.32}$$

//The Newton's algorithm

Else

$$\mathbf{h}^{(i)} = -\nabla F_{p_i,\Omega_i}\left(\mathbf{x}^{(i)}\right) \tag{4.33}$$

//The steepest descent direction

Endif

3. Compute the step length by line search. Find the smallest integer k satisfying

$$F_{p_i,\Omega_i}\left(\mathbf{x}^{(i)} + \beta^k \mathbf{h}^{(i)}\right) - F_{p_i,\Omega_i}\left(\mathbf{x}^{(i)}\right) \leq -\alpha\beta^k \left\|\mathbf{h}^{(i)}\right\|^2.$$

//make sure the objective function is decreased

and

$$F_{p_i,\Omega_i}\left(\mathbf{x}^{(i)} + \beta^k \mathbf{h}^{(i)}\right) - F\left(\mathbf{x}^{(i)} + \beta^k \mathbf{h}^{(i)}\right) \geq -10\log N/p_i.$$

//make sure the point reduction by ε-active set does not influence the solution seriously.

4. Update the solution:

$$\mathbf{x}^{(i+1)} = \mathbf{x}^{(i)} + \beta^k \mathbf{h}^{(i)}$$

$$\Omega_{i+1} = \Omega_i \cup \Omega_{\varepsilon_i}\left(\mathbf{x}^{i+1}\right)$$

5. Adjust smoothing parameter.

If $\|\nabla F_{p_i, \Omega_i}(\boldsymbol{x}^{(i+1)})\| \leq 2\varepsilon_i$, the solution stagnates for this configuration, then set

$\mathbf{x}^*_{(j)}=\mathbf{x}^{(i+1)}$, $p_{i+1}=\xi p_i$, *and* $\varepsilon_{i+1}=\varepsilon_i/\varsigma$.

//accept the solution as optimum for this configuration, enlarge smoothing parameter p, and reduce point number in Ω_{ε_i}.

$i \leftarrow i+1$ and $j \leftarrow j+1$

Else set

$p_{i+1}=p_i$, $\varepsilon_{i+1}=\varepsilon_i$, and $i \leftarrow i+1$

Endif

If termination criterion is satisfied

Break

Else

Go to **Step 2**.

Endif

Without losing generality the nominal geometric shape is represented in an implicit form as $f(\mathbf{q};\mathbf{s})=0$. Here, $\mathbf{q}=[x,y,z]^T$ is the coordinate of an arbitrary point on the surface, and **s** is the shape parameter (intrinsic characteristics) of the surface, for example, the half axes lengths of an ellipsoid and half vertex angle of a cone.

The Chebyshev fitting problem is equivalent to

$$\min_{\mathbf{x}} \max_{1\le j\le N} \left\| \mathbf{p}_j - \mathbf{q}_j \right\|^2$$

This formula is applied in this paper hereafter. The calculation of the gradient and Hessian matrices is discussed in detail in the succeeding text.

Set $d_j=\|\mathbf{p}_j-\mathbf{q}_j\|^2=(\mathbf{p}_j-\mathbf{q}_j)^T(\mathbf{p}_j-\mathbf{q}_j)$, then

$$\begin{cases} \nabla d_j = 2\left(\mathbf{p}_j-\mathbf{q}_j\right)^T \dfrac{\partial\left(\mathbf{p}_j-\mathbf{q}_j\right)}{\partial \mathbf{x}} \\ \nabla^2 d_j = 2\left[\dfrac{\partial\left(\mathbf{p}_j-\mathbf{q}_j\right)}{\partial \mathbf{x}}\right]^T \dfrac{\partial\left(\mathbf{p}_j-\mathbf{q}_j\right)}{\partial \mathbf{x}} + 2\left(\mathbf{p}_j-\mathbf{q}_j\right)^T \dfrac{\partial^2\left(\mathbf{p}_j-\mathbf{q}_j\right)}{\partial \mathbf{x}^2} \end{cases} \tag{4.34}$$

For the sake of clarity, the subscript j is omitted. Now, we consider the term $(\mathbf{p}-\mathbf{q})^T\partial(\mathbf{p}-\mathbf{q})/\partial\mathbf{x}$. The matrix $\partial\mathbf{q}/\partial\mathbf{x}$, that is, the dependency between the projection points and the motion/shape parameters, is very difficult to be calculated. Most researchers ignore it in their programs, but consequentially the convergence rate will be seriously reduced [10]. We investigate calculating it in a tactful way.

The variables **x** are divided into two groups, the motion parameters **m** and shape parameters **s**. From $\begin{cases} \dfrac{\partial f}{\partial \mathbf{q}} \times (\mathbf{p}-\mathbf{q}) = 0 \\ f(\mathbf{q};\mathbf{s}) = 0 \end{cases}$, it can be obtained that

$$\begin{cases} \dfrac{\partial f}{\partial \mathbf{q}} \times \left(\dfrac{\partial \mathbf{p}}{\partial m_i} - \dfrac{\partial \mathbf{q}}{\partial m_i} \right) = 0 \\ \dfrac{\partial f}{\partial \mathbf{q}} \dfrac{\partial \mathbf{q}}{\partial m_i} = 0 \end{cases} \text{ and } \begin{cases} \dfrac{\partial f}{\partial \mathbf{q}} \times \dfrac{\partial \mathbf{q}}{\partial s_i} = 0 \\ \dfrac{\partial f}{\partial \mathbf{q}} \cdot \dfrac{\partial \mathbf{q}}{\partial s_i} + \dfrac{\partial f}{\partial s_i} = 0 \end{cases}.$$

Therefore

$$\begin{cases} \dfrac{\partial (\mathbf{p}-\mathbf{q})}{\partial m_i} = \dfrac{\dfrac{\partial \mathbf{p}}{\partial m_i} \cdot (\mathbf{p}-\mathbf{q})}{\|\mathbf{p}-\mathbf{q}\|^2} (\mathbf{p}-\mathbf{q}) \\ \dfrac{\partial (\mathbf{p}-\mathbf{q})}{\partial s_i} = \dfrac{\dfrac{\partial f}{\partial s_i}}{\left\| \dfrac{\partial f}{\partial \mathbf{q}} \right\|^2} \dfrac{\partial f}{\partial \mathbf{q}} \end{cases}. \tag{4.35}$$

The second term of $\nabla^2 d_j$ is in practice very small compared with the first one; thus it is ignored for the convenience of calculation.

The flowchart of the Chebyshev fitting program is shown in Fig. 4.9.

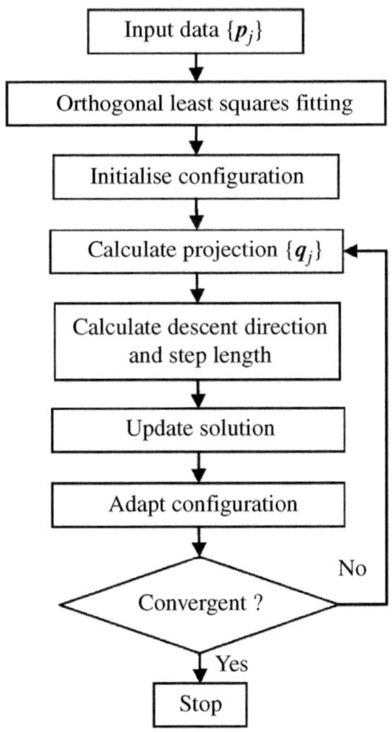

Fig. 4.9 Flowchart of the Chebyshev fitting program.

4.5.3.1 Experimental validation

To demonstrate the validity of this algorithm, three examples are applied: an ellipsoid $ax^2 + by^2 + cz^2 = 1$, a hyperbolic paraboloid $ax^2 - bz^2 - y = 0$, and an aspheric surface

$$z = \frac{r^2/R}{1 + \sqrt{1 - (1+k)r^2/R^2}} + a_4 r^4 + a_6 r^6$$

with $r = \sqrt{x^2 + y^2}$.

The exponential penalty function (EPF) method, the differential evolution (DE) algorithm [29], and the primal-dual interior point (PDIP) method are applied for the Chebyshev fitting. The obtained results are listed in Table 4.6.

In the table, it can be seen that the EPF method performs best among the three. In fact the DE program terminates when it has run a specified number of iterates. The result of DE fitting can be improved by allowing the program to run further, subsequently spending more time. The situation of PDIP fitting program is similar. The running time of EPF is much less than that of DE for ellipsoid and hyperbolic paraboloid fitting and even less than DE fitting with alpha-shape point reduction for the aspheric surface. If carrying out orthogonal distance least squares fitting direct to these three point sets, the obtained PV parameters will be 86.81, 22.47, and 3.79 μm, respectively, which are 44.9%, 26.2%, and 18.0% greater than the PV parameters obtained by the EPF Chebyshev fitting.

These examples effectively demonstrated the necessity and validity of Chebyshev fitting to obtain the correct PV form error parameters. Furthermore, they also revealed the superiority of EPF minimum zone fitting of complex-shaped surfaces over the differential evolution algorithm and the primal-dual interior point method in terms of computational efficiency and numerical stability.

4.6 Summary

Geometrical fitting is a key step to working out the intrinsic parameters and forming quality metrics, which are usually directly linked to the functionalities of free-form surfaces. Algebraic fitting based on the least squares method is commonly used in commercial software, but bias can occur in the fitting results. Geometrical fitting is recommended in consideration of parameter fidelity. Robust estimators are preferred

Table 4.6 Comparison of DE, PDIP, and EPF methods.

	DE results		PDIP results		EPF results	
Data sets	**PV/μm**	**Time/s**	**PV/μm**	**Time/s**	**PV/μm**	**Time/s**
Ellipsoid	63.69	56.17	61.68	14.30	59.89	3.11
Hyper parab	17.80	38.56	17.85	15.26	17.80	3.73
Asphere	3.21	5.40	3.21	4.67	3.21	2.34

when the measured data contain a lot of noise and outliers, while the minimum zone fitting is a proper fitting error metric to obtain a narrow tolerance band. These nondifferentiable optimization problems can be converted into continuous surrogate functions or constrained optimization problems. Then they can be solved straightforwardly using some trusted region methods or interior point methods. Specific choices should be made in accordance with the fabrication process, functionality requirement, computational cost, and technical regulations.

References

[1] ISO17450-1:2011. Geometrical product specifications (GPS)—General concepts Part 1: Model for geometrical specification and verification. Geneva: ISO; 2011.
[2] ISO 10110-12:2007. Optics and photonics—Preparation of drawings for optical elements and systems—Part 12: Aspheric surfaces. Geneva: ISO; 2007.
[3] Gross H. Handbook of Optical Systems. In: Aberration Theory and Correction of Optical Systems. Volume 3. Wiley; 2006.
[4] Zhao C, Burge J. Orthogonal curvature polynomials over a unit circle: basis set derived from curvatures of Zernike polynomials. Opt Express 2013;21:31430.
[5] Forbes G. Fitting freeform shapes with orthogonal bases. Opt Express 2013;21:19061.
[6] Gross H, Brömel A, Beier M, Steinkopf R, Hartung J, Zhong Y, Oleszko M, Ochse D. Overview on surface representations for freeform surfaces. In: Optical Systems Design 2015: Optical Design and Engineering VI. Vol. 9626. International Society for Optics and Photonics; 2015 Sep 23. p. 96260U.
[7] Zhang X, Zhang H, He X, Xu M. Bias in parameter estimation of form errors. Surf Topogr: Metrol Prop 2014;2(3) 035006.
[8] Ahn SJ. Least squares orthogonal distance fitting of curves and surfaces in space. Springer Science & Business Media; 2004 Dec 7.
[9] Boggs PT, Byrd RH, Schabel RB. A stable and efficient algorithm for nonlinear orthogonal distance regression. SIAM J Stat Comput 1987;8(6):1052–78.
[10] Jiang X, Zhang X, Scott PJ. Template matching of freeform surfaces based on orthogonal distance fitting for precision metrology. Meas Sci Technol 2010;21(4). 045101.
[11] Rey WJ. Introduction to robust and quasi-robust statistical methods. Springer Science & Business Media; 2012 Dec 6.
[12] Anderson R. Modern methods for robust regression. SAGE Inc; 2007.
[13] Björck A. Numerical methods for least squares problems. SIAM; 1996.
[14] Huber PJ. Robust estimation of a location parameter. Ann Math Stat 1964;35(1):73–101.
[15] Mosteller F, Tukey JW. Date analysis and regression. Reading, MA: Addison-Wesley; 1977.
[16] Rousseeuw PJ. Least median of squares regression. J Am Stat Assoc 1984;79(388):871–80.
[17] Gonin R, Money AH. Nonlinear lp-norm estimation. Marcel Dekker Ltd; 1989.
[18] Cooper P, Mason JC. Rational and lp approximations-renovating existing algorithms. In: Proc Conf on Math Methods for Curves and Surf. Tromso, Norway; 2004.
[19] Hunter DR, Lange K. Quantile regression via an MM Algorithm. J Comput Graph Stat 2000; 9(1):60–77.
[20] ISO 1101:2017. Geometrical product specifications-geometrical tolerancing—tolerances of form, orientation, location and run-out. Geneva: ISO; 2017.
[21] Venkaiah N, Shunmugam MS. Evaluation of form data using computational geometric techniques—Part II: Cylindricity error. Int J Mach Tool Manuf 2007;47:1237–45.
[22] Malyscheff AM, Trafalis TB, Raman S. Support vector machine learning to the determination of the minimum enclosing zone. Comput Ind Eng 2002;42:59–72.
[23] Kanada T. Evaluation of spherical for errors: computation of sphericity by means of minimum zone method and some examination with using simulated data. Proc Eng 1995;17:281–9.

[24] Rice JR, Usow KH. The Lawson algorithm and extensions. Math Comput 1968;22(101):118–27.
[25] Al-Subaihi I, Watson GA. Fitting parametric curves and surfaces by l_∞ distance regression. BIT Numer Math 2005;45(3):443–61.
[26] Wang F, Wang Y. Nonmonotone algorithm for minimax problems. Appl Math Comput 2011;217:6296–308.
[27] Zhang X, Jiang X, Forbes AB, Minh HD, Scott PJ. Evaluating the form errors of spheres, cylinders and cones using the primal–dual interior point method. Proc Inst Mech Eng B 2013;227(5):720–5.
[28] Zhang X, Xu M, Zhang H, He X, Jiang X. Chebyshev fitting of complex surfaces for precision metrology. Measurement 2013;46(9):3720–4.
[29] Zhang X, Jiang X, Scott PJ. Minimum zone evaluation of the form errors of quadric surfaces. Precis Eng 2011;35:383–9.

CHAPTER 5

Free-form surface reconstruction

5.1 Introduction

Free-form surface reconstruction refers to the process of transforming a set of points (a point cloud) or a polygonal mesh into an analytical surface that may be represented as an analytical function, usually expressed by means of a parametric equation. The goal of the surface fitting is to represent the measured points using a continuous model. Surface reconstruction can be divided in two categories: interpolation and approximation. A surface reconstructed using an interpolation technique passes exactly through the measured point, while the approximated surface passes near the points [1, 2]. Surface reconstruction has a key role in the conversion from a point cloud to a format that can be used by a computer-aided design (CAD) software. CAD readable format is the base form of free-form components manufactured using computer-aided manufacturing (CAM), such as single-point diamond turning, ultraprecision polishing, electrolytic in-process dressing, and plasma chemical vaporization machining [3].

To convert a mesh to a CAD readable surface, the first preprocess step is called parametrization. An abstract coordinate system is computed using parametrization techniques that are reviewed in Section 5.2. The B-spline approximation, which is discussed in Section 5.3, is a technique that generates a uniform grid of control points; this may lead to some poor approximation of local features if the control net is not dense enough, or it could lead to some wave effects in the flat portion of the surface if the grid is too fine. To overcome these issues some local refinement methods were recently proposed; these are detailed and discussed in Section 5.4. An interpolation method based on triangular Bezier surfaces is presented in Section 5.5, and two methods to perform the reconstruction of volume data sets are discussed in Section 5.6. In the last section, conversion from a gray scale volume to a triangular mesh is presented using one of the previous methods to convert it to a CAD readable surface.

5.2 Triangular mesh parametrization

Triangular mesh parametrization may be described as "attaching a coordinate system" to the triangular mesh [4]. This process provides a description of each measured 3D point,

* For more examples and practical implementation, check the website: www.digitalsurf.com/freeform.

Advanced Metrology
https://doi.org/10.1016/B978-0-12-821815-0.00005-8

$\boldsymbol{x}=(x, y, z)^T \in \mathbb{R}^3$, as a function of two abstract parameters $\boldsymbol{u}=(u, v)^T \in \mathbb{R}^2$. It is a crucial step for a good free-form surface reconstruction.

Only open meshes with one border will be investigated here, for parametrization of complex meshes with holes refer to Gu et al. [5].

One of the most famous parametrization techniques is based on Tutte mapping theorem [6], which assures that, if the triangulated surface is homeomorphic to a disk, the parametric coordinates of the boundary lie in a convex polygon. If the coordinates of the internal vertices are a convex combination of their neighbors, the parametrization is valid; that is, there are no self-intersections.

The results of this theorem were used by Floater [7] to parametrize a triangular mesh. The vertices of the boundary are first mapped to a convex polygon, usually a square or a circle, using the arch length parametrization. From this the parameters of the internal vertices are finally computed solving a linear system of equations:

$$\boldsymbol{A} \cdot \boldsymbol{U} = \boldsymbol{0}_n$$

where $\boldsymbol{A} \in \mathbb{R}^{n \times n}$ is a matrix of known coefficients and $\boldsymbol{U}=[\boldsymbol{u}_1^{int}, \boldsymbol{u}_2^{int}, \ldots, \boldsymbol{u}_{n_{int}}^{int}, \boldsymbol{u}_1^b, \boldsymbol{u}_2^b, \ldots, \boldsymbol{u}_{n_b}^b]^T \in \mathbb{R}^{n \times n}$ represents the matrix of the abstract unknown parameters. The parameters of the internal vertices are stored before the boundary ones. The entries of the matrix $\boldsymbol{A}$ must correctly utilize Tutte's theorem:

$$a_{ij} = \begin{cases} >0 & \text{if } i=j \text{ and } \boldsymbol{x}_i \text{ and } \boldsymbol{x}_j \text{ are connected by an edge,} \\ -\sum_{i \neq j} a_{ij} & \text{if } i=j, \\ 0 & \text{otherwise.} \end{cases}$$

Since the value of the boundary vertices is already set, the linear system can be solved as

$$[\boldsymbol{A}_{int}, \boldsymbol{A}_b] \begin{bmatrix} \boldsymbol{U}_{int} \\ \boldsymbol{U}_b \end{bmatrix} = \boldsymbol{0}_n$$

$$\boldsymbol{A}_{int} \cdot \boldsymbol{U}_{int} = -\boldsymbol{A}_b \cdot \boldsymbol{U}_b$$

where the unknown matrix to be found is $\boldsymbol{U}_{int}$. The last step is to choose the value of the matrix $\boldsymbol{A}$. One possible solution is to use the values of the discrete Laplace-Beltrami computed using the cotangent formula [7]:

$$a_{ij} \begin{cases} \frac{1}{2A_i}\left(\cot \alpha_{ij} + \cot \beta_{ij}\right) & \text{if } i=j \\ -\sum_{i \neq j} a_{ij} & \text{otherwise} \end{cases}$$

where α_{ij} and β_{ij} are the angles opposite to the edge $(\boldsymbol{x}_i - \boldsymbol{x}_j)$ and A_i is the Voronoi area of the vertex $\boldsymbol{x}_i$. A different definition of weights was proposed by Floater [8], based on the mean value theorem:

$$a_{ij}\begin{cases} \dfrac{1}{\left\|\boldsymbol{x}_i - \boldsymbol{x}_j\right\|}\left(\tan\left(\dfrac{\delta_{ij}}{2}\right) + \tan\left(\dfrac{\gamma_{ij}}{2}\right)\right) & \text{if } i=j \\ -\sum_{i\neq j} a_{ij} & \text{otherwise} \end{cases}$$

where δ_{ij} and γ_{ij} are the angles around $\boldsymbol{x}_i$ formed by the edge $(\boldsymbol{x}_i - \boldsymbol{x}_j)$. Both the parametrizations proposed by Floater are fast, but they create some distortion on the parameter space; that is, distance between two points on the parameters' space is not proportional to the distance of the points computed along the mesh. Yoshizawa et al. [9] proposed a method to compute a stretch-minimizing parametrization starting from Floater's mean value parametrization and minimizing a quadratic function. Let $\boldsymbol{J}(\boldsymbol{u})$ the Jacobian matrix of the parametrized surface:

$$\boldsymbol{J}(\boldsymbol{u}) = \left[\frac{\partial \boldsymbol{x}(\boldsymbol{u})}{\partial u}, \frac{\partial \boldsymbol{x}(\boldsymbol{u})}{\partial v}\right].$$

The local stretch can be computed as [10].

$$\sigma(\boldsymbol{u}) = \sqrt{\frac{\lambda_1^2(\boldsymbol{u}) + \lambda_2^2(\boldsymbol{u})}{2}}$$

where $\lambda_1(\boldsymbol{u})$ $(\lambda_2(\boldsymbol{u}))$ is the maximum (minimum) eigenvalue of the metric matrix $\boldsymbol{J}^T(\boldsymbol{u}) \cdot \boldsymbol{J}(\boldsymbol{u})$. The authors compute the per-vertex stretch as the average of the neighborhood stretches weighted by their areas:

$$\sigma(\boldsymbol{u}_i) = \sqrt{\frac{\Sigma_{j\in N(\boldsymbol{x}_i)} A_j \cdot \sigma_j}{\Sigma_{j\in N(\boldsymbol{x}_i)} A_j}}$$

where A_j and σ_j are the area and the stretch of the jth triangle, while $N(\boldsymbol{x}_i)$ represents the triangle around the ith vertex. The parametrization is achieved by minimizing, for each vertex i, the quadratic function:

$$E(\boldsymbol{u}_i) = \sum_{j\in N(\boldsymbol{x}_i)} w_{ij}\left(\boldsymbol{u}_j - \boldsymbol{u}_i\right)^2.$$

The weight is updated as $w_{ij}^{new} = \frac{w_{ij}^{old}}{\sigma_j}$, where w_{ij}^0 is the weight of the shape-preserving parametrization proposed by Floater, and $\sigma_j = \sigma(\boldsymbol{u}_j)$. The stretch-minimizing parametrization is implemented in the open source C++ library OpenGI [11].

5.2.1 Case studies

Two meshes, a blade of an impeller (see Fig. 5.1A) and a portion of a lattice surface (see Fig. 5.2), are parametrized with the presented algorithms. After the parametrization a checkboard image is added to the interpolated surface. If each square of the checkboard image has a constant area, we can conclude that the parametrization has not introduced any distortion. It is possible to observe that both shape-preserving (Fig. 5.1B) and mean value parametrizations

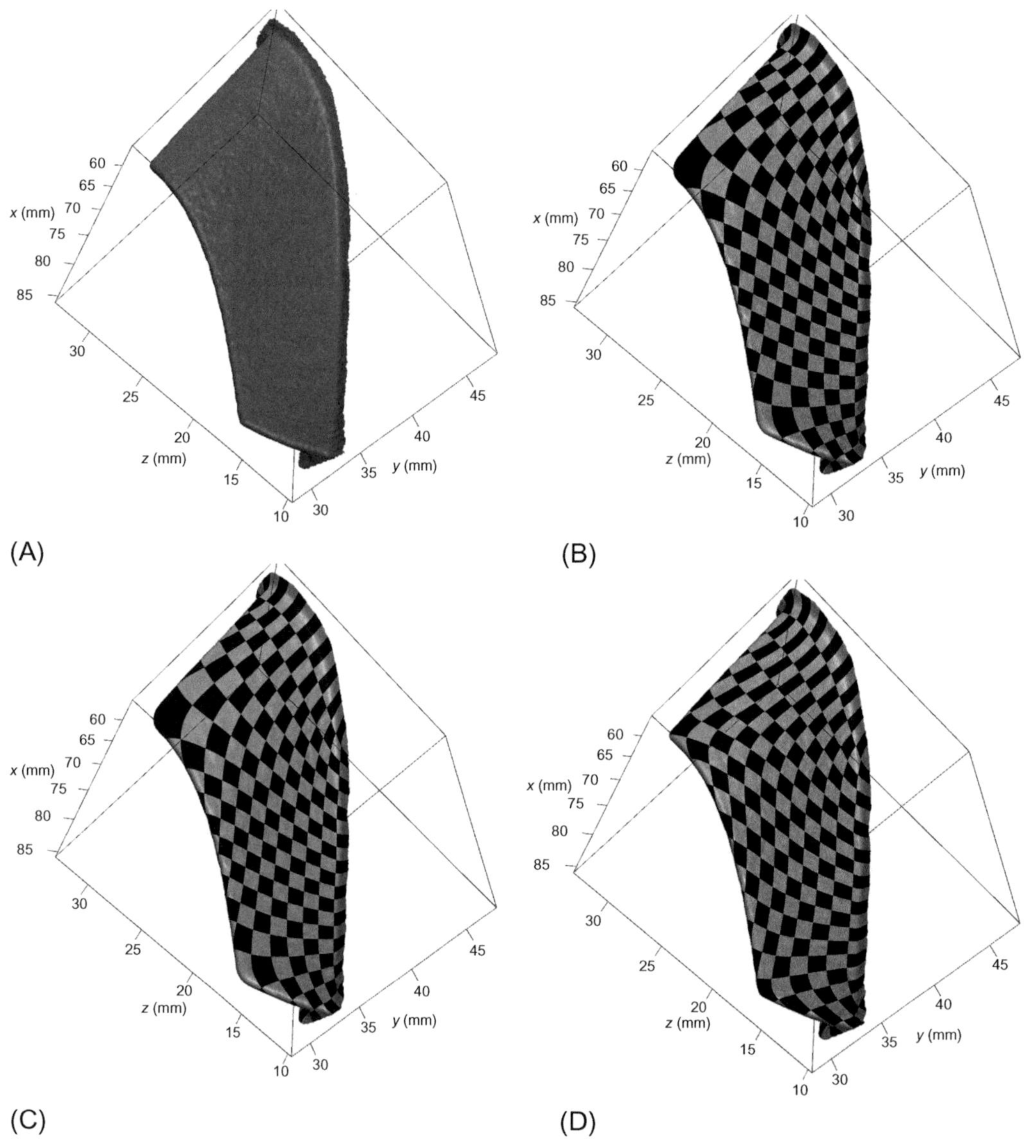

Fig. 5.1 Parametrization of a portion of an impeller blade: analyzed mesh (A), shape-preserving parametrization (B), mean value parametrization (C), and stretch-minimizing parametrization (D).

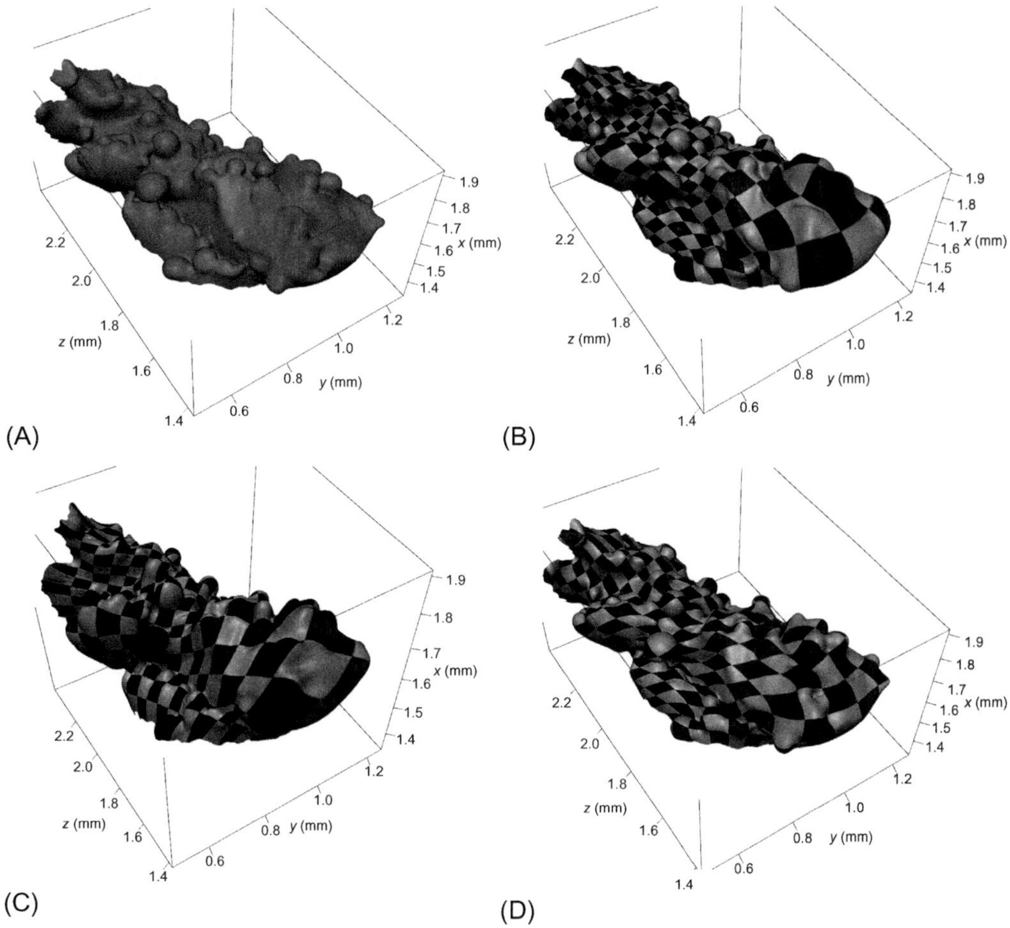

Fig. 5.2 Parametrization of a portion of lattice surface: analyzed mesh (A), shape-preserving parametrization (B), mean value parametrization (C), and stretch-minimizing parametrization (D).

(Fig. 5.1C) create a distortion on the edges of the blade. When applying the stretch-minimizing parametrization, you can see the distortion is smaller (Fig. 5.1D).

The distortion introduced by the shape-preserving and mean value parametrization is higher in the portion of a lattice mesh (see Fig. 5.2B and C, respectively). In contrast the stretch-minimizing algorithm is able to reduce the distortion introduced by the mean value parametrization (see Fig. 5.2D).

A good parametrization is important in the fitting stage because the estimation of the coefficients is performed in the parameters' space, so a parametrization technique that reduces any distortions of the area allows for better results.

5.3 Surface reconstruction with B-splines and NURBS

B-spline curve and surfaces are defined by a linear combination of a B-spline basis function and a set of control points. The base of a B-spline curve of degree d is defined by a nondecreasing sequence of knots $\boldsymbol{t} = (t)_{i=0}^{n+d}$ through the De Boor's algorithm [12]:

$$B_{i,d}(t) = \frac{t - t_i}{t_{i+d} - t_i} B_{i,d-1}(t) + \frac{t_{i+d} - t}{t_{i+d+1} - t_{i+1}} B_{i+1,d-1}(t)$$

with

$$B_{i,0}(t) = \begin{cases} 1 & \text{if } t_i \le t \le t_{i+1} \\ 0 & \text{otherwise.} \end{cases}$$

The De Boor's algorithm can lead to the division $\frac{0}{0}$; this is set by convection to 0. Given a set of $n+1$ control points $\boldsymbol{p}_i \in \mathbb{R}^3$, a B-spline curve can be computed by

$$\boldsymbol{r}(t) = \sum_{i=0}^{n} B_{i,d}(t) \cdot \boldsymbol{p}_i \quad a \le t < b$$

A B-spline curve is therefore simply a weighted, linear combination of the control points, with the weight represented by the B-spline basis functions.

The properties of a B-spline curve include the following [12]:

- $\boldsymbol{r}(t)$ is a piecewise polynomial curve.
- The curve interpolates the first and the last control point.
- Affine invariance: an affine transformation (scaling and rotation) is applied to the curve, thereby applying it to the control points.
- A strong convex hull property: the curve is contained in the convex hull of the control polygon.
- A local modification scheme: moving control point $\boldsymbol{p}_i$ the curve changes only in the interval $[t_i, t_{i+p+1}]$.
- The control polygon represents a piecewise linear approximation of the curve.
- Moving along the curve along the parameter space, the basis function acts like a switch; that is, each base function is switched on or off depending on the parameter interval.
- Variation diminishing property: no plane has more intersections with the curve than with the control polygon.
- Continuity and differentiability of the curve: $\boldsymbol{r}(t)$ is infinitely differentiable in the interior knot interval and at least $d-k$ times continuously differentiable at a knot multiplicity k.
- It is possible to use multiple coincidental control points.

The extension to the surface can be achieved by the cross product of two B-spline basis function as follows:

$$\boldsymbol{r}(u, v) = \sum_{i=0}^{n_u} \sum_{j=0}^{n_v} B_{i,p}(u) \cdot B_{j,q}(v) \cdot \boldsymbol{p}_{ij}, \quad a_u \le u < b_u, a_v \le v \le b_v \tag{5.1}$$

where p and q are degrees of the basis function in u and v direction and $\boldsymbol{p}_{ij}$ is the control point corresponding to the product of the basis function $B_{i,p}(u)$ and $B_{j.q}(v)$.

The reconstruction of a B-spline surface from an observed set of available points $\boldsymbol{x}_i$, $i=1, \ldots, n$ requires finding the control points of the surface given a parameter space $\boldsymbol{u}=(u, v)$ and two sets of basis function $B_{i,p}(u)$ and $B_{j,q}(v)$.

In this book, it is assumed that the surface is (or it can be easily converted to) a triangular mesh; for the reconstruction from point cloud to triangular mesh, refer to [13–16].

Before describing the algorithm used to estimate the control points, the matrix notation of the B-spline surface is introduced. From Eq. (5.1) the surface can be converted in matrix notation as

$$\boldsymbol{r}(u, v)=\boldsymbol{B}_u \cdot \boldsymbol{P} \cdot \boldsymbol{B}_v$$

where $\boldsymbol{B}_u=(B_{0,p}(u), B_{0,p}(u), \ldots, B_{n_u,p}(u))^T$ is the vector of the B-spline basis functions in u direction, $\boldsymbol{B_v}=(B_{0,q}(v), B_{0,q}(v), \ldots, B_{n_v,q}(v))^T$ is the vector of B-spline basis function in v direction, and

$$\boldsymbol{P}=\begin{bmatrix} \boldsymbol{p}_{00} & \cdots & \boldsymbol{p}_{0n_v} \\ \vdots & \ddots & \vdots \\ \boldsymbol{p}_{n_u0} & \cdots & \boldsymbol{p}_{n_un_v} \end{bmatrix}$$

is the matrix of control points.

5.3.1 Surface reconstruction

Surface reconstruction means finding the unknown control points of a B-spline surface given a finite number of sampled points $\boldsymbol{x}_i$, $i=1, \ldots, n$.

The first step consists of computing the space of the abstract parameters u-v through one of the methods described in Section 5.2. After this the control points can be estimated.

To compute the unknown coefficients, it is useful to rewrite the evaluation of a point in a standard model that is used in linear regression [17].

Let $\boldsymbol{P}_{\cdot i}$ be the ith row of the matrix $\boldsymbol{P}$, $\mathbf{B}=(\boldsymbol{P}_{\cdot 1}, \boldsymbol{P}_{\cdot 2}, \ldots, \boldsymbol{P}_{\cdot (n_u+1)})^T \in \mathbb{R}^{p=(n_u+1)\cdot(n_v+1)}$ be the vector with the control points, and $\boldsymbol{f}^T(u, v)=\boldsymbol{B}_u^T \otimes \boldsymbol{B}_v^T=[B_{0,\,p}(u)\boldsymbol{B}_v, B_{0,\,p}(u)\boldsymbol{B}_v, \ldots, B_{n_u,\,p}(u)\boldsymbol{B}_v]^T \in \mathbb{R}^p$, where $\otimes$ represents the Kronecker product.

The evaluation of a point on the surface at parameter value (u, v) can be computed as

$$\boldsymbol{x}^T(u, v)=\boldsymbol{f}^T(u, v)\cdot\mathbf{B}.$$

In matrix form, it can be written as

$$\boldsymbol{X}=\boldsymbol{F}\cdot\mathbf{B}$$

where $\boldsymbol{X} = [\boldsymbol{x}_1, \boldsymbol{x}_2, \dots, \boldsymbol{x}_n]^T \in \mathbb{R}^{n,3}$ is the matrix of the observed points and $\boldsymbol{F} = [f_1, f_2, \dots, f_n]^T \in \mathbb{R}^{n,p}$ is the model matrix.

For each coordinate the coefficients are estimated, therefore minimizing the sum of square of the residuals

$$\min_{\boldsymbol{\beta}_i} \boldsymbol{\epsilon}_i^T \cdot \boldsymbol{\epsilon}_i = (\boldsymbol{F} \cdot \boldsymbol{\beta}_i)^T (\boldsymbol{F} \cdot \boldsymbol{\beta}_i) = \boldsymbol{\beta}_i^T \left(\boldsymbol{F}^T \boldsymbol{F}\right) \boldsymbol{\beta}_i$$

where $\boldsymbol{\beta}_i$ is the ith column of the matrix $\mathbf{B}$. The coefficients can then be computed using the least squares method:

$$\hat{\mathbf{B}} = \left(\boldsymbol{F}^T \boldsymbol{F}\right)^{-1} \boldsymbol{F} \boldsymbol{X}.$$

Due to the high number of coefficients estimated in the results, the previous equation can lead to some unwanted oscillation due to the overfitting, or the matrix $\boldsymbol{F}^T \boldsymbol{F}$ maybe unable to be inverted.

To overcome this issue a penalty term is usually added to the minimization problem:

$$\min_{\boldsymbol{\beta}_i} \boldsymbol{\beta}_i^T \left(\boldsymbol{F}^T \boldsymbol{F}\right) \boldsymbol{\beta}_i + \lambda \boldsymbol{\beta}_i^T \boldsymbol{E} \boldsymbol{\beta}_i$$

where the entries of the matrix $\boldsymbol{E}$ are $e_{ij} = \iint_{\mathbb{R}^2} \frac{\partial^2 f(u,v)}{\partial u^2} + 2\frac{\partial^2 f(u,v)}{\partial u\, \partial v} + \frac{\partial^2 f(u,v)}{\partial v^2}\, du\, dv$; that is, the roughness of the surface is penalized by the second derivative. The matrix of unknown coefficients can be estimated as

$$\hat{\mathbf{B}} = \left(\boldsymbol{F}^T \boldsymbol{F} + \lambda \boldsymbol{E}\right)^{-1} \boldsymbol{F} \boldsymbol{X} \tag{5.2}$$

Due to the limited domain of each basis function, the matrices $\boldsymbol{F}$ and $\boldsymbol{E}$ are sparse; that is, the majority of the entries of the matrices are zeros. The system can therefore be estimated using the numerical algorithm based on sparse matrices. The C++ code can be downloaded from GitHub [18].

5.3.2 Case studies

In this section the mesh that was parametrized in Section 5.2.1 will be reconstructed. Different factors will be analyzed:

- The parametrization of the mesh
- The smoothing term λ
- The isotropy (anisotropy) of the measured triangle mesh

Fig. 5.3 shows the stretching of the different parametrizations, a magnification of the reconstructed surface, and the measured points. For all the reconstructions a number of control points were used equal to 150 both in u and v direction and λ equal to 0.01. It is possible to observe that the shape-preserving and mean mesh parametrizations do not allow a good reconstruction because the portion of the area in the u-v space is not proportional to the area of the mesh. If the parametrization preserves the area, it is possible to achieve better reconstruction; see Fig. 5.3E and F.

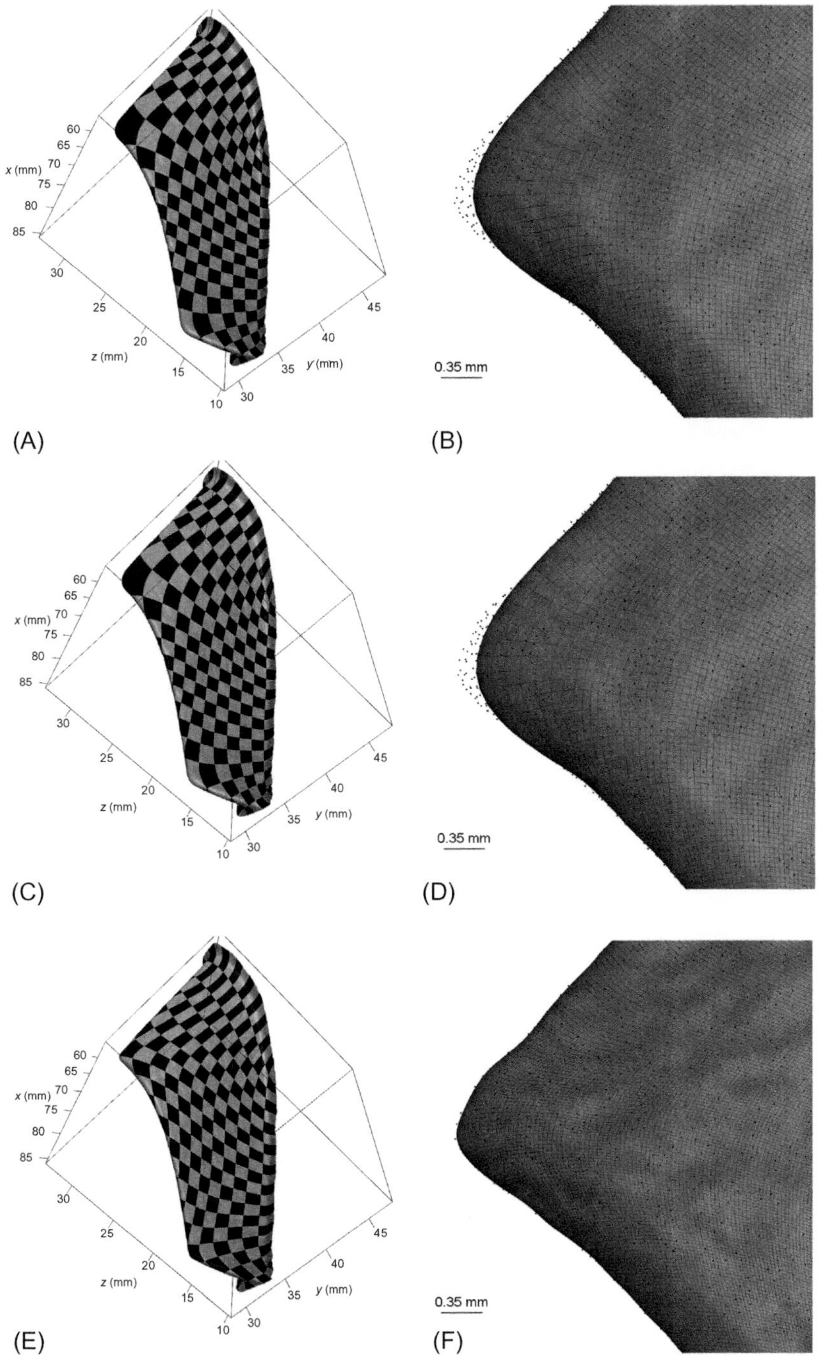

Fig. 5.3 Effect of parametrization: (A, B) shape-preserving, (C, D) mean value, and (E, F) stretch-minimizing parametrizations.

The effect of the smoothing parameters is analyzed using the stretch-minimizing parametrization with the estimated parameters because it leads to better results. The different reconstructions, setting λ equal to 0.01, 0.1, and 1, are shown in Fig. 5.4. It is possible to observe that, by increasing the value of the parameters, the surface is smoother, but some details are lost.

Fig. 5.5 shows the measured mesh and the reconstructed mesh using a number of 150 control points in u and v directions and using smoothing parameters equal to 0.01.

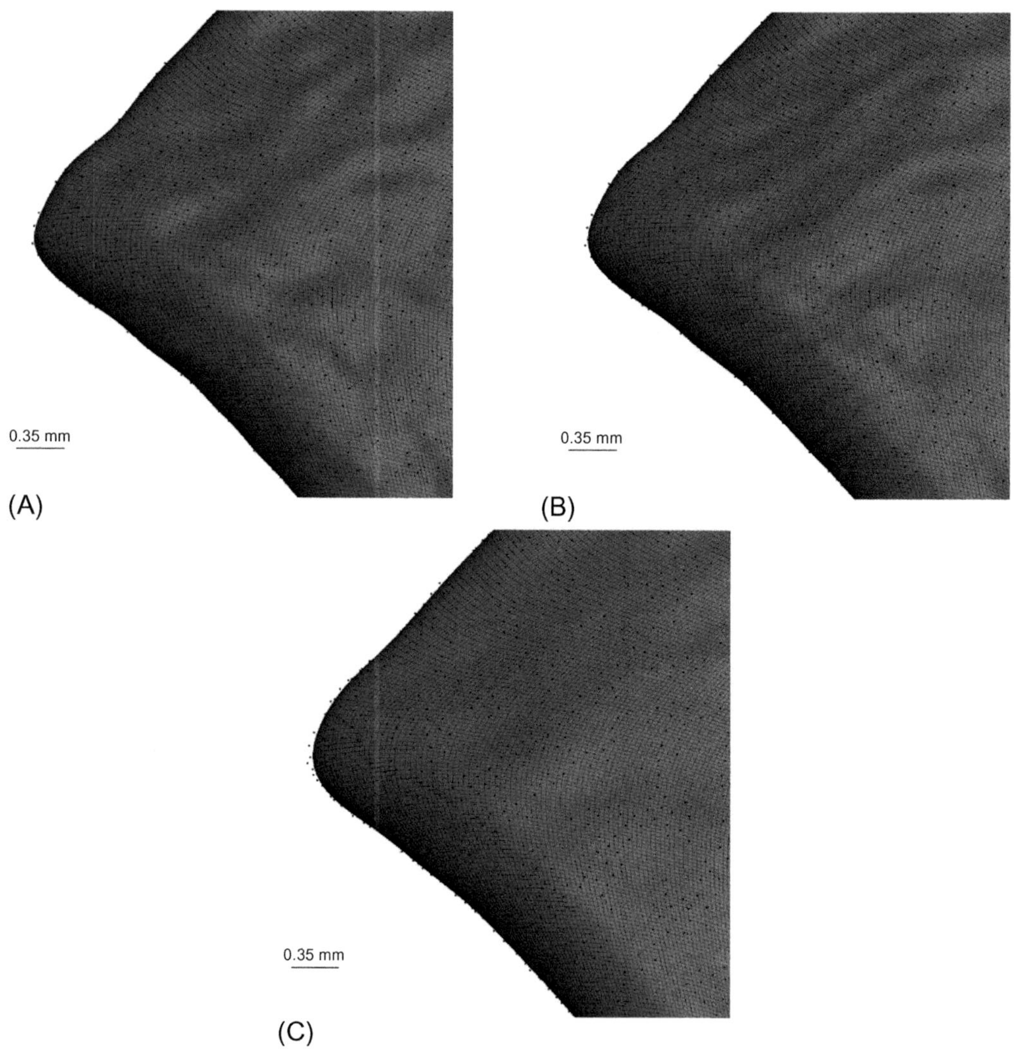

Fig. 5.4 Effect of λ: (A) 0.01, (B) 0.1, and (C) 1.

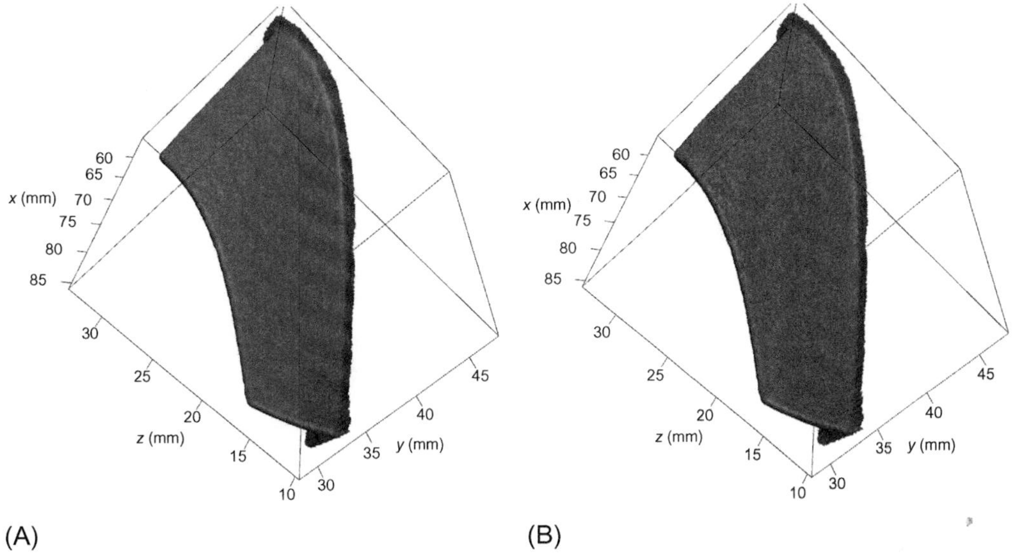

Fig. 5.5 Measured mesh (A) and reconstructed (B).

The second analyzed mesh is the portion of lattice structure with isotropic remeshing. Two hundred control points in u and v direction and a λ parameter equal to 1 were used to reconstruct the surface. In Fig. 5.6 the effect of the mesh parametrization on a portion of the surface can be seen. Both the shapes preserving and mean parametrization do not reconstruct the globules on the surface. Using the stretch-minimizing parametrization, it is possible to reçonstruct the globules, but some smaller details cannot be estimated because there still is a distortion in the parametrization phase. The effectiveness of the reconstruction using the stretch-minimizing parametrization can be observed in Fig. 5.7.

The effect of the anisotropic mesh is shown in Fig. 5.8. It is possible to observe that there are some oscillations at the base of some globules.

5.4 Surface reconstruction with local B-splines models

In the B-spline model described in the previous section, the number of coefficients was defined by the product of the number of coefficients in u $(n_u + 1)$ and v $(n_v + 1)$ direction. This procedure may lead to the introduction of control points that are not needed. Let us suppose that the surface to reconstruct is a free-form surface with some features on an almost planar base, such as the dental bracket in Fig. 5.9. The ideal reconstruction, using some kind of B-spline technique, would be to adaptively add the control points only in the locations where there are the features and minimize the number of control points in the other part of the surface. To overcome this issue hierarchical (H), truncated hierarchical (TH), and locally refined (LR) B-splines were introduced [19]. The construction

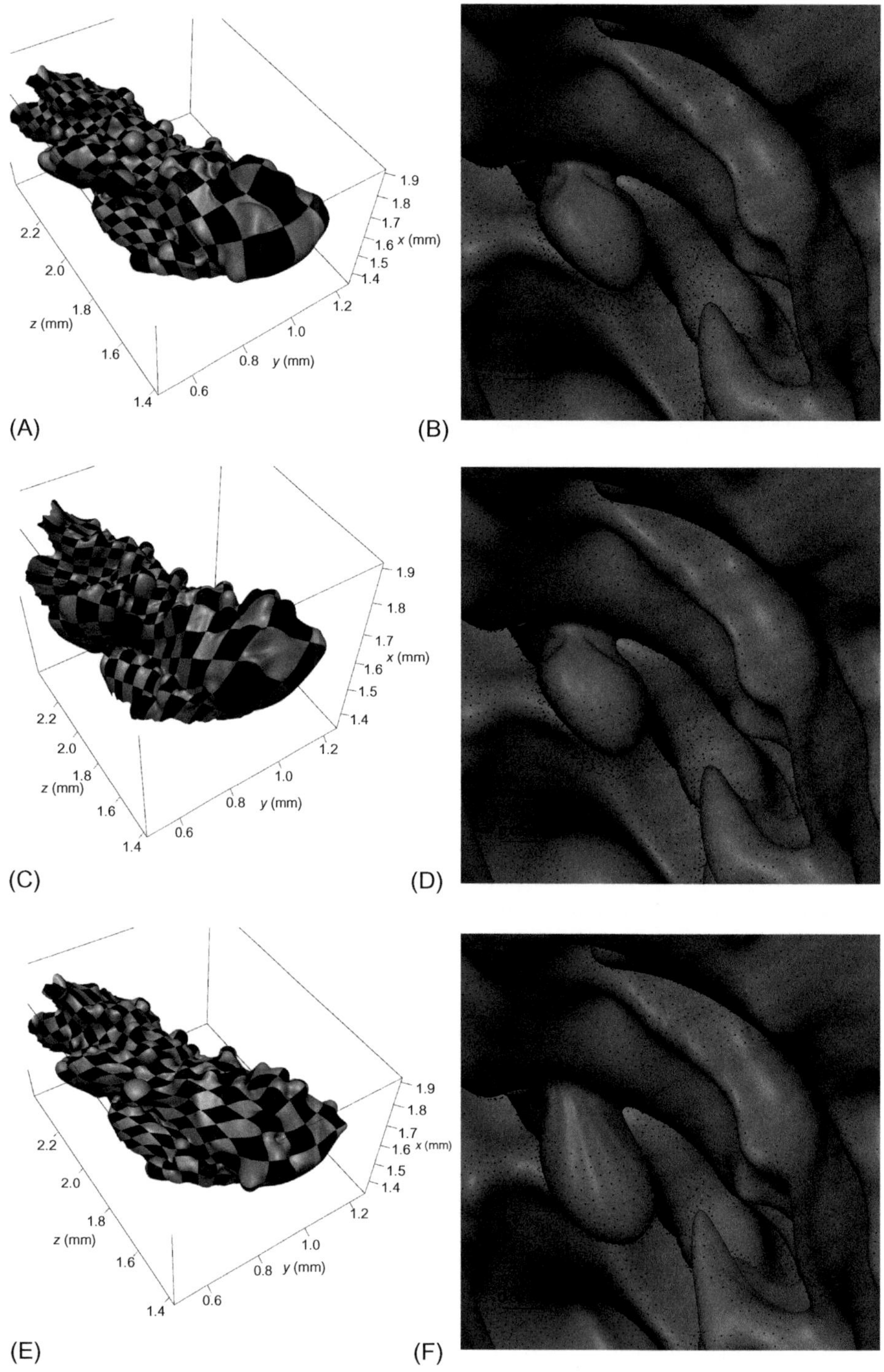

Fig. 5.6 Effect of parametrization: (A, B) shape preserving, (C, D) mean value, and (E, F) stretch minimizing.

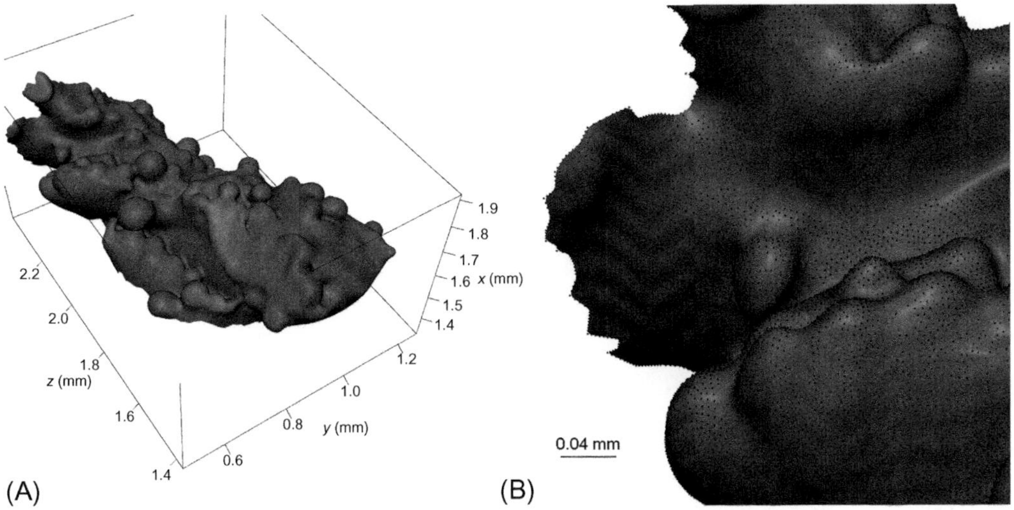

Fig. 5.7 Reconstructed mesh (A) and magnification (B).

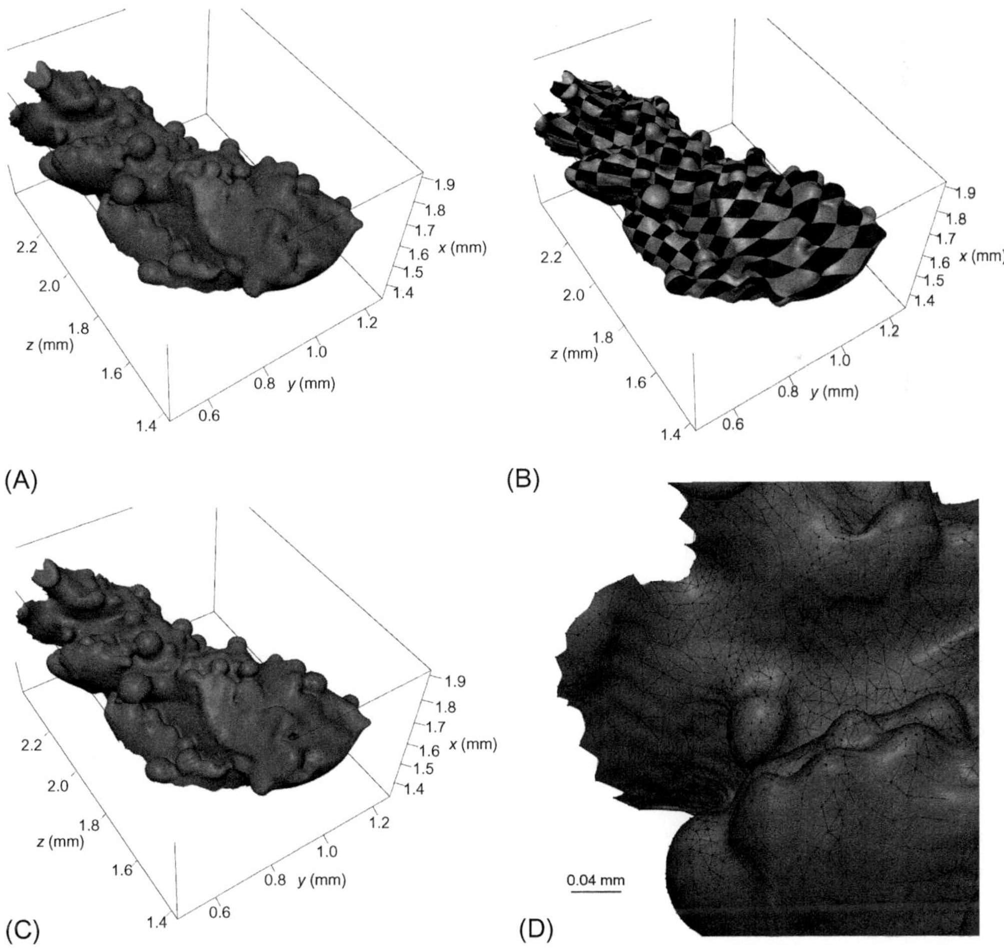

Fig. 5.8 Effect of anisotropic meshing: (A) measured mesh, (B) parametrization, (C) reconstructed surface, and (D) magnification of the reconstructed surface.

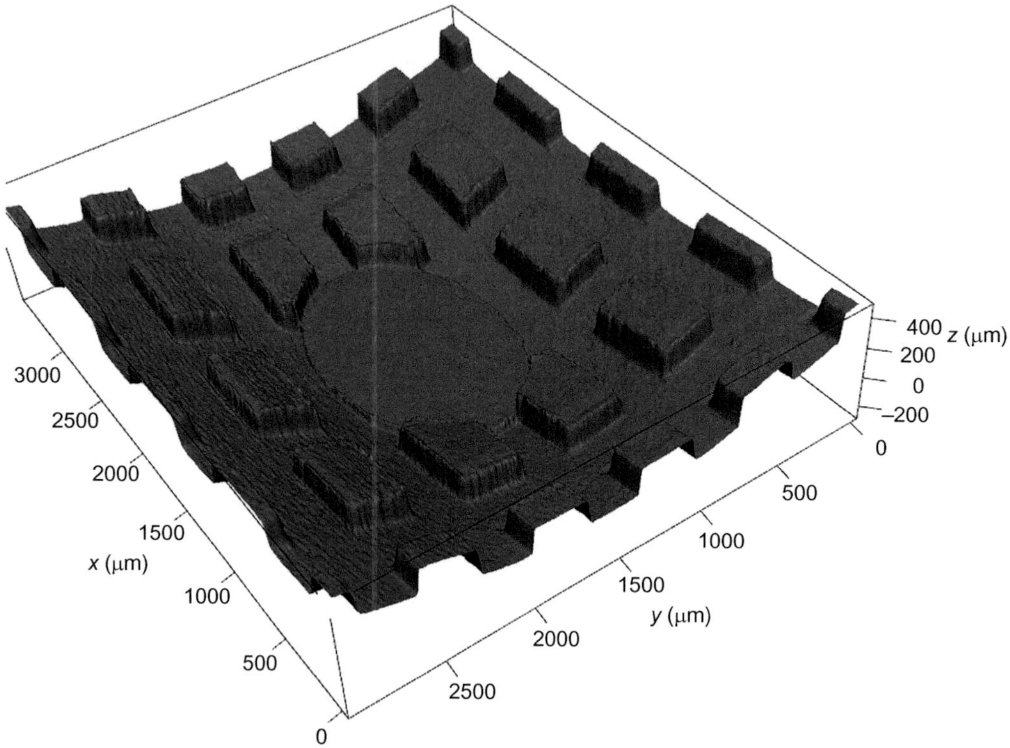

Fig. 5.9 Dental bracket mesh.

and the properties of the TH and LR B-splines are first analyzed, and then, some test cases are presented.

5.4.1 Truncated hierarchical B-splines

Before introducing the TH B-splines, we must first explore H B-splines, which were developed by Forsey and Bartels [20]. Starting from an initial set of knots, only some subsets of the space are refined, while the coarse areas are substituted by a finer basis function. The construction of quadratic H B-splines is shown in Fig. 5.10 (bottom row). The basic functions of refined parts are contained in the coarse mesh, and the missing blue points in the second image of the last row show which basis functions are deleted. In the last image shown is a further iteration of the hierarchical process. It should be noted that, although the basis functions are removed, there is still an overlapping between different hierarchical levels as seen in Fig. 5.11. If the refinement process is evaluated multiple times, the number of overlapping functions can rapidly increase. This behavior has a negative impact on the solution of linear systems. To remove the overlapping

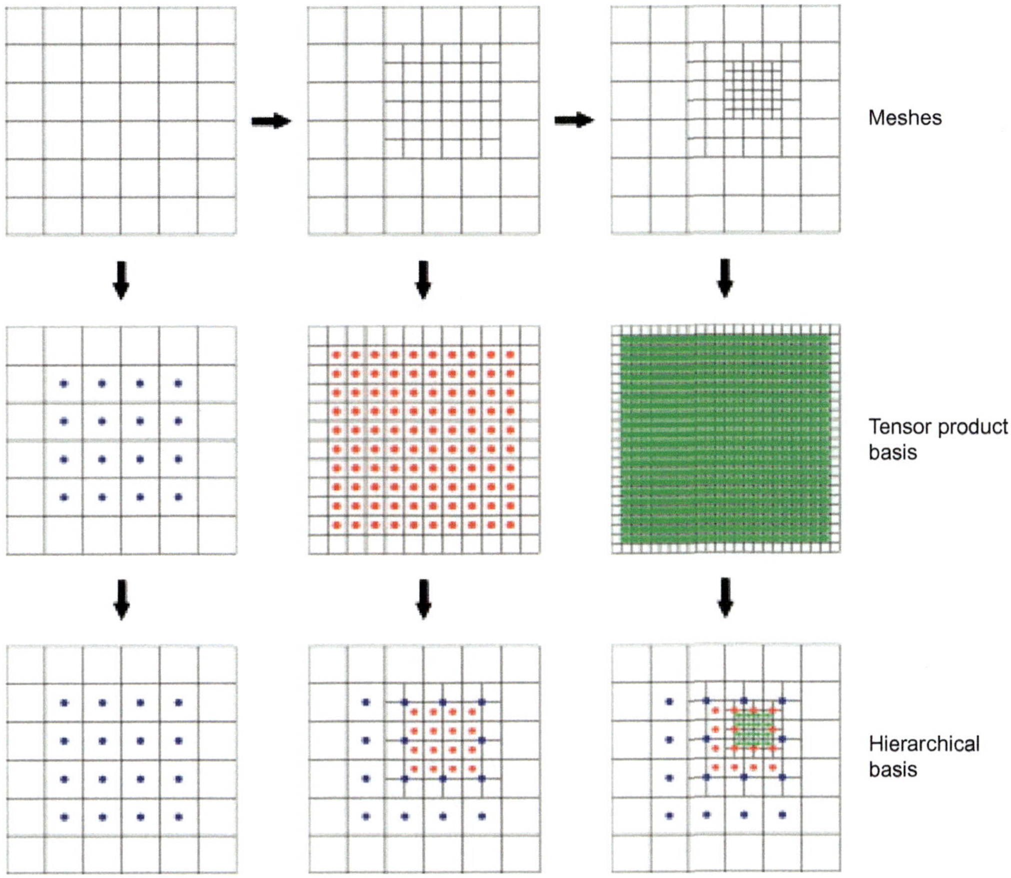

Fig. 5.10 2D H B-spline basis. Top row: hierarchical refinement. Middle row: tensor product basis functions. Bottom row: Basis function of the H B-splines at each level of accuracy. *(Credit: Forsey DR, Bartels RH. Hierarchical B-spline refinement. SIGGRAPH Comput Graph 1988;22(4):205–12.)*

between basis functions at different levels, the TH B-splines were introduced by Giannelli et al. [21] (Fig. 5.12). The idea is to truncate the basis function of the coarse mesh to reduce the overlapping area. The properties of the TH B-splines are as follows:

- The basis functions are linearly independent.
- The spaces spanned by the basis are nested.
- The basis maintains a partition of unity.

The control points are estimated using Eq. (5.2) starting with a tensor product spline with few internal knots. At each iteration the portions of the surface, whose distance to the points is higher than a set tolerance, are refined. The C++ code can be downloaded from the G+Smo page [22].

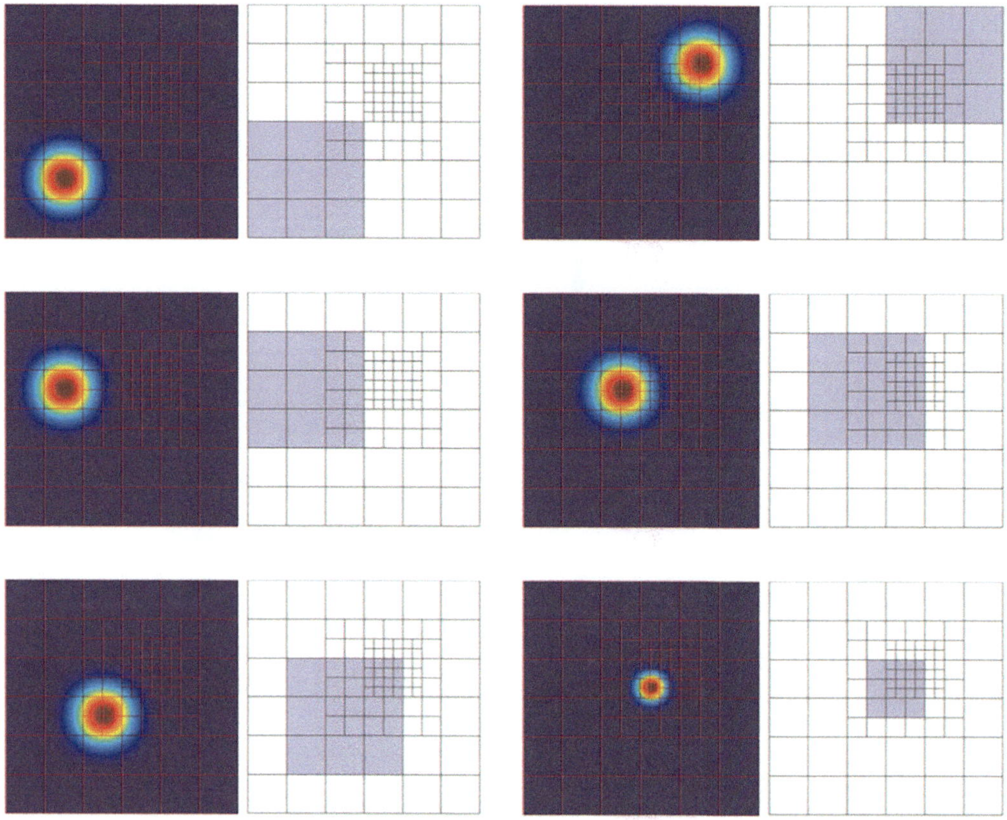

Fig. 5.11 H B-spline basis: on the left shown is the evaluated point while on the right the support. Note that, for the evaluation of each points in the example, the basis at different hierarchical level has to be evaluated. *(Credit: Forsey DR, Bartels RH. Hierarchical B-spline refinement. SIGGRAPH Comput Graph 1988;22(4):205–12.)*

5.4.2 Locally refined B-splines

LR B-spline was introduced by Dokken et al. [23]. The refinement procedure of the LR B-splines is based on the knot insertion procedure. A new knot is inserted for each refinement step; the old B-spline knots are then split into two new ones.

The properties of the LR B-splines are as follows:

- The basis functions are linearly independent.
- The spaces are nested.
- The LR B-splines defined on a mesh are not affected by the order in which each refinement is performed.
- Partition to unity.

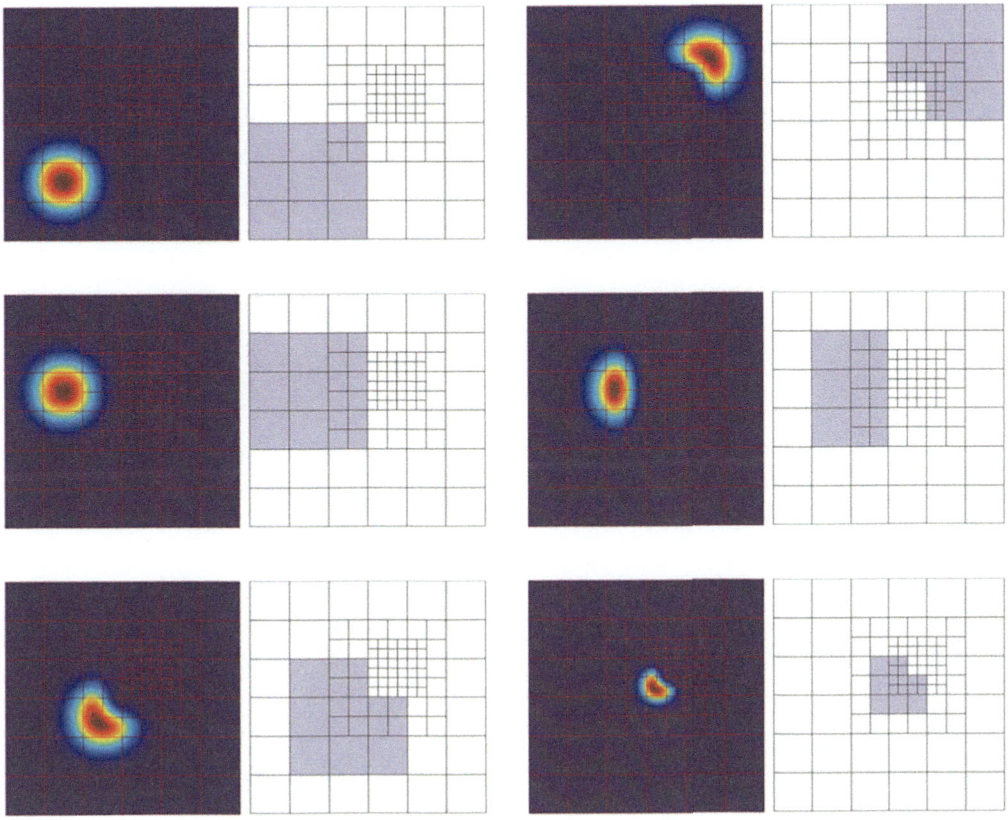

Fig. 5.12 TH B-spline basis: on the left shown is the evaluated point while on the right the support. The number of overlapping basis is reduced compared with the H B-splines. *(Credit: Forsey DR, Bartels RH. Hierarchical B-spline refinement. SIGGRAPH Comput Graph 1988;22(4):205–12.)*

Fig. 5.13 shows the refined meshes and the basis active on a point evaluation. The overlapping between different levels is lower than the H B-splines but higher compared with the TH B-splines.

The computation of the coefficients is computed using the multilevel B-spline algorithm (MBA) [24]. Starting from a coarse grid, the portion of the surface where the distance is not below a tolerance is split. The new control point is computed as

$$\boldsymbol{p}_{ij} = \frac{\sum_{c \in N_{ij}} \left[\gamma_{ij} B_i(u_c) B_j(v_c)\right]^2 \boldsymbol{\phi}_c}{\sum_{c \in N_{ij}} \left[\gamma_{ij} B_i(u_c) B_j(v_c)\right]^2}$$

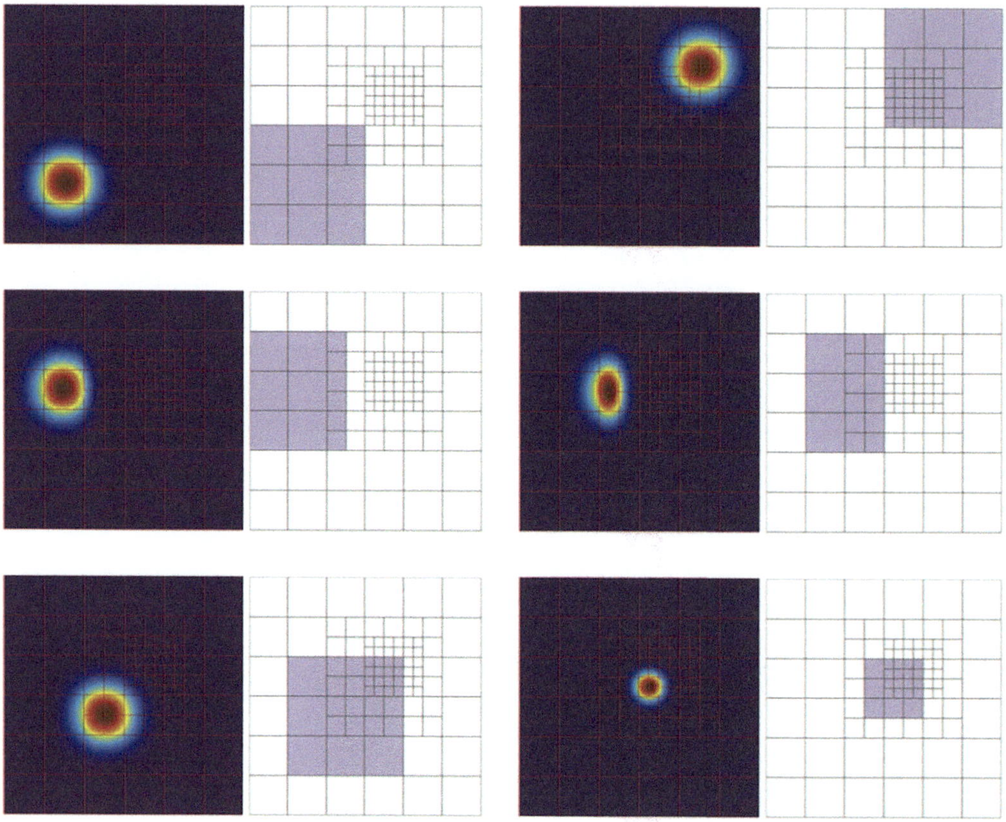

Fig. 5.13 LR B-spline basis: on the left shown is the evaluated point while on the right the support. *(Credit: Forsey DR, Bartels RH. Hierarchical B-spline refinement. SIGGRAPH Comput Graph 1988;22(4): 205–12.)*

where N_{ij} refers to all the points in the support of $B_i(u_c)B_j(v_c)$ and γ_{ij} is a scaling factor proper of the LR B-splines and

$$\boldsymbol{\phi}_c = \frac{\gamma_{ij} B_i(u_c) B_j(v_c) \mathbf{x}(u_c, v_c)}{\Sigma_{(k,\, l) \in N_c} [\gamma_{kl} B_k(u_c) B_l(v_c)]^2}$$

where N_c refers to all the B-spline basis that have the point (u_c, v_c) in their support.

The MBA algorithm was proposed by Lee et al. [25]; it was designed to maximize the speed of computation for the approximation function. It does not require the inversion of a linear system, but the coefficients are computed using the point in the splines support. The C++ code can be downloaded from GitHub [26].

5.4.3 Test cases

In this section the surfaces are reconstructed using the local refinement algorithm. Fig. 5.14 shows a magnification of the impeller blade reconstructed using different mesh parametrizations. A degree 3 B-splines in both u and v direction with a tolerance of 0.03 mm and a λ equal to 10^{-7} (for the TH B-splines algorithm) is used to reconstruct the mesh. The refinement methods allow a better approximation in the portions of the surface where there is a high distortion, but the best results are obtained with the stretch-minimizing parametrization.

The reconstruction of the portion of the lattice mesh with both the TH and LR B-spline algorithm has the same behavior of the least squares B-spline algorithm (see Fig. 5.15). Degree 3 B-splines with a tolerance of 0.001 mm, λ equal to 10^{-7}, and a maximum number of 10 iterations were used as input parameters for the algorithms.

The last test case regards the reconstruction of the dental bracket. Shown in Fig. 5.16 is the measured mesh and the reconstruction with the B-spline method. To compute the reconstruction, since there is a high difference between the range of the three coordinates, the values are first coded between 0 and 1. The surface is approximated as a z function of the other coordinates: $z = z(x, y)$. A total of 100×100 control points with a λ equal to 0.001 was used in the LS algorithm, while degree 2 B-splines with a maximum number of 10 iterations, a tolerance of 5%, and a λ equal to 10^{-7} were all set for the local methods. From a magnification of the reconstructed meshes (see Fig. 5.17), it is possible to conclude that the LS algorithm added some oscillations on the reconstruction that can be avoided using the local refining algorithms.

5.5 Surface reconstruction with triangular Bezier surfaces

Reconstruction with triangular Bezier patches is different compared with the methods presented in the previous sections because the aim is to fit a Bezier triangular patch to each triangle of the mesh. The goal is to interpolate the points, but the resultant surfaces have a higher degree of accuracy and arguably better properties. A triangular Bezier patch is defined as a parametric polynomial surface in $\mathbb{R}^3$ whose parameter space is defined on a triangular domain T [27]. Let T be a triangle with vertices $\boldsymbol{v}_0$, $\boldsymbol{v}_1$, and $\boldsymbol{v}_2$ and $\boldsymbol{x} \in T$; a point on the Bezier patch of degree n is computed as

$$\boldsymbol{r}(x) = \sum_{|\boldsymbol{i}|=n} \boldsymbol{p}_i B_i^n(\boldsymbol{u})$$

where $\boldsymbol{i} = (i, j, k)^T$ is a vector of indices, $|\boldsymbol{i}| = i + j + k$, $\boldsymbol{u} = (u, v, w)^T = (u_0, u_1, u_2)^T$ is a vector of barycentric coordinates, $\boldsymbol{p}_i \in \mathbb{R}^3$ are the control points, and

$$B_{\boldsymbol{i}}^n(\boldsymbol{u}) = \frac{n!}{i!j!k!} u^i v^j w^k, i + j + k = n$$

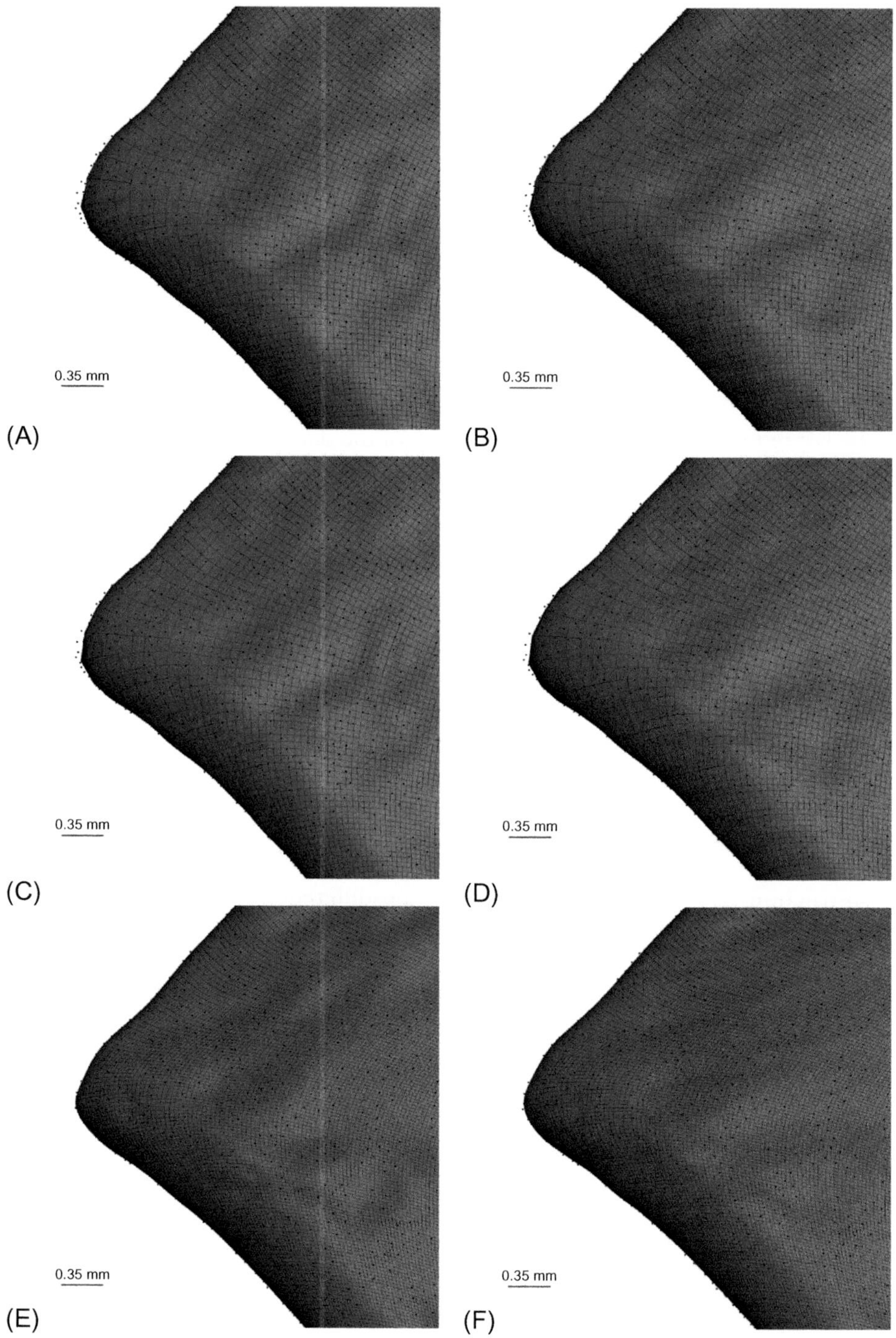

Fig. 5.14 Magnification of the reconstruction using the local refinement B-spline algorithm: left, TH B-splines; right, LR B-spline; first row, shape-preserving parametrization; second row, mean value parametrization; and third row, stretch-minimizing parametrization.

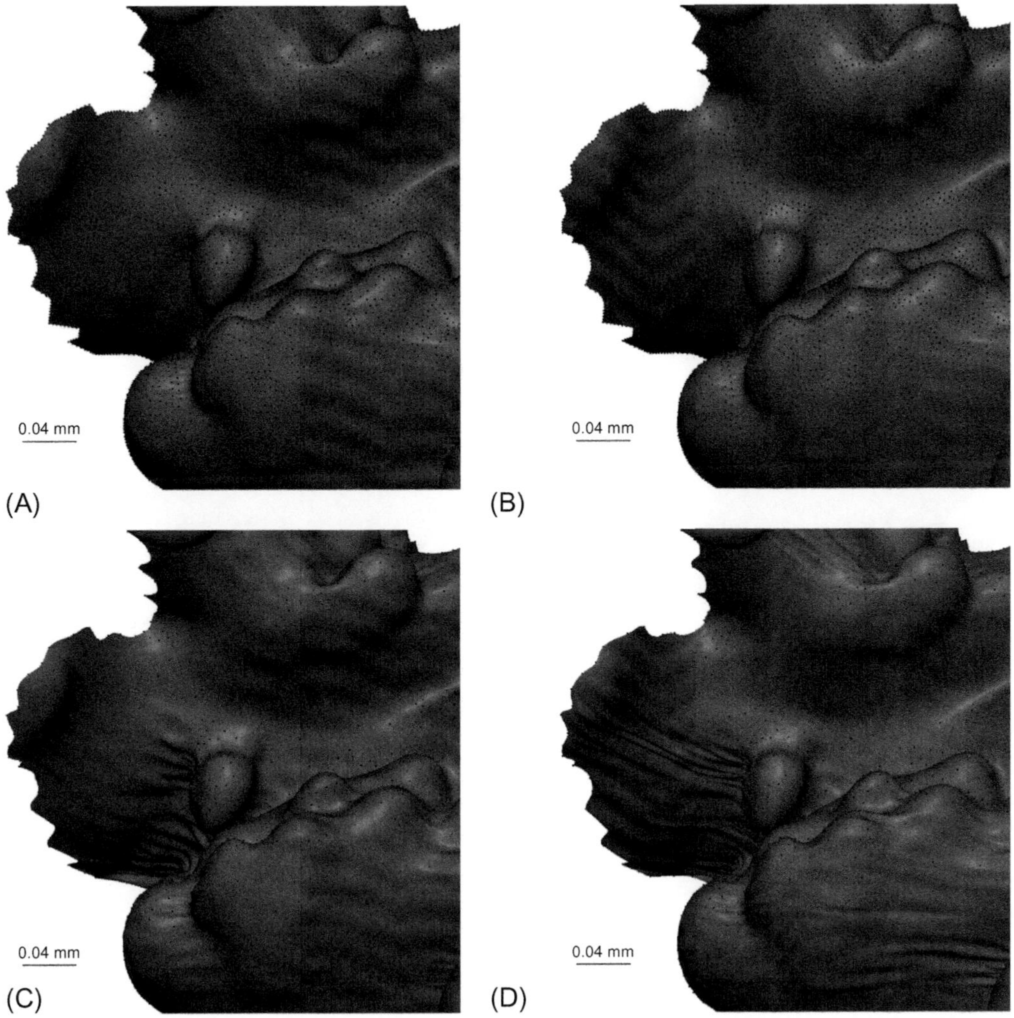

Fig. 5.15 Magnification of the reconstruction of the portion of lattice mesh using the local refinement B-spline algorithm: left, TH B-splines; right, LR B-spline; first row, isotropic mesh; and second row, anisotropic mesh.

is the Bernstein polynomial of degree n. The evaluation of a point as a function of the parametric domain is performed through the de Casteljau algorithm; for a detail explanation, see [27]. To compute the control points, the directional derivatives are needed; thus the computation of these quantities is now introduced.

The derivative of a Bezier triangular patch is defined on a direction $\boldsymbol{d} = (u_d, v_d, w_d)^T$ in the parameters' space as

$$\boldsymbol{r}_d(\boldsymbol{u}) = D_d\boldsymbol{r}(\boldsymbol{u}) = \sum_{|\boldsymbol{i}|=n} \boldsymbol{p}_i D_d B_i^n(\boldsymbol{u})$$

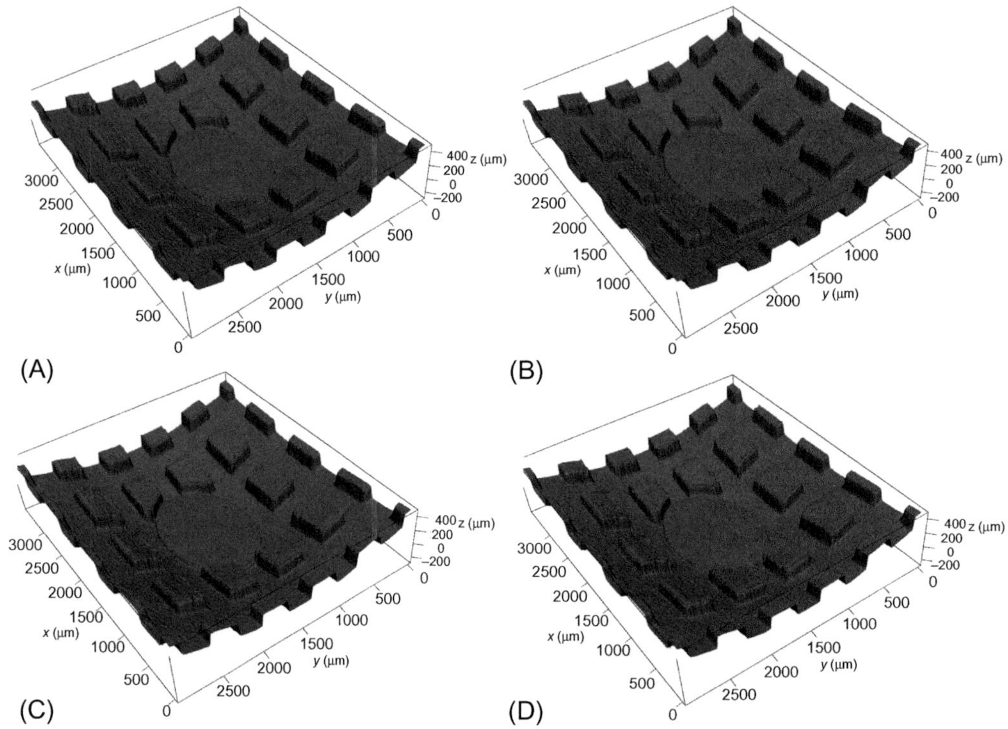

Fig. 5.16 Reconstruction of the dental bracket. Measured mesh (A), least squares B-splines (B), TH B-splines (C), and LR B-splines (D).

and applying the chain rule

$$
\begin{aligned}
D_d B_{\boldsymbol{i}}^n(\boldsymbol{u}) &= \sum_{k=0}^{2} \frac{\partial B_{\boldsymbol{i}}^n(\boldsymbol{u})}{\partial u_k} D_d u_k \\
&= n \sum_{k=0}^{2} B_{\boldsymbol{i}-\boldsymbol{e}_k}^{n-1}(\boldsymbol{u}) D_d u_k
\end{aligned}
$$

where $\boldsymbol{e}_k$ is a vector of zeros except for the kth element that is equal to one. The directional derivative is

$$
\boldsymbol{r}_d(\boldsymbol{u}) = n \sum_{|\boldsymbol{i}|=n-1} B_{\boldsymbol{i}}^{n-1} \sum_{k=0}^{2} D_d u_k \boldsymbol{p}_{\boldsymbol{i}+\boldsymbol{e}_k}.
$$

5.5.1 Degree 2 interpolation

A degree 2 Bezier patch allows a second-order approximation of the surface. It is defined by six control points as shown in Fig. 5.18.

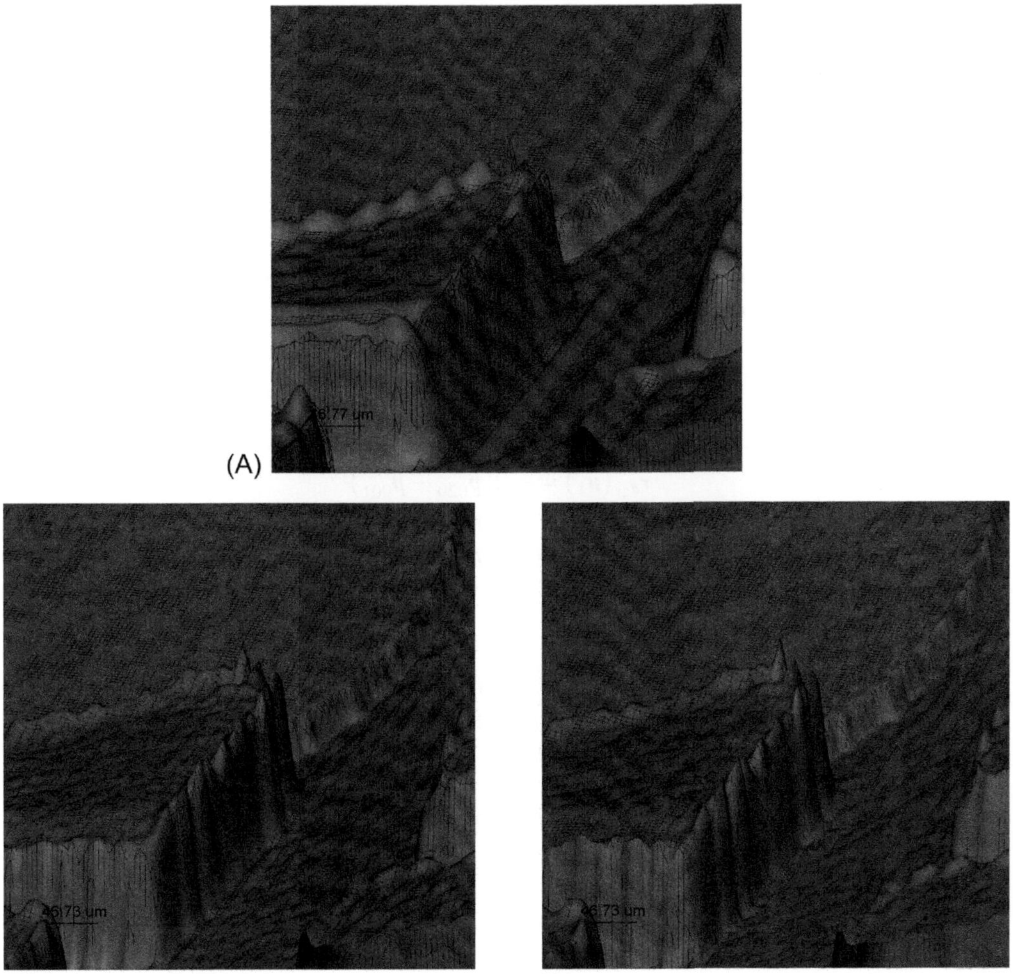

(A)

(B) (C)

Fig. 5.17 Magnification of the reconstruction of the dental bracket: least squares B-splines (A), TH B-splines (B), and LR B-splines (C).

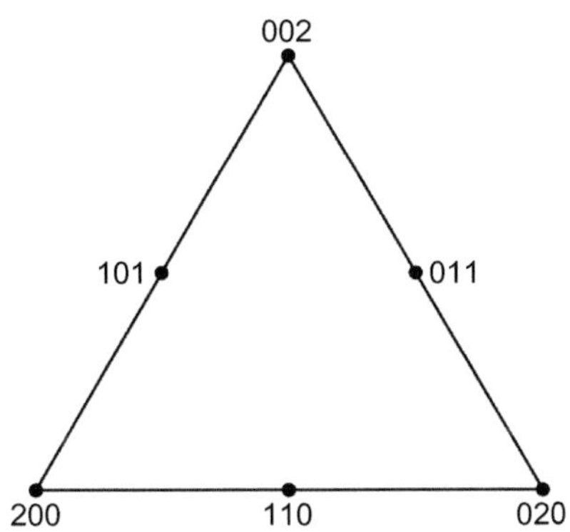

Fig. 5.18 Control points of a triangular Bezier patch of degree 2.

To interpolate the triangular mesh, the values of the control points must coincide with the vertices of the triangle:

$$\begin{aligned} \boldsymbol{r}(\boldsymbol{u}_0) &= \boldsymbol{p}_{200} \\ \boldsymbol{r}(\boldsymbol{u}_1) &= \boldsymbol{p}_{020} \\ \boldsymbol{r}(\boldsymbol{u}_2) &= \boldsymbol{p}_{002} \end{aligned}$$

where $\boldsymbol{u}_0 = (1, 0, 0)^T$, $\boldsymbol{u}_1 = (0, 1, 0)^T$, and $\boldsymbol{u}_2 = (0, 0, 1)^T$ are the vertices of the triangle.

The directional derivatives at the vertices are

$$\begin{aligned} \boldsymbol{r}_{\boldsymbol{d}_{10}}(\boldsymbol{u}_0) &= -2(\boldsymbol{p}_{200} - \boldsymbol{p}_{110}) \\ \boldsymbol{r}_{\boldsymbol{d}_{20}}(\boldsymbol{u}_0) &= -2(\boldsymbol{p}_{020} - \boldsymbol{p}_{101}) \\ \boldsymbol{r}_{\boldsymbol{d}_{01}}(\boldsymbol{u}_1) &= -2(\boldsymbol{p}_{020} - \boldsymbol{p}_{110}) \\ \boldsymbol{r}_{\boldsymbol{d}_{21}}(\boldsymbol{u}_1) &= -2(\boldsymbol{p}_{020} - \boldsymbol{p}_{001}) \\ \boldsymbol{r}_{\boldsymbol{d}_{02}}(\boldsymbol{u}_2) &= -2(\boldsymbol{p}_{002} - \boldsymbol{p}_{101}) \\ \boldsymbol{r}_{\boldsymbol{d}_{12}}(\boldsymbol{u}_2) &= -2(\boldsymbol{p}_{002} - \boldsymbol{p}_{011}) \end{aligned}$$

where

$$\begin{aligned} \boldsymbol{d}_{10}(\boldsymbol{u}) &= -\boldsymbol{d}_{01}(\boldsymbol{u}) = \boldsymbol{v}_1 - \boldsymbol{v}_0 = (-1, 1, 0)^T \\ \boldsymbol{d}_{20}(\boldsymbol{u}) &= -\boldsymbol{d}_{02}(\boldsymbol{u}) = \boldsymbol{v}_2 - \boldsymbol{v}_0 = (-1, 0, 1)^T \\ \boldsymbol{d}_{21}(\boldsymbol{u}) &= -\boldsymbol{d}_{12}(\boldsymbol{u}) = \boldsymbol{v}_2 - \boldsymbol{v}_1 = (0, -1, 1)^T. \end{aligned}$$

As it is possible to observe, the middle control points are used in the computation of directional derivatives at different vertices; for example, $\boldsymbol{p}_{110}$ is present both in the computation of $\boldsymbol{r}_{\boldsymbol{d}_{10}}(\boldsymbol{u}_0)$ and $\boldsymbol{r}_{\boldsymbol{d}_{01}}(\boldsymbol{u}_1)$.

To estimate the middle control points, it is possible to compute the average of the values using the following terms to estimate:

$$\begin{aligned} \boldsymbol{p}_{110} &= \frac{1}{2}\left(\boldsymbol{p}_{200} + \frac{\boldsymbol{r}_{\boldsymbol{d}_{10}}(\boldsymbol{u}_0)}{2} + \boldsymbol{p}_{020} + \frac{\boldsymbol{r}_{\boldsymbol{d}_{01}}(\boldsymbol{u}_1)}{2}\right) \\ \boldsymbol{p}_{101} &= \frac{1}{2}\left(\boldsymbol{p}_{200} + \frac{\boldsymbol{r}_{\boldsymbol{d}_{20}}(\boldsymbol{u}_0)}{2} + \boldsymbol{p}_{002} + \frac{\boldsymbol{r}_{\boldsymbol{d}_{02}}(\boldsymbol{u}_2)}{2}\right) \\ \boldsymbol{p}_{011} &= \frac{1}{2}\left(\boldsymbol{p}_{020} + \frac{\boldsymbol{r}_{\boldsymbol{d}_{21}}(\boldsymbol{u}_1)}{2} + \boldsymbol{p}_{002} + \frac{\boldsymbol{r}_{\boldsymbol{d}_{12}}(\boldsymbol{u}_2)}{2}\right). \end{aligned}$$

It should be noted that, while using this method to compute the middle control points, if the directional derivatives are recomputed, they do not correspond to the ones used to estimate the points. Since the middle control points depend on the numerical order of directional derivatives at different vertices, it is not possible to reconstruct a C^1 surface. C^1 surfaces are surfaces with a continuous first derivative.

Using the orthogonality between the first-order directional derivatives and the vertex normal, a possible estimation can be

$$r_{d_{ji}}(u) = R_{ji} n_i$$

where n_i is the normal of the ith vertex and R_{ji} is a $\frac{\pi}{2}$ rotation matrix of the vector n_i toward the vector $(r(u_j) - r(u_i))$.

The vertex normal can be computed as the weighted average of the incident face normal [4]:

$$n_i = \frac{\sum_{t \in N_i} \alpha_t n_t}{\sum_{t \in N_i} \alpha_t}$$

where N_i represents the face neighborhood of the ith vertex and α_t is the incident angle of.

5.5.2 Degree 5 interpolation

To get an approximation of a surface of class C^1, each of the control points must be included in only one vertex directional derivative computation, that is, the terms involved in the first order of directional derivative computation are independent. To obtain this result a degree 5 Bezier patch is needed. It has to estimate a total of 21 control points (see Fig. 5.19).

Computing the first and second order of directional derivatives is possible to estimate the control points. Only the estimated values of the bottom left corner are reported

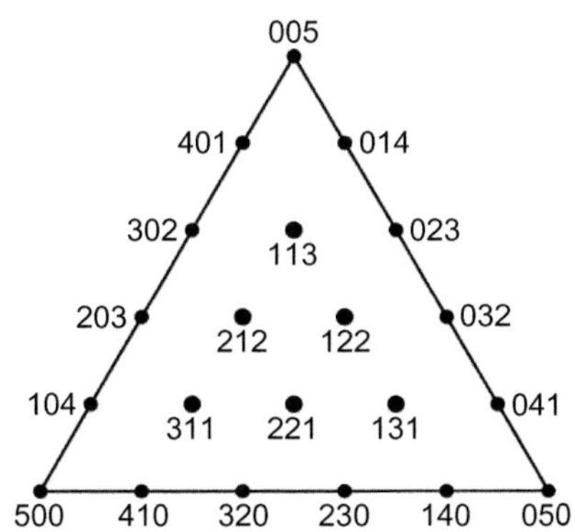

Fig. 5.19 Control points of a triangular Bezier patch of degree 5.

$$
\begin{aligned}
\boldsymbol{b}_{500} &= \boldsymbol{r}(\boldsymbol{u}_0) \\
\boldsymbol{b}_{410} &= \boldsymbol{r}(\boldsymbol{u}_0) + \frac{\boldsymbol{r}_{\boldsymbol{d}_{10}}(\boldsymbol{u}_0)}{5} \\
\boldsymbol{b}_{401} &= \boldsymbol{r}(\boldsymbol{u}_0) + \frac{\boldsymbol{r}_{\boldsymbol{d}_{20}}(\boldsymbol{u}_0)}{5} \\
\boldsymbol{b}_{320} &= \boldsymbol{r}(\boldsymbol{u}_0) + \frac{2}{5}\boldsymbol{r}_{\boldsymbol{d}_{10}}(\boldsymbol{u}_0) + \frac{\boldsymbol{r}_{\boldsymbol{d}_{10}\boldsymbol{d}_{10}}(\boldsymbol{u}_0)}{20} \\
\boldsymbol{b}_{302} &= \boldsymbol{r}(\boldsymbol{u}_0) + \frac{2}{5}\boldsymbol{r}_{\boldsymbol{d}_{20}}(\boldsymbol{u}_0) + \frac{\boldsymbol{r}_{\boldsymbol{d}_{20}\boldsymbol{d}_{20}}(\boldsymbol{u}_0)}{20} \\
\boldsymbol{b}_{311} &= \boldsymbol{r}(\boldsymbol{u}_0) + \frac{1}{5}\boldsymbol{r}_{\boldsymbol{d}_{10}}(\boldsymbol{u}_0) + \frac{1}{5}\boldsymbol{r}_{\boldsymbol{d}_{20}}(\boldsymbol{u}_0) + \frac{\boldsymbol{r}_{\boldsymbol{d}_{10}\boldsymbol{d}_{20}}(\boldsymbol{u}_0)}{20}
\end{aligned}
$$

while the central control points can be computed as [28].

$$
\begin{aligned}
\boldsymbol{b}_{221} = & \frac{8}{15}\boldsymbol{r}_{\boldsymbol{d}_{\eta_2}}(\boldsymbol{\eta}_{01}) - \frac{1}{6}(\boldsymbol{b}_{401} + 4\boldsymbol{b}_{311} + 4\boldsymbol{b}_{131} + \boldsymbol{b}_{041}) \\
& + \frac{1}{12}(\boldsymbol{b}_{500} + 4\boldsymbol{b}_{410} + 6\boldsymbol{b}_{320} + 4\boldsymbol{b}_{230} + \boldsymbol{b}_{140}) \\
& + \frac{1}{12}(\boldsymbol{b}_{410} + 4\boldsymbol{b}_{320} + 6\boldsymbol{b}_{230} + 4\boldsymbol{b}_{140} + \boldsymbol{b}_{050}) \\
\boldsymbol{b}_{122} = & \frac{8}{15}\boldsymbol{r}_{\boldsymbol{d}_{\eta_0}}(\boldsymbol{\eta}_{12}) - \frac{1}{6}(\boldsymbol{b}_{140} + 4\boldsymbol{b}_{131} + 4\boldsymbol{b}_{113} + \boldsymbol{b}_{104}) \\
& + \frac{1}{12}(\boldsymbol{b}_{050} + 4\boldsymbol{b}_{041} + 6\boldsymbol{b}_{032} + 4\boldsymbol{b}_{023} + \boldsymbol{b}_{014}) \\
& + \frac{1}{12}(\boldsymbol{b}_{041} + 4\boldsymbol{b}_{032} + 6\boldsymbol{b}_{023} + 4\boldsymbol{b}_{014} + \boldsymbol{b}_{005}) \\
\boldsymbol{b}_{122} = & \frac{8}{15}\boldsymbol{r}_{\boldsymbol{d}_{\eta_1}}(\boldsymbol{\eta}_{20}) - \frac{1}{6}(\boldsymbol{b}_{014} + 4\boldsymbol{b}_{113} + 4\boldsymbol{b}_{311} + \boldsymbol{b}_{410}) \\
& + \frac{1}{12}(\boldsymbol{b}_{500} + 4\boldsymbol{b}_{401} + 6\boldsymbol{b}_{302} + 4\boldsymbol{b}_{203} + \boldsymbol{b}_{104}) \\
& + \frac{1}{12}(\boldsymbol{b}_{401} + 4\boldsymbol{b}_{302} + 6\boldsymbol{b}_{203} + 4\boldsymbol{b}_{104} + \boldsymbol{b}_{005})
\end{aligned}
$$

where

$$
\begin{aligned}
\boldsymbol{\eta}_{01} &= \frac{\boldsymbol{u}_0 + \boldsymbol{u}_1}{2} = (0.5, 0.5, 0)^T \\
\boldsymbol{\eta}_{12} &= \frac{\boldsymbol{u}_1 + \boldsymbol{u}_2}{2} = (0, 0.5, 0.5)^T \\
\boldsymbol{\eta}_{20} &= \frac{\boldsymbol{u}_2 + \boldsymbol{u}_0}{2} = (0.5, 0, 0.5)^T \\
\boldsymbol{d}_{\eta_0} &= \boldsymbol{u}_0 - \boldsymbol{\eta}_{12} = (1, -0.5, -0.5)^T \\
\boldsymbol{d}_{\eta_1} &= \boldsymbol{u}_1 - \boldsymbol{\eta}_{20} = (-0.5, 1, -0.5)^T \\
\boldsymbol{d}_{\eta_2} &= \boldsymbol{u}_2 - \boldsymbol{\eta}_{01} = (-0.5, -0.5, 1)^T.
\end{aligned}
$$

For the estimation of the control points, both the first- and the second-order derivatives are needed. To compute the first-order derivative, it is possible to use the method presented before, while to compute the second-order ones, the surface is locally approximated with a second-order degree polynomial [29]. For each vertex $\boldsymbol{v}_i$ a height function, a paraboloid, is used to approximate a smooth surface:

$$h(u, v) = \frac{1}{2}\left(\alpha u^2 + 2\beta uv + \gamma v^2\right).$$

The height value of the vertex corresponds to 0. To estimate the parameters a local coordinate system must be introduced. A versor can be represented by the vertex normal; a second versor ($\boldsymbol{a}$) can be computed rotating the normal by $\frac{\pi}{2}$ in the direction of one of the edges of the vertex and the last one as the cross product of the first two ($\boldsymbol{b} = \boldsymbol{n} \times \boldsymbol{a}$). Let $\boldsymbol{T} = [\boldsymbol{a}, \boldsymbol{b}] \in \mathbb{R}^{3\times 2}$, the local coordinates can be computed as

$$\left(u_j, v_j\right) = \boldsymbol{T}^T\left(\boldsymbol{v}_j - \boldsymbol{v}_i\right)$$

and the heights values as

$$\boldsymbol{h}_j = \boldsymbol{v}_j - \boldsymbol{v}_i.$$

The least squares system to solve to compute the coefficients is

$$\begin{bmatrix} \frac{u_0^2}{2} & u_0 v_0 & \frac{v_0^2}{2} \\ \frac{u_1^2}{2} & u_1 v_1 & \frac{v_0^2}{2} \\ \vdots & \vdots & \vdots \\ \frac{u_{n-1}^2}{2} & u_{n-1} v_{n-1} & \frac{v_{n-1}^2}{2} \end{bmatrix} \begin{bmatrix} \alpha_x & \beta_x & \gamma_x \\ \alpha_y & \beta_y & \gamma_y \\ \alpha_z & \beta_z & \gamma_z \end{bmatrix} = \begin{bmatrix} \boldsymbol{h}_0^T \\ \boldsymbol{h}_1^T \\ \vdots \\ \boldsymbol{h}_{n-1}^T \end{bmatrix}.$$

After computing the Hessian as

$$\boldsymbol{H}_x = \begin{bmatrix} \alpha_x & \beta_x \\ \beta_x & \gamma_x \end{bmatrix}$$

$$\boldsymbol{H}_y = \begin{bmatrix} \alpha_y & \beta_y \\ \beta_y & \gamma_y \end{bmatrix}$$

$$\boldsymbol{H}_z = \begin{bmatrix} \alpha_z & \beta_z \\ \beta_z & \gamma_z \end{bmatrix}$$

the second-order partial derivatives can be estimated as

$$\boldsymbol{r}_{\boldsymbol{d}_{ij}\boldsymbol{d}_{kl}}(\boldsymbol{u}) = \left(\boldsymbol{d}_{ij}^T \boldsymbol{H}_x \boldsymbol{d}_{kl}, \boldsymbol{d}_{ij}^T \boldsymbol{H}_y \boldsymbol{d}_{kl}, \boldsymbol{d}_{ij}^T \boldsymbol{H}_z \boldsymbol{d}_{kl}\right)^T.$$

5.5.3 Case studies

Interpolation with a triangular Bezier surface is needed to approximate triangles with a higher approximation order compared with linear ones. In this section the reconstruction of the portion of a lattice surface, the anisotropic mesh, is reconstructed with the triangular Bezier surfaces of degree 2 (Fig. 5.20) and 5 (Fig. 5.21). Since the reconstructed surfaces are of a degree greater than 1, if a magnification is performed, they appear smoother. The difference between a linear interpolation and second or quintic degree can be observed in Fig. 5.20D and Fig. 5.21D, respectively. The two reconstructed

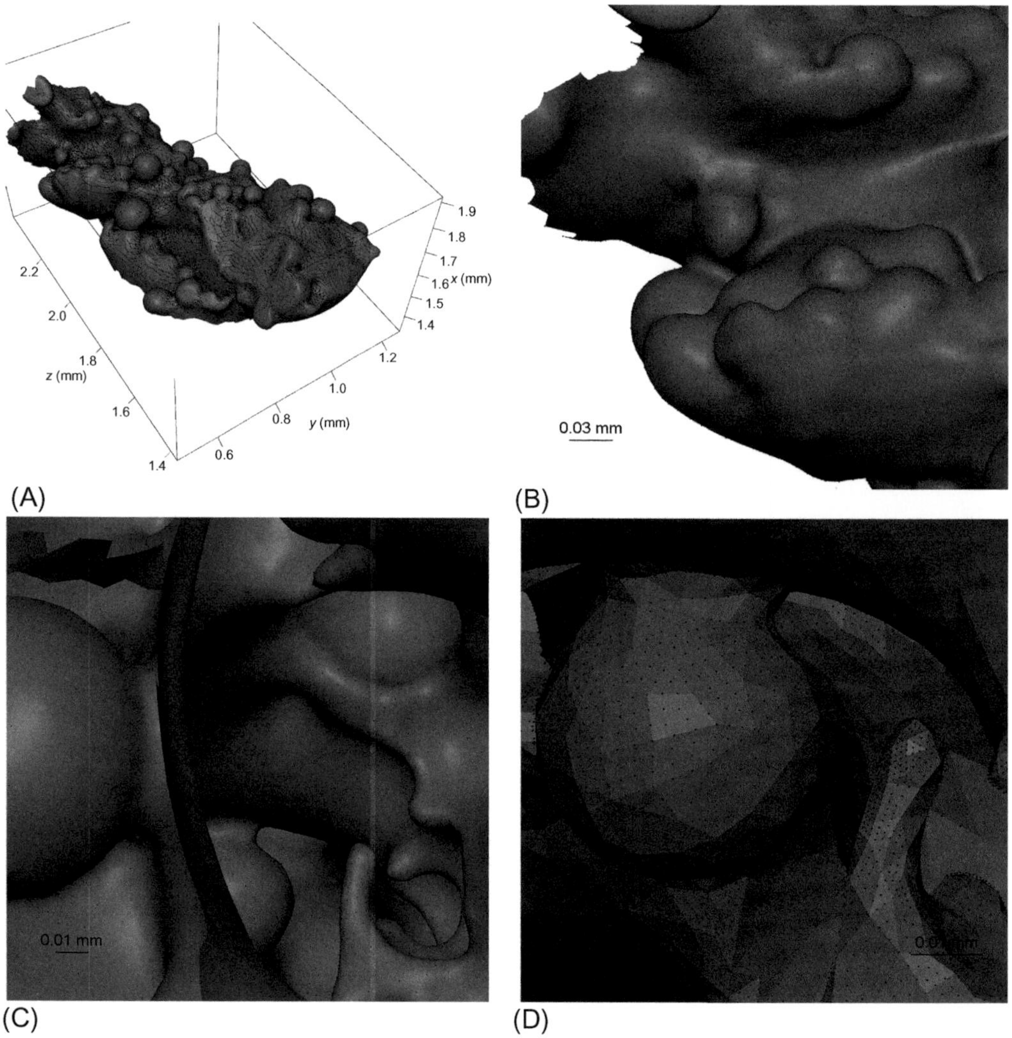

Fig. 5.20 Reconstruction of the portion of lattice surface with a triangular Bezier surface of degree 2: reconstructed surface (A), magnifications of the reconstructed surface (B, C), and magnification of the surface to reconstruct with some of the reconstructed points (D).

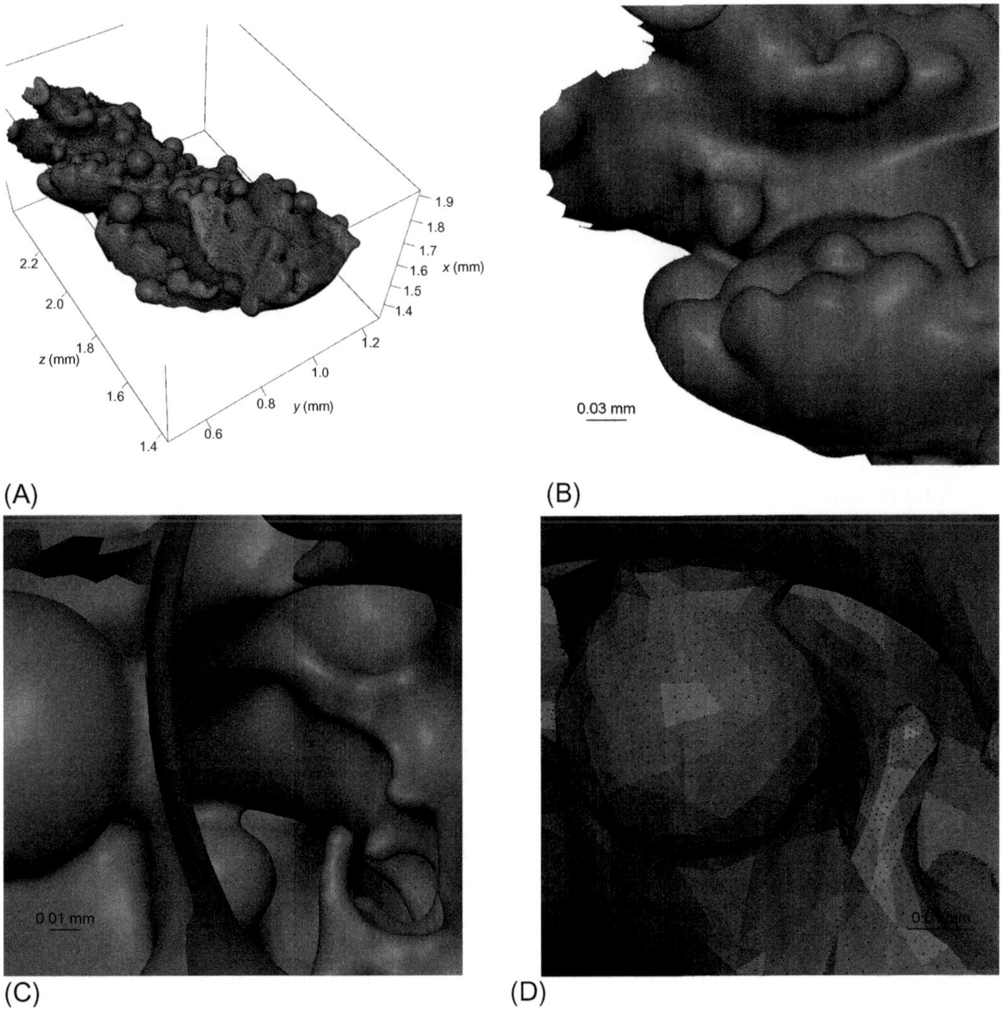

Fig. 5.21 Reconstruction of the portion of lattice surface with a triangular Bezier surface of degree 2: reconstructed surface (A), magnifications of the reconstructed surface (B, C), and magnification of the surface to reconstruct with some of the reconstructed points (D).

surfaces are close to each other, but if a C^1 surface is needed, the most complex surface of degree 5 must be used.

5.6 Implicit function surface reconstruction

The surface reconstruction of implicit functions $\phi(\boldsymbol{x}) = k$ consists of converting a volume of scalar values to a triangular mesh where k is the isovalue (scalar) representing the surface to be found. Implicit functions may be the measurement results of computed tomography (CT) scans [30] or an optical coordinate measuring machine (CMM). The output of these systems is an image, 3D for a CT scan, while 2-D for the optical CMM.

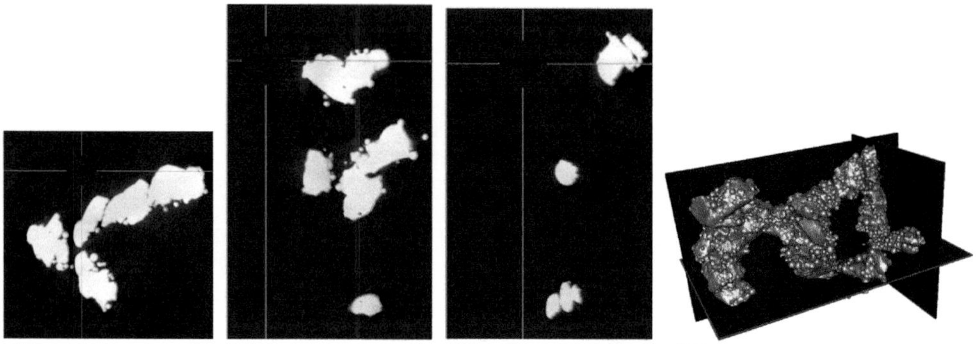

(A) (B) (C) (D)

Fig. 5.22 Reconstruction of a CT scan of a portion of lattice mesh. Axial (A), sagital (B), coronal (C), and views and volume rendering (D).

The identification of the surface (the profile) is represented by the edges of the image, where the gradient of the gray level image reaches its maximum. An example of the volume reconstructed using a CT scan is shown in Fig. 5.22. To use medical imaging linguistics, the images represent some slices of the volume along the axial (z constant), sagittal (y constant), and coronal (x constant) views, while in Fig. 5.22D the volume rendering is shown.

The identification of the contours is usually performed through level set functions, that is, starting from a gray level supplied by the user; the level set (surface) evolves according to a partial differential equation to find the border of the object to reconstruct. For more detail about level set functions, refer to [31]. The output of the level set method is an image representing a signed distance function, that is, an implicit function such that

$$\phi(\boldsymbol{x}) = \begin{cases} -d(\boldsymbol{x}, \Omega) & \text{if the point is inside the object} \\ 0 & \text{if the point is on the boundary of the object} \\ d(\boldsymbol{x}, \Omega) & \text{if the point is outside the object} \end{cases}$$

where $d(\boldsymbol{x}, \Omega)$ is the distance between a point in the space and the boundary of the object (Ω) to reconstruct. The goal of the implicit function reconstruction is to approximate the unknown surface $\phi(\boldsymbol{x}) = 0$. To compute the final signed distance function, the geodesic active contour, implemented in ITK [32], used only the advection term [33]. The computed signed distance function with the extracted surface is shown in Fig. 5.23.

5.6.1 Marching cubes

Among the algorithms used to reconstruct an implicit function, the most famous is the marching cubes [34] (Fig. 5.24). The algorithm was initially developed in 1987 to allow a fast visualization of an isosurface. Let's suppose, without loss of generality, that the surface to extract is represented by the 0 level set function, the algorithm search, for each voxel, if

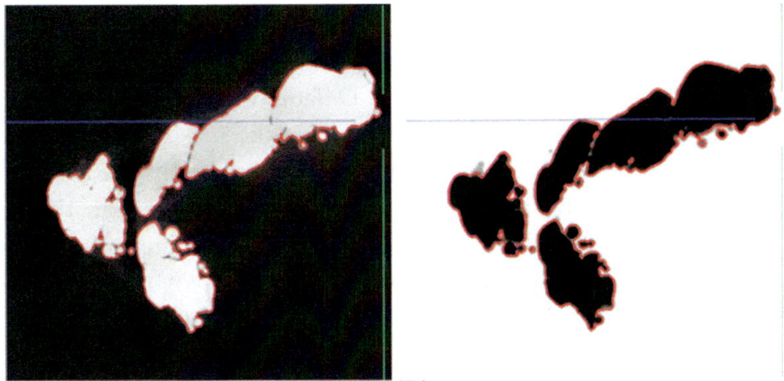

(A) (B)

Fig. 5.23 Reconstruct image of a CT scan (A) and signed distance function (B) with the isosurface at level set equal to 0.

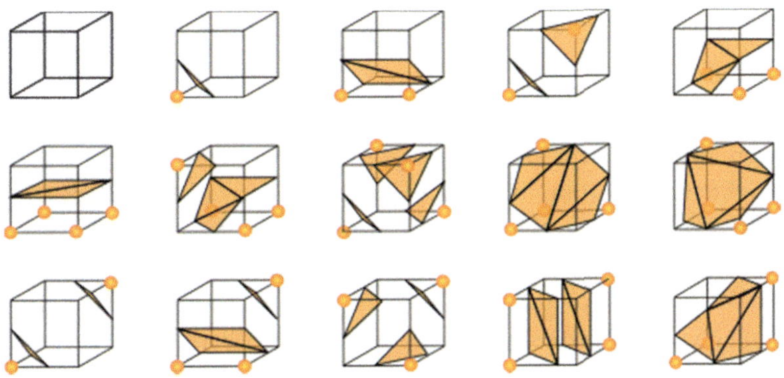

Fig. 5.24 Original marching cubes lookup table [37]. *Credit: By Jmtrivial—Own work, GPL, https://commons.wikimedia.org/w/index.php?curid=1282165.*

there is a zero crossing point in the edges of its neighborhood (a cube). All the possible crossing configurations are shown in Fig. 5.23. For example, in the second cube (row 1, column 2), there are only three crossing points forming a triangle, while in the last image, there are multiple crossing points forming a polygon that can be easily converted to a triangular mesh. The final step is to merge all the duplicated points created by the algorithm. An implementation of the algorithm can be found in VTK [35] or the improvement of the algorithm using Hermite interpolation [36]. After the mesh extraction a mesh simplification can be performed due to the high number of triangles produced. A drawback however is that elongated triangles, triangles with two edges much longer compared with the third one, are produced.

5.6.2 Adaptive surface reconstruction

Another algorithm able to extract the surface information was proposed by Boissonnat and Oudot [38]. Their algorithm uses a local Delaunay triangulation, which starts from an initial set of points inside the surface, and updates the mesh as a new point inside the surface. The algorithm needs the initial points that should lie inside the surface as input parameters. This refers to the maximum allowed distance between the reconstructed mesh and the isosurface and the value representing the radius of a sphere, centered at provided starting point. An example of the evolution of the surface is shown in Fig. 5.25. The surface incrementally adds triangles to better approximate the implicit surface that is shown as a torus in the example. The code can be downloaded as part of the CGAL [39] library. The algorithm produces a mesh with a lower number of triangles compared with the marching cubes, so the simplification step is not needed. A drawback of this method is the time required, which is higher compared with the marching cubes. Additionally, if there are some pores inside the surface, occasionally the shape is not reconstructed.

5.6.3 Test case

The reconstructed lattice mesh with the described algorithms is shown in Fig. 5.26. The reconstruction using the marching cube algorithm produces a mesh with 4,814,392 triangles. From the magnification, it's clear to see that there is a visual effect due to the voxel representation. Using the adaptive surface reconstruction with a maximum distance set to 1 μm, Fig. 5.26E and F, the surface appears visually smoother thanks to a smaller number of triangles (431,948 to be precise). In Fig. 5.26C and D, the mesh is reconstructed using the marching cube algorithm using the quadratic edge collapse decimation implemented in Meshlab [40], producing a number of triangles equal to the mesh of the adaptive surface generator.

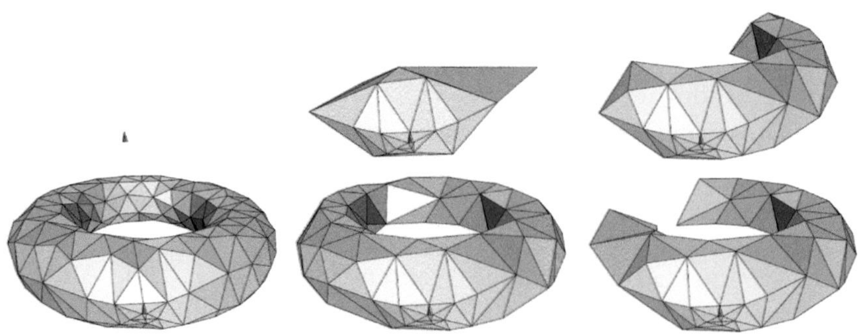

Fig. 5.25 Evolution of the algorithm proposed in [38]. *Credit: https://www.sciencedirect.com/science/article/pii/S1524070305000056.*

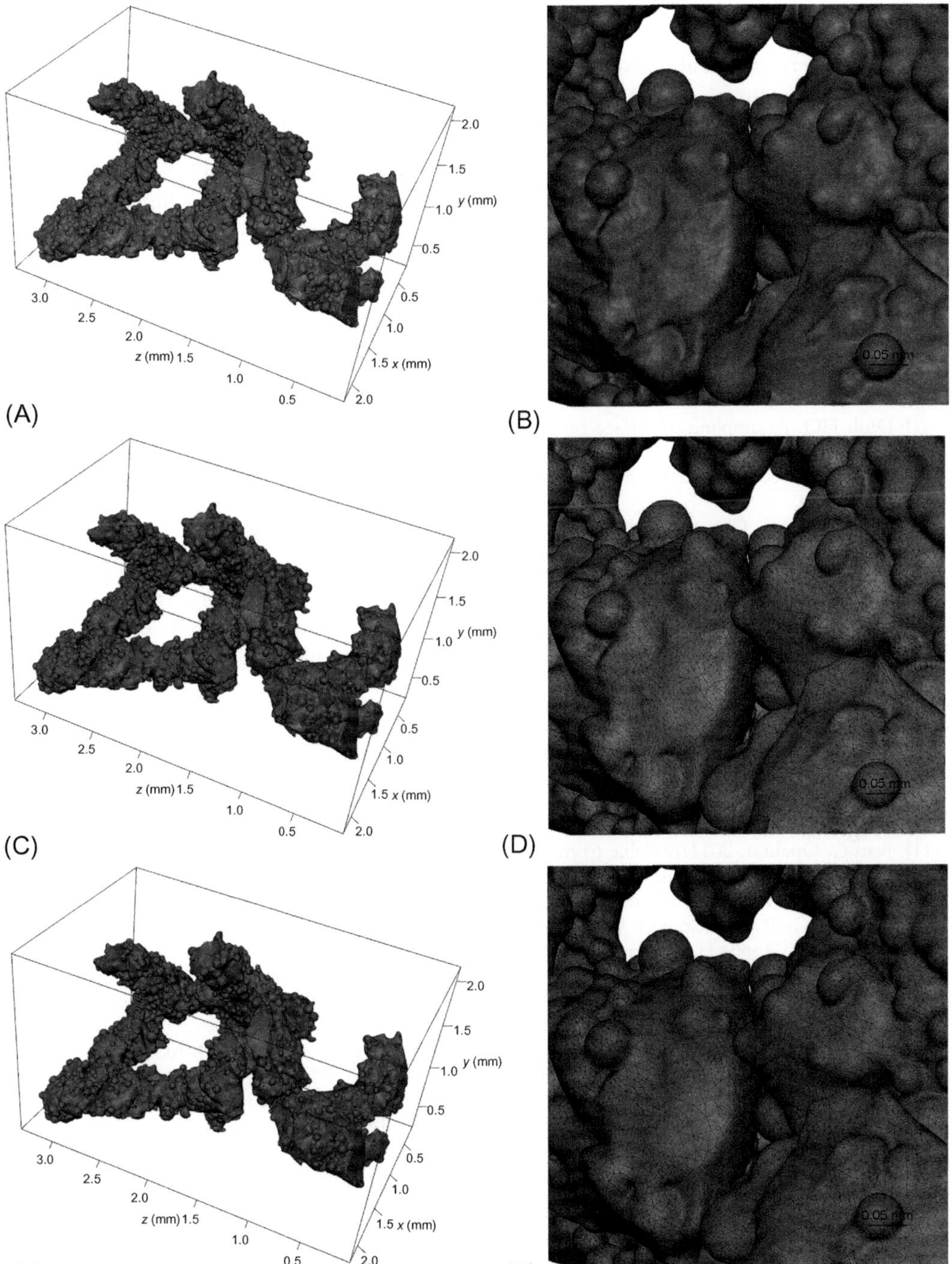

Fig. 5.26 Reconstruction of the portion of lattice mesh and magnification. Marching cubes (A, B), decimation after marching cubes (C, D), and adaptive surface reconstruction (E, F).

5.7 Summary

In this chapter, surface approximation methods have been presented. Surface reconstruction using B-spline basis, global and local, requires an initial parametrization of the measured mesh. They allow the conversion from a set of vertices to a parametric model that can then be used during the design phase. Triangular Bezier surfaces are also presented; they can be used to approximate a triangular mesh with a higher degree of accuracy. Finally, two methods to reconstruct an implicit function are described; they are useful to reconstruct a surface of the computed tomography scan.

References

[1] Dinh HQ. A sampling of surface reconstruction techniques; 2000Georgia Institute of Technology; 2000.
[2] Zhang X. Free-form surface fitting for precision coordinate metrology; 2009University of Huddersfield; 2009.
[3] Lee W, To S, Cheung C. Design and advanced manufacturing technology for freeform optics; 2005Hong Kong Polytechnic University; 2005.
[4] Botsch M, Kobbelt L, Pauly M, Alliez P, Lévy B. Polygon mesh processing. Natick: A K Peters; 2010.
[5] Gu X, Gortler SJ, Hoppe H. Geometry images. ACM Trans Graph 2002;21(3):355–61.
[6] Tutte W. Convex representation of graphs. Proc London Math Soc 1960;10:381–9.
[7] Floater MS. Parametrization and smooth approximation of surface triangulations. Comput Aided Geom Des 1997;14(3):231–50.
[8] Floater MS. Mean value coordinates. Comput Aid Geometr Des 2003;20(1):19–27.
[9] Yoshizawa S, Belyaev A, Seidel H-P, editors. A fast and simple stretch-minimizing mesh parameterization. Proceedings Shape Modeling ApplicationsIEEE; 2004;
[10] Sander PV, Snyder J, Gortler SJ, Hoppe H. Texture mapping progressive meshes. In: Proceedings of the 28th annual conference on Computer graphics and interactive techniques. 383307: ACM; 2001. p. 409–16.
[11] Rau C. OpenGI, 2011Available from:http://opengi;sourceforge.net.
[12] Piegl L, Tiller W. The NURBS book. New York: Springer-Verlag New York, Inc.; 1997.
[13] Hoppe H, DeRose T, Duchamp T, McDonald J, Stuetzle W. Surface reconstruction from unorganized points. SIGGRAPH Comput Graph 1992;26(2):71–8.
[14] Kazhdan M, Hoppe H. Screened Poisson surface reconstruction. ACM Trans Graph 2013;32(3):1–13.
[15] Amenta N, Choi S, Dey TK, Leekha N. A simple algorithm for homeomorphic surface reconstruction. In: Proceedings of the sixteenth annual symposium on Computational geometry; Clear Water Bay, Kowloon, Hong Kong. 336207: ACM; 2000. p. 213–22.
[16] Amenta N, Choi S, Kolluri RK. The power crust. In: Proceedings of the sixth ACM symposium on Solid modeling and applications; Ann Arbor, Michigan, USA. 376986: ACM; 2001. p. 249–66.
[17] Johnson RA, Wichern DW. Applied multivariate statistical analysis. 6th ed. New Jersey: Pearson; 2007.
[18] Pagani L. Least squares B-spline approximation (LSBA). Available from: https://github.com/lucapagani/LSBA; 2015.
[19] Johannessen KA, Remonato F, Kvamsdal T. On the similarities and differences between classical hierarchical, truncated hierarchical and LR B-splines. Comput Methods Appl Mech Eng 2015;291:64–101.
[20] Forsey DR, Bartels RH. Hierarchical B-spline refinement. SIGGRAPH Comput Graph 1988; 22(4):205–12.
[21] Giannelli C, Jüttler B, Speleers H. THB-splines: The truncated basis for hierarchical splines. Comput Aid Geometr Des 2012;29(7):485–98.

[22] Mantzaflaris A, et al. G+Smo (Geometry plus Simulation modules) v0.8.1 2017 [Available from: http://gs.jku.at/gismo].
[23] Dokken T, Lyche T, Pettersen KF. Polynomial splines over locally refined box-partitions. Comput Aid Geometr Des 2013;30(3):331–56.
[24] Skytt V, Barrowclough O, Dokken T. Locally refined spline surfaces for representation of terrain data. Comput Graph 2015;49(C):58–68.
[25] Lee S, Wolberg G, Shin SY. Scattered data interpolation with multilevel B-splines. IEEE Trans Vis Comput Graph 1997;3(3):228–44.
[26] SINTEF-Geometry: GoTools. Available from: https://github.com/SINTEF-Geometry/GoTools; 2018.
[27] Farin G. Triangular Bernstein-Bézier patches. Comput Aid Geometr Des 1986;3(2):83–127.
[28] Lai M-J, Schumaker LL. Spline functions on triangulations. Cambridge University Press; 2007.
[29] Bærentzen JA, Gravesen J, Anton F, Aanæs H. Curvature in Triangle Meshes. {Guide to Computational Geometry Processing: Foundations, Algorithms, and Methods}. London: Springer London; 2012. p. 143–58.
[30] Carmignato S, Dewulf W, Leach R. Industrial X-ray computed tomography. Springer International Publishing; 2018. VII, 369 p.
[31] Osher S, Fedkiw R. Level set methods and dynamic implicit surfaces. New York: Springer-Verlag; 2003. XIII, 273 p.
[32] Yoo TS, Ackerman MJ, Lorensen WE, Schroeder W, Chalana V, Aylward S, et al. Engineering and algorithm design for an image processing Api: A technical report on ITK—the insight toolkit. Stud Health Technol Inform 2002;85:586–92.
[33] Caselles V, Kimmel R, Sapiro G. Geodesic active contours. Int J Comput Vision 1997;22(1):61–79.
[34] Lorensen WE, Cline HE. Marching cubes: A high resolution 3D surface construction algorithm. ACM Comput Graph 1987;21(4):163–9.
[35] Schroeder W, Martin K, Lorensen B. The visualization toolkit; 2006.
[36] Fuhrmann S, Kazhdan M, Goesele M. Accurate Isosurface Interpolation with Hermite Data. In: 2015 International Conference on 3D Vision; 2015. 19–22 Oct. 2015.
[37] Jmtrivial. The originally published 15 cube configurations. Available from: https://commons.wikimedia.org/w/index.php?curid=1282165.
[38] Boissonnat J-D, Oudot S. Provably good sampling and meshing of surfaces. Graph Model 2005;67:405–51.
[39] Project TC. CGAL User and Reference Manual, 4.9 ed. CGAL Editorial Board. 2016.
[40] Cignoni P, Callieri M, Corsini M, Dellepiane M, Ganovelli F, Ranzuglia G. Meshlab: An open-source mesh processing tool, In: Eurographics Italian Chapter Conference. 2008. p. 129–36.

CHAPTER 6

Free-form surface filtering using the diffusion equation

6.1 Introduction

In this chapter, a linear filter based on the well-known heat equation, a partial differential equation (PDE), is analyzed. In Euclidean geometries, this method is equivalent to linear Gaussian filtering, but it is possible to utilize the theory of differential geometry to solve the PDE on a non-Euclidean surface. This theoretically provides a tool for performing "Gaussian filtering" on a surface that is appropriate for and dependent on the geometry of that surface and should theoretically yield results that are free of distortion. In areal surfaces (2.5D), nominal surfaces are mathematically described as a continuous function that defines a height value over a planar domain of interest. If the domain of interest is signified as $A \in \mathbb{R}^2$, the surface height is then calculated using the function $g: A \rightarrow \mathbb{R}$, and the continuous surface is described by the graph of g:

$$\{(u_1, u_2, g(u_1, u_2)) : (u_1, u_2) \subset A\}.$$

If the domain of g is Euclidean, well-established methods for analyzing the surface data are available. Some examples are Fourier analysis, wavelet decomposition, and linear filtering. The filtering operation is performed on a scalar field defined on a plane. If the reference surface is not simple, the metric tensor has to be taken into account.

In this chapter the parametric surface is used as surface representation. A general parametric surface is denoted by $M \subset \mathbb{R}^3$ and parametrized by a function $X: U \rightarrow \mathbb{R}^3$ such that.

$$M = \{X(u_1, u_2) \subset \mathbb{R}^3 : (u_1, u_2) \in U \tag{6.1}$$

where $U \subset \mathbb{R}^2$ is the parameter space.

In the field of metrology, the filtration of surface data has been performed using a variety of techniques including wavelets and Gaussian filters. These techniques assume that the underlying geometry of the surface portion being analyzed is planar. In the case of the measurement of free-form surfaces, this assumption is no longer valid. Standard band-pass filters based on Fourier analysis, which are valid for Euclidean geometries, are not immediately applicable for geometries that are non-Euclidean. A generalization of the Fourier analysis on free-form surfaces and some test cases will be presented in the rest of this chapter.

* For more examples and practical implementation, check the website: www.digitalsurf.com/freeform.

Advanced Metrology
https://doi.org/10.1016/B978-0-12-821815-0.00006-X

6.2 Linear Gaussian filters

Before presenting methods for filtering on free-form surfaces, the linear Gaussian filter is briefly discussed. The linear areal Gaussian filter is the standard method for analyzing the measurement data of surface textures for nominal planar surfaces. Each linear areal filter has an associated weight calculation function $s(x, y)$. For the areal Gaussian filter, the calculation is given by

$$s(x, y) = \frac{1}{\alpha^2 \lambda_c^2} \exp\left[-\frac{\pi}{\alpha^2}\left(\frac{x^2 + y^2}{\lambda_c^2}\right)\right] \tag{6.2}$$

where x and y are the distances from the center of the weight function in the x and y directions, respectively. λ_c is the cutoff wavelength, and α is given by

$$\alpha = \sqrt{\frac{\log 2}{\pi}} \approx 0.4697.$$

The value of α is chosen to provide a 50% transmission characteristic at the cutoff wavelength λ_c. Applying a filter to a scalar field means substituting the value of each point with a weighted average of its neighborhood, where the weights are defined by the function $s(x, y)$. The filter is applied by performing a discrete convolution with the surface data and discrete values sampled from the weight function. Repeated application of the Gaussian filter using an increasing sequence of standard deviation parameters yields a set of increasingly smooth representations of the initial data, known as the Gaussian scale space [1]. This provides a means of analyzing data at a number of increasingly coarse resolutions.

6.3 Diffusion filtering and relationship with Gaussian filters

6.3.1 Diffusion equation

Diffusion equations are used to model changes in concentration of a quantity of interest inside a specified region with respect to spatial and temporal variables. If the value of the quantity of interest at a particular point in space and time is described by a continuous function $f: V \times \mathbb{R}^+ \rightarrow \mathbb{R}$ (where $V \in \mathbb{R}^3$ denotes the volume of interest), then the evolution of this quantity is described by the partial differential equation:

$$\frac{\partial f(\boldsymbol{x}, t)}{\partial t} - \nabla f(\boldsymbol{x}, t) = 0$$

where $\nabla = \frac{\partial^2}{\partial u^2} + \frac{\partial^2}{\partial v^2}$ denotes the Laplacian operator with respect to the spatial variables $\boldsymbol{x} = (x, y, z)^T$. This equation is more commonly known, in the partial differential equation literature, as the heat equation and is an offshoot of the more general anisotropic diffusion equation (6.2). As mentioned in the previous section, a general continuous surface defined over a planar domain can be represented by Eq. (6.1). If a boundary condition of the form $f(x, y, 0) = g(x, y)$ is imposed on the diffusion equation, we have seen that the solution, $f(x, y, t)$ at time t, is given by a continuous convolution of the function $g(x, y)$ with a Gaussian function of standard deviation $\sigma = \sqrt{2t}$. The analytical solution of the diffusion equation is expressed as a continuous convolution, but in general practice a measured surface is in a discrete format, and the function values to be diffused are discretely sampled. In the discrete setting, a common approach is to approximate the convolution process by performing a discrete convolution using sampled values from the Gaussian kernel [2]. Furthermore, performing diffusion filtering on a free-form surface can be achieved through a generalizing Eq. (6.3) on non-Euclidean manifolds with the use of the Laplace-Beltrami operator [3–5].

6.3.2 The relationship between the diffusion time parameter and the Gaussian cutoff wavelength

The relationship between the time parameter t of the diffusion process and the standard deviation σ of the equivalent Gaussian function is given by $\sigma = \sqrt{2t}$. From this relationship and the definition of the areal Gaussian weight function, the required value of σ can be calculated for a given cutoff wavelength λ_c according to

$$\lambda_c = \pi\sigma\sqrt{\frac{2}{\log 2}}.$$

From this relationship the appropriate diffusion time parameter t is given by

$$t \approx 0.0175\,\lambda_c^2.$$

In the Euclidean case the cutoff wavelength provides a 50% transmission characteristic of a sine wave. In the case of free-form surfaces since there is currently no clear equivalent to a sine wave on a free-form surface, this interpretation requires further exploratory research. It is sensible though to interpret the value of λ_c as a nesting index for data smoothing until a fully developed interpretation is available.

6.4 Laplace-Beltrami operator

The Laplace-Beltrami operator is the generalization of the Laplacian operator to functions defined on surfaces, or more generally Riemannian manifolds. When the manifold in question is a Euclidean space, the Laplace-Beltrami operator simplifies to the standard Laplacian operator. Although the definitions apply to general manifolds, for areal surface filtering, it is the two-dimensional manifolds embedded in $\mathbb{R}^3$ that are of interest, particularly parametric surfaces such as spline surfaces. To define the Laplace-Beltrami operator, some basic definitions from differential geometry are required. To allow a more compact description, partial derivatives are denoted by

$$\boldsymbol{x}_{u_i} = \frac{\partial \boldsymbol{x}}{\partial u_i}$$

$$\boldsymbol{x}_{u_i u_j} = \frac{\partial^2 \boldsymbol{x}}{\partial u_i \, \partial u_j}$$

where $i=1, 2, j=1, 2, u_1=u$ and $u_2=v$. Given a parametric surface $M \subset \mathbb{R}^3$ and a function $f: M \rightarrow \mathbb{R}$, such that $f \in C^1$, the tangent vectors at a point $\boldsymbol{x} \in M$ that are given in u and v directions are given by $\boldsymbol{x}_u$ and $\boldsymbol{x}_v$, respectively. The surface normal $\boldsymbol{n}$ at $\boldsymbol{x}$ can be evaluated as

$$\boldsymbol{n} = \frac{\boldsymbol{x}_u \times \boldsymbol{x}_v}{\|\boldsymbol{x}_u \times \boldsymbol{x}_v\|}$$

where $\|\cdot\|$ is the Euclidean norm and $\times$ is the cross product. The first fundamental form of M is denoted by the matrix $\mathbf{G} \in \mathbb{R}^{2\times 2}$; the components are given by

$$g_{ij} = \boldsymbol{x}_{u_i} \cdot \boldsymbol{x}_{u_j}, \ (i,j) \in (1,2) \times (1,2)$$

where $\cdot$ is the usual Euclidean scalar product. We can now define the tangential gradient operator on the manifold as M, ∇_M, acting on a function $f(\boldsymbol{u})$ by

$$\nabla_M f = [\boldsymbol{x}_u, \boldsymbol{x}_v]^T \mathbf{G}^{-1} [f_u, f_v]^T \in \mathbb{R}^3$$

and the tangential divergence operator as div_M acting on a C^1 continuous vector field $\boldsymbol{r}(\boldsymbol{u})$ on M by

$$\mathrm{div}_M \boldsymbol{r} = \frac{1}{\sqrt{\det(\mathbf{G})}} \left(\frac{\partial}{\partial u}, \frac{\partial}{\partial v} \right) \left(\sqrt{\det(\mathbf{G})} \, \mathbf{G}^{-1} [\boldsymbol{x}_u, \boldsymbol{x}_v]^T \boldsymbol{r} \right).$$

The Laplacian operator of a function f defined on a surface M can then be computed as

$$\Delta_M f = \mathrm{div}_M (\nabla_M f)$$

which can be represented explicitly in terms of coordinates by

$$\Delta_M f = \frac{1}{\sqrt{\det(\mathbf{G})}} \left(\frac{\partial}{\partial u}, \frac{\partial}{\partial v} \right) \left(\sqrt{\det(\mathbf{G})} \, \mathbf{G}^{-1} [\boldsymbol{x}_u, \boldsymbol{x}_v]^T (f_u, f_v)^T \right).$$

The isotropic diffusion of a function f defined on a surface M can be found solving the following equation:

$$\frac{\partial f}{\partial t} - \Delta_M f = 0. \qquad (6.3)$$

6.5 Mean curvature motion

It is possible to prove, from the theory of differential geometry [6], that the Laplace-Beltrami operator applied directly to the vertices of the measured surface is related to the mean curvature as

$$\Delta_M \boldsymbol{x} = -2H(\boldsymbol{x})\,\boldsymbol{n}(\boldsymbol{x})$$

where $H(\boldsymbol{x})$ and $\boldsymbol{n}(\boldsymbol{x})$ are, respectively, the mean curvature and surface normal vector at $\boldsymbol{x}$. This can also be thought of as applying the Laplace-Beltrami operator to the identity function on the surface M. The mean curvature flow describes the evolution of a continuous surface with a velocity, normal to the surface at each point, equal to the mean curvature. In the discrete setting where data take the form of a mesh, solving the diffusion equation at increasing time steps will result in a sequence of increasingly smoothed surface meshes. Thus mean curvature motion is seen to be the application of the diffusion process to the geometry itself, rather than the diffusion of function values defined over the geometry. One of the shortcomings of using diffusion filtering in its basic form for surface texture purposes is that it has a tendency to smooth out sharp edges and the general noise. Perona and Malik [7] proposed a method to overcome this by introducing nonlinearity into Eq. (6.3) in the form of a diffusivity function. The diffusivity function is usually chosen to identify small values for regions that have large differences between local surface height values. The diffusivity function acts like a weight function for the diffusion process, resulting in reduced diffusion around edges, while general noisy regions are smoothed normally. Another development is the application of anisotropic diffusion. This allows the diffusion to be directed in a specific area via the introduction of a diffusivity tensor [2].

6.6 Discrete differential geometry

6.6.1 Discrete Laplace-Beltrami operators

For most practical applications the geometry is represented by a three-dimensional triangular mesh formed from discrete data points. In such cases, there are a number of numerical methods available for approximating both the Laplace-Beltrami operator and solutions of the diffusion equation at discrete time values.

Most discrete approximations to the Laplace-Beltrami operator involve triangular meshes, although Liu et al. [8] describe a method for use with quadrilateral meshes. In this chapter, triangular mesh approximations proposed by Meyer et al. [9] and by Desbrun et al. [5] will be presented. As shown by Xu [10], these schemes show the true value of the Laplace-Beltrami operator when the triangulation satisfies certain criteria, described later. Some notation required for the description of discrete Laplace-Beltrami approximations is now introduced.

A triangular mesh M is considered to be composed of N measured data points $\boldsymbol{x}_i \in \mathbb{R}^3$, which provide a discrete approximation of the continuous but unknown underlying surface M. Edges are defined as straight line segments joining two neighboring vertices $[\boldsymbol{x}_i, \boldsymbol{x}_j]$. Triangles are denoted by the triplet $[\boldsymbol{x}_i, \boldsymbol{x}_j, \text{and } \boldsymbol{x}_k]$, where $\boldsymbol{x}_i$, $\boldsymbol{x}_j$, and $\boldsymbol{x}_k$ are the vertices of the triangle. $N(i)$ is the set of all points $\boldsymbol{x}_j$ that are vertices of triangles containing $\boldsymbol{x}_i$ (members of $N(i)$ are called one-ring neighbors of $\boldsymbol{x}_i$). The number of elements, of the set $N(i)$, is referred to as the valence of vertex $\boldsymbol{x}_i$. Using this notation the approximations for Meyer et al. [9] and Desbrun et al. [5] for the Laplace-Beltrami operator applied to a function $f: M \rightarrow \mathbb{R}$ are, respectively, given by

$$\Delta_M f(\boldsymbol{x}_i) = \frac{3}{2\,A_M(\boldsymbol{x}_i)} \sum\nolimits_{j \in N(i)} \left(\cot\alpha_{ij} + \cot\beta_{ij}\right)\left(f(\boldsymbol{x}_j) - f(\boldsymbol{x}_i)\right) \tag{6.4}$$

and

$$\Delta_M f(\boldsymbol{x}_i) = \frac{3}{2\,A(\boldsymbol{x}_i)} \sum\nolimits_{j \in N(i)} \left(\cot\alpha_{ij} + \cot\beta_{ij}\right)\left(f(\boldsymbol{x}_j) - f(\boldsymbol{x}_i)\right). \tag{6.5}$$

Here, $A(\boldsymbol{x}_i)$ is the area of all triangles that contain vertex $\boldsymbol{x}_i$, and the angles α_{ij} and β_{ij} are the two angles opposing the edge $[\boldsymbol{x}_i, \boldsymbol{x}_j]$ of the two triangles sharing the edge. $A_m(\boldsymbol{x}_i)$ is the area of the region formed by joining the circumcenters of all triangles containing vertex $\boldsymbol{x}_i$. In cases where a triangle is obtuse, the midpoint of the edge opposite the obtuse angle is used instead of the circumcenter. Fig. 6.1 provides a visual representation of the angles α_{ij} and β_{ij}.

6.6.2 Criteria for convergence

Xu [10] has shown that the Meyer (Eq. 6.4) and Desbrun (Eq. 6.5) approximations are shown to converge to the Laplace-Beltrami operator, subject to certain conditions on the positioning of each point and its one-ring neighbor in the immediate vicinity. The convergence is defined in the sense that the theoretical Laplace-Beltrami operator is the limiting value of these approximations calculated on increasingly fine subdivisions of the initial triangular mesh. The first condition for convergence is that the vertex, which we are calculating the approximation for, has a valence of 6. The second condition is that in parameter space, each vertex and its one-ring neighbor forms a sheared hexagon.

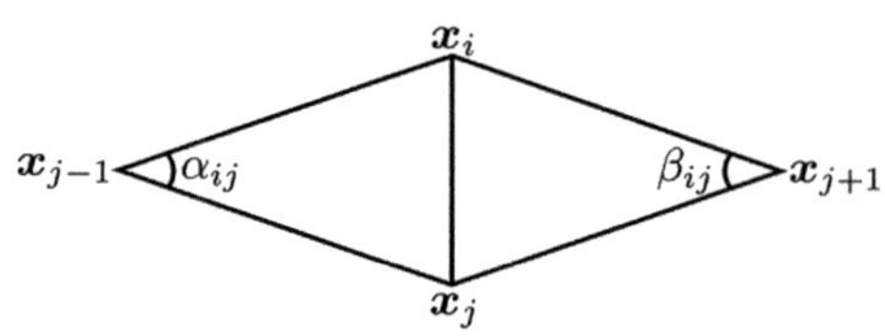

Fig. 6.1 Angles α_{ij} and β_{ij}.

Usually if the calculation point $x_c \in \mathbb{R}^3$ has six one-ring neighbors $x_i \in \mathbb{R}^3$, $i = 1, \ldots, 6$, each with corresponding parameter values u_c, u_1, ..., u_6, then convergence will occur if

$$q_i = q_{i+1} + q_{i-1} - q_c, \; i = 1, \ldots, 6$$

where index arithmetic is performed modulo 6. Further details and proof of the convergence for these approximations can be found in Xu [10].

6.6.3 Numerical solutions of the diffusion equation

To solve the diffusion equation at a point in time, we have used the semiimplicit Euler scheme as described by Desbrun et al. [5]. Assuming that the surface is evolving in time, we denote the vector of the scalar field values at time t by $\boldsymbol{f}^t \in \mathbb{R}^N$. The ith entry of this matrix contains the function value at vertex $\boldsymbol{x}_i$. For a time stamp of δt, the time derivative is approximated using a finite difference scheme:

$$\frac{\partial f}{\partial t} \approx \frac{\boldsymbol{f}^{t+\delta t} - \boldsymbol{f}^t}{\delta t}$$

where the initial value of $\boldsymbol{f}^{t_0}$ is the initial value of the scalar field. Now the value of the function as a result of the diffusion process has an approximate solution $\boldsymbol{f}^{t+\delta t}$ at time $t + \delta t$ that is given by the solution of the linear system:

$$(\boldsymbol{I} - \delta t\, \boldsymbol{\Delta})\boldsymbol{f}^{t+\delta t} = \boldsymbol{f}^t$$

where the matrix $\boldsymbol{\Delta} \in \mathbb{R}^{N \times N}$ represents the discrete approximation of the Laplace-Beltrami operator using either Eq. (6.4) or (6.5). For the Meyer et al. approximation (6.4), $\boldsymbol{\Delta}$ has element Δ_{ij} given by

$$\Delta_{ij} = \begin{cases} \frac{3}{2A_M(\boldsymbol{x}_i)} \sum_{l \in N(i)} (\cot\alpha_{il} + \cot\beta_{il})(f(\boldsymbol{x}_j) - f(\boldsymbol{x}_l)), i = j \\ -\frac{1}{2A_M(\boldsymbol{x}_i)} (\cot\alpha_{il} + \cot\beta_{il}), j \in N(i), i \in N(j) \\ 0, \quad \notin N(j) \end{cases}$$

while, for the Desbrun et al. approximation (6.5)

$$\Delta_{ij} = \begin{cases} \frac{3}{2A(\boldsymbol{x}_i)} \sum_{l \in N(i)} (\cot\alpha_{il} + \cot\beta_{il})(f(\boldsymbol{x}_j) - f(\boldsymbol{x}_l)), i = j \\ -\frac{1}{2A(\boldsymbol{x}_i)} (\cot\alpha_{il} + \cot\beta_{il}), j \in N(i), i \in N(j) \\ 0, \notin N(j) \end{cases}.$$

From these matrices, it can be seen that the discrete Laplace-Beltrami operator for vertex i at time t is given by the ith element of the vector $\boldsymbol{\Delta f}^t$. The system can be solved using standard techniques for linear systems; although given the sparse nature of the matrix $\mathbf{\Delta}$, methods for solving sparse linear systems may be more appropriate. This is of particular relevance for the analysis of a large number of data points. Desbrun et al. [5] propose the use of a preconditioned biconjugate gradient descent method to provide an iterative solution to the sparse linear system.

6.7 Application approach

Application of mean curvature motion as described earlier using both Meyer and Desbrun forms of the discrete Laplace-Beltrami operator was found to be unsatisfactory for the purposes of surface metrology [11]. Results obtained from these methods using simulated and experimental data in the form of surface meshes with boundary were found to have an unacceptable amount of shrinkage and drift in space. Owing to the nature of surface metrology, precision is of major importance, and as a consequence the unpredictable results of mean curvature motion were considered to be unsuitable for data analysis.

It is important that the filtered data remain close to the initial set of data points. Bearing this in mind, then, an alternative approach is used. Instead of performing a diffusion of the geometry itself, a least squares approximation is fitted to the surface data to provide a reference surface, and diffusion filtering of the approximation residuals is performed on the geometry of the reference surface. Firstly a least squares NURBS surface approximation is fitted to the initial set of N measurement points $\boldsymbol{r}_i$, $i=1, \ldots, N$. This reference surface is denoted by M. Points $\boldsymbol{x}_i=\boldsymbol{x}_i(\boldsymbol{u}_i) \in M$ on the surface are then calculated for each measurement point $\boldsymbol{r}_i$. The least squares approximation residual e_i at each point is also calculated. In this way, samples of a function $f: M \rightarrow \mathbb{R}$ defined by $f(\boldsymbol{x}_i)=e_i$ are generated. The idea is to then perform the diffusion filtering of the residuals e_i on the new mesh obtained from the reference points $\boldsymbol{x}_i$ in previously described methods. The triangulation of the point $\boldsymbol{x}_i$ to form an approximate rectangular grid in parameter space is performed as follows: Form two triangles from each quadrilateral in the mesh by joining its lower left and top right vertices; this ensures a valence of six for all interior vertices and very closely approximates the convergence condition. By choosing to diffuse the least squares residuals on the reference surface, the data are unable to drift too far from its initial configuration. Additionally, it is reasonable to assume (in the case of a good surface approximation) from statistical theory that the least squares residuals e_i will be approximately normally distributed with mean zero, that is, $e_i \approx N(0, \sigma)$. In such a case, it would be expected that over time, the diffusion process would result in the residuals converging to their mean value, 0. Assuming the validity of this statistical reasoning and provided that surface M is a good approximation to the initial data, it is reasonable to expect e_i to be close to zero. As a consequence the limiting surface of this diffusion process will be very close to the reference surface M.

6.7.1 Surfaces without boundaries

Most of the literature on diffusion filtering deals with the case of Euclidean geometry or more general geometries where the data in question are perceived as measurements from an underlying manifold that has no boundary. In cases where the manifold has no boundary, the triangulation can be closed, and so there are no issues with evaluation of Laplace-Beltrami as every vertex has a complete set of one-ring neighbors. Use of Eqs. (6.4) and (6.5) is not particularly satisfactory at edge vertices, and experimentation with these methods has not yielded good results at boundaries in such cases. At such a time when surface measurement data take the form of a closed mesh without a boundary, this issue will be mitigated. Currently, most surface measurement data take the form of surface sheets with a clear boundary, and so a method of dealing with boundary vertices is required for a satisfactory data analysis.

6.7.2 Surfaces without boundaries

Because of the undesirable results that we obtain for the Laplace-Beltrami operator at edge vertices, they are used only at interior vertices for the calculations (6.4) and (6.5) in the Laplace-Beltrami operator. They are included in the calculations, but their actual diffused residual values are disregarded for analysis purposes. This ensures that only vertices that have a valence of six are actually filtered, to ensure the reliability and accuracy of the filtered function values. Providing that the measured surface is initially sampled with a sufficiently large number of points, this approach should not result in the loss of a large percentage of data. Xu [10] suggests that as a solution to the diffusion equation on surfaces with a boundary, a possible way of dealing with the problem is to add extra surface points around the boundary to provide an outer layer of triangles. This results in the initial boundary vertices having an artificial valence of six. For surface fairing, where high accuracy may not be the chief consideration, this may be an acceptable approach, but in a field such as surface metrology where data accuracy and reliability are of high importance, this seems to be an unreliable way to proceed. For filtering residuals on a reference surface, this problem would be exacerbated by the need to simulate a residual value for each new artificial surface point. This approach provides no means of assessing how much inaccuracy is introduced by the simultaneous extrapolation of geometry and simulation of function values. For these reasons, at this point in time, the exclusion of boundary vertices for filtering purposes is proposed.

6.7.3 Simulation test case

For the purposes of testing these methods, a simple degree 4 NURBS surface [12] with 6×6 control points was created and is shown in Fig. 6.2. For the surface points a uniform rectangular mesh was created in the parameter domain, and this was then triangulated. To simulate a realistic surface roughness sample, the reference surface residuals were selected

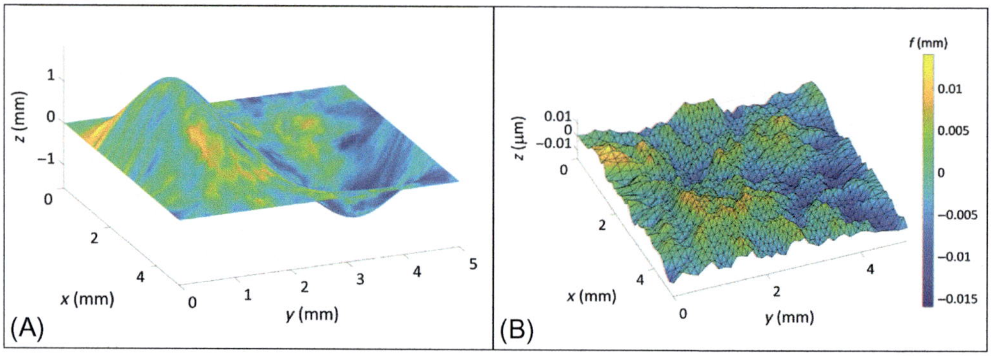

Fig. 6.2 Simulated surface and residuals: (A) Simulated surface, (B) residual dataset.

as a subset of a stochastic model that generated a roughness surface (derived from the Internet-based surface metrology algorithm testing system of the National Institute of Standards and Technology [13]). These residuals were then filtered using the diffusion equation as described previously using the Meyer form of the discrete Laplace-Beltrami operator (6.4).

Fig. 6.3 illustrates the results of the diffusion process at increasing values of the time parameter t that correspond to common choices of the cutoff wavelength λ_c. Due to the

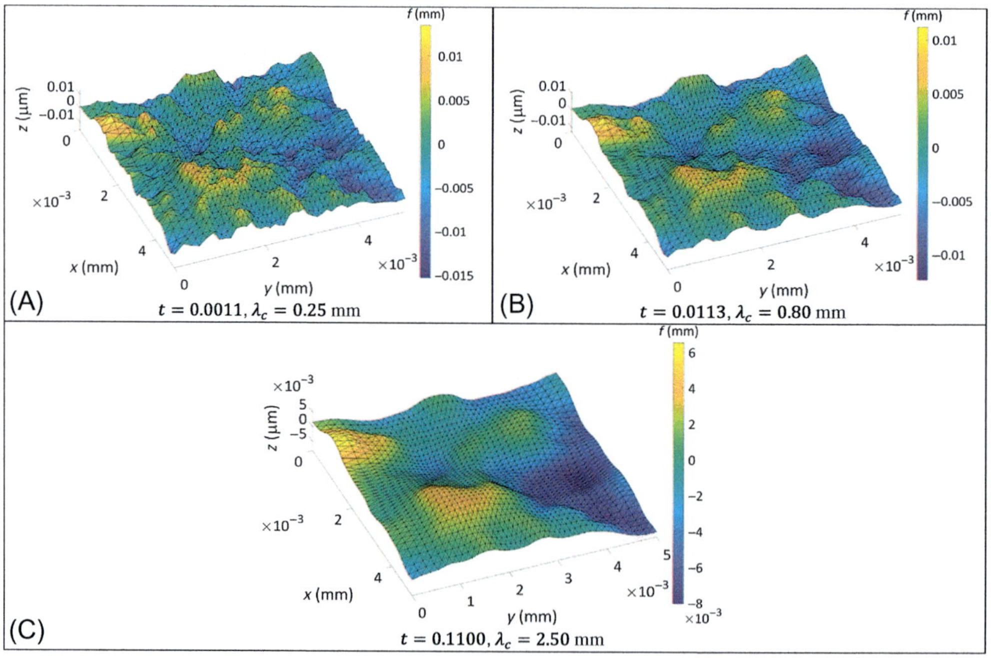

Fig. 6.3 Residual of the simulated surface after applying the diffusion equation with different cutoff values: (A) $t = 0.0011$, $\lambda_c = 0.25$ mm, (B) $t = 0.0113$, $\lambda_c = 0.80$ mm, (C) $t = 0.1100$, $\lambda_c = 2.50$ mm.

fact that the surface form change is much larger than the residuals, the surface residuals are plotted against their values in parameter space to enhance the visibility of the filtration effects. The boundary residuals are not diffused, and so only the internal vertex residuals are illustrated.

6.7.4 Experimental test case

After the initial application of these methods to simulated data, diffusion filtering was performed on real surface measurement data. The data were obtained from a coordinate measuring machine (CMM) producing a measurement for a portion of a knee replacement component. The data are composed of 60×35 points taken on an approximately rectangular grid. The approximating reference surface is a cubic NURBS surface with 14×17 control points and is illustrated in Fig. 6.4. As with the simulated data, the diffusion of the fitted surface residuals was implemented using the Meyer approximation of the Laplace-Beltrami operator. Fig. 6.5 show the results of diffusion of the residuals at increasing values of the time parameter. As with the simulated data result, the residuals are plotted against the parameter values of the surface reference points so that the effects of filtration are clearly visible. Boundary vertices are again omitted.

6.7.5 Mean curvature flow: Simulation test case

The results show that a potential way forward for filtration of free-form surface metrology data is to use the established field of PDE filtering. In particular, diffusion filtering can be applied to obtain a Gaussian scale space for free-form surfaces owing to its theoretical link with the linear Gaussian filter. As PDE filtering is a wide and active area of research, the theory of these methods is well discussed in other research fields, such as image analysis. This chapter is intended to provide a useful starting point for solving free-form filtration difficulties by using PDEs. The proposed use of a reference surface to provide a manifold on which to apply the diffusion equation provides a method of ensuring that the filtered data do not deviate excessively from the initial data. This is an important requirement for

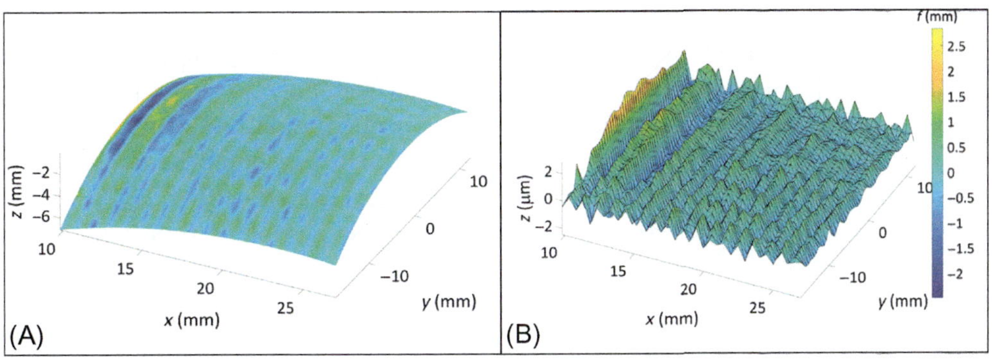

Fig. 6.4 Measurement surface data obtained with CMM and residuals: (A) Measured surface, (B) residual dataset.

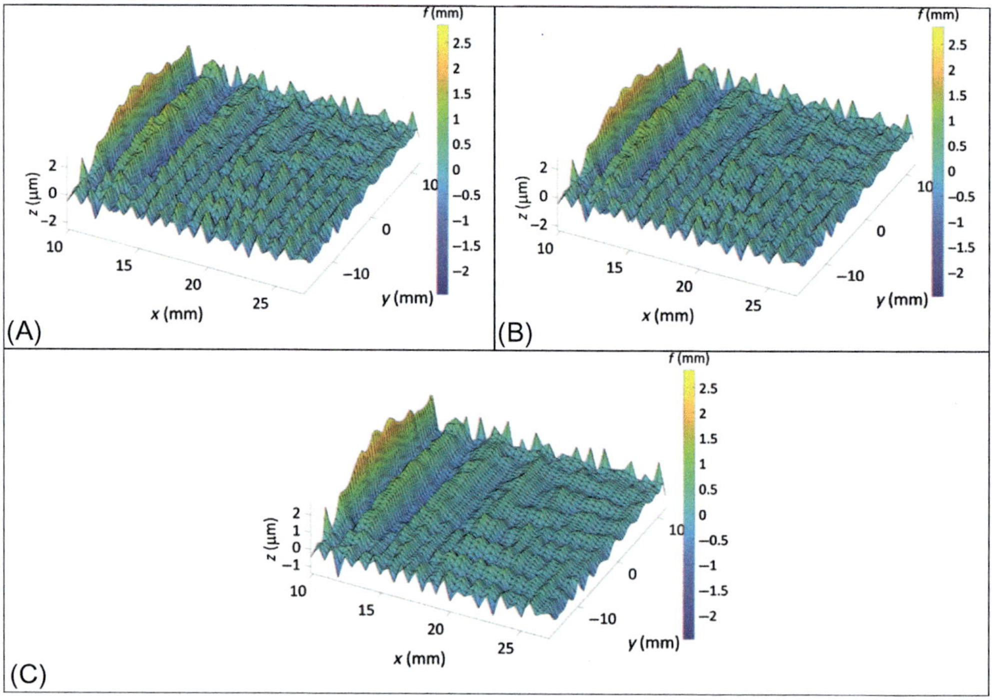

Fig. 6.5 Residual of the measured surface after applying the diffusion equation with different cutoff values: (A) $t=0.0011$, $\lambda_c=0.25\,\text{mm}$, (B) $t=0.0113$, $\lambda_c=0.80\,\text{mm}$, (C) $t=0.1100$, $\lambda_c=2.50\,\text{mm}$.

analyzing surface measurement data and overcomes the undesirable mesh area shrinkage and spatial drift effects that were exhibited in experimentation with mean curvature motion. To illustrate the behavior of this method, the smoothing was applied to an artificially created surface. A structured planar surface roughness (formed of a subset of a stochastic model derived from the internet-based surface metrology algorithm testing system of NIST, Bui et al. 2007 [13]) was augmented and superimposed onto a real data set of CMM measurements from a knee replacement joint. Although the resulting surface is not representative of true roughness, there is a need to amplify the roughness profile for the purposes of visibility. If the true roughness obtained using the residuals of a least squares surface fits to the measurement data, the roughness would be dominated by the surface form, and all plots would be indistinguishable from the other. As the filtering is performed on a non-Euclidean surface, it is not simply possible to project the surface points to a plane as each surface point is able to evolve along a general vector, unlike in the Euclidean setting where only the height value (z coordinate) is changing. Fig. 6.6 shows the initial surface mesh and the results using increasing values $t=0.01$, 0.1, 1, 10 for the diffusion time parameter, respectively.

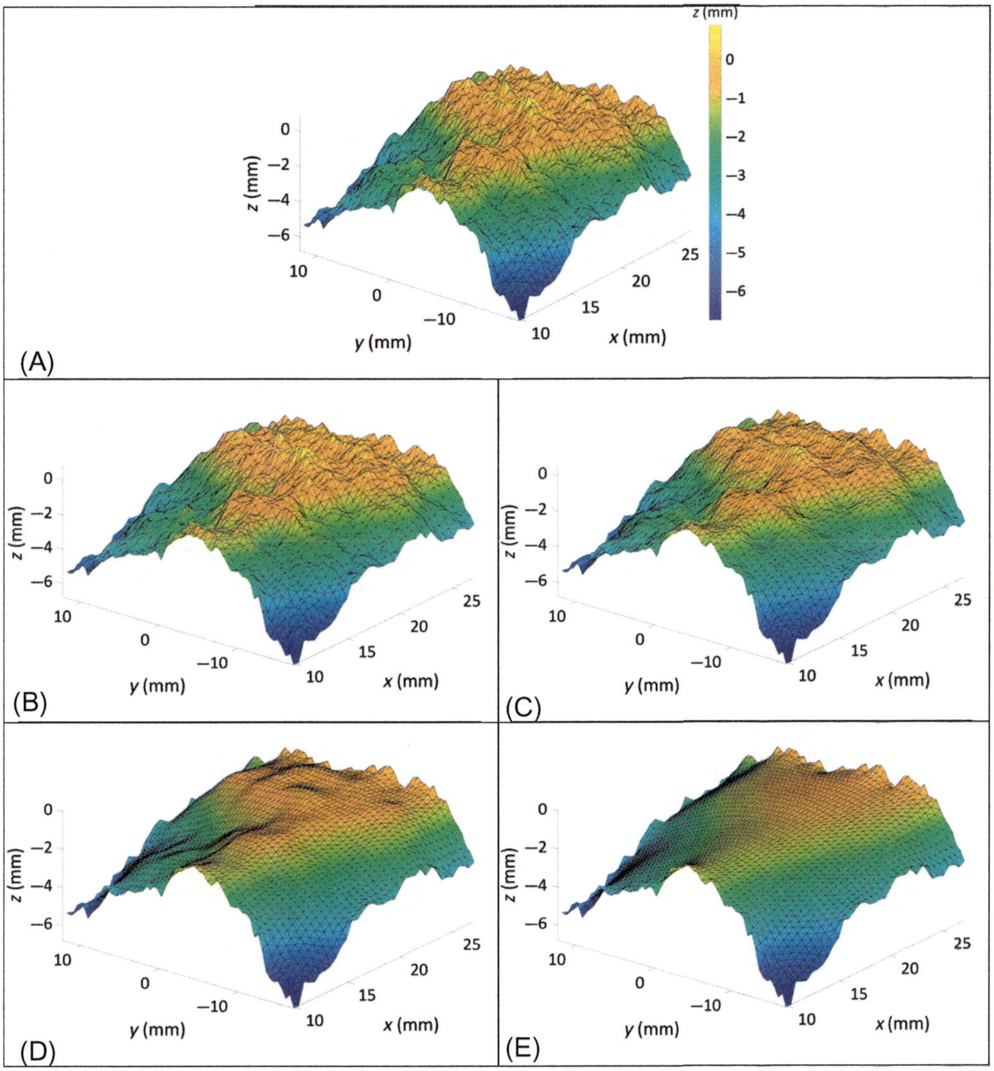

Fig. 6.6 Initial surface and surface smoothed using the mean curvature flow: (A) Simulated surface, (B) $t = 0.01$, (C) $t = 0.1$, (D) $t = 1$, (E) $t = 10$.

6.8 Summary

A key area of future research will be the application of this methodology to a large number of different types of free-form surface data and to establish whether there are any particular surfaces for which the method yields very good or poor results. Other areas of interest will be the development of ways of speeding up the algorithms and an investigation into the propagation of uncertainty through the filter. Other definitions for the discrete Laplace-Beltrami operator are available, and a deeper investigation of these would seem appropriate, particularly mesh-free definitions. Experimentation with different types of (and higher order) geometric flow is another area that is likely to be worthy of attention. So far, this chapter has only considered linear isotropic diffusion as a means of filtering the data, but experimentation with anisotropic diffusion or other edge-enhancing types of diffusion may lead to superior results, particularly in the analysis of structured surfaces. Finally, it would be desirable to try and establish a theoretical link between the cutoff wavelengths of the linear areal Gaussian filter and the time step in the evolution of the diffusion equation.

References

[1] Witkin AP. Scale-space filtering. In: Readings in computer vision. Amsterdam: Elsevier; 1987. p. 329–32.
[2] Weickert J. Anisotropic diffusion in image processing. Stuttgart: Teubner; 1998.
[3] Bajaj CL, Xu G. Anisotropic diffusion of surfaces and functions on surfaces. ACM Trans Graph (TOG) 2003;22(1):4–32.
[4] Clarenz U, Diewald U, Rumpf M. Anisotropic geometric diffusion in surface processing. IEEE; 2000.
[5] Desbrun M, Meyer M, Schröder P, Barr AH, editors. Implicit fairing of irregular meshes using diffusion and curvature flow. In: Proceedings of the 26th annual conference on Computer graphics and interactive techniques. Citeseer; 1999.
[6] do Carmo MP. Differential geometry of curves and surfaces. New Jersey: Prentice-Hall; 1976.
[7] Perona P, Malik J. Scale-space and edge detection using anisotropic diffusion. IEEE Trans Pattern Anal Mach Intell 1990;12(7):629–39.
[8] Liu D, Xu G, Zhang Q. A discrete scheme of Laplace–Beltrami operator and its convergence over quadrilateral meshes. Comput Math Appl 2008;55(6):1081–93.
[9] Meyer M, Desbrun M, Schröder P, Barr AH. Discrete differential-geometry operators for triangulated 2-manifolds. In: Visualization and mathematics III. Berlin: Springer; 2003. p. 35–57.
[10] Xu G. Convergence of discrete Laplace-Beltrami operators over surfaces. Comput Math Appl 2004;48(3–4):347–60.
[11] Jiang X, Cooper P, Scott PJ. Freeform surface filtering using the diffusion equation. Proc Roy Soc A: Math Phys Eng Sci 2010;467(2127):841–59.
[12] Piegl L, Tiller W. The NURBS book. New York: Springer-Verlag, Inc. 1997.
[13] Bui SH, Vorburger TV. Surface metrology algorithm testing system. Precis Eng 2007;31(3):218–25.

CHAPTER 7

Morphological filtering of free-form surfaces

7.1 Introduction

The surface governs the functional behaviors of the product, whether that be a mechanical, tribological, hydrodynamic, optical, thermal, chemical, or biological property, all of which are of tremendous importance to product performance [1, 2]. The controlling behaviors often involve surface details at micrometer and nanometer scale. Many emerging products and devices are based on achieving surfaces with special functionalities. Manufactured items such as micro- and nanometer-scale transistors, microelectromechanical systems (MEMS), and nanoelectromechanical systems (NEMS); microfluidic devices; optic components with free-form geometry; and structured surface products are clear evidence of products where the surface plays the functional role [3].

Surfaces and their measurement provide a link between the manufacture of these engineering components and their use [4]. On the one hand, it can help to control the manufacturing process, monitoring changes in the surface texture, and indicating issues with that process such as machine tool vibration and tool wear [5, 6]. On the other hand, it can help with functional prediction such as characterizing geometrical features that will directly impact on tribology and physical properties of the whole system [7–9]. An example of which is the friction of two contact surfaces and the optical fatigue of one reflecting surface. Controlling the manufacturing process helps to increase repeatability and hence improve overall quality of conformance. Functional prediction helps performance and assists in its optimization.

The early use of surface measurement was mainly to control the manufacturing process. The surface texture is a fingerprint of the stages of a manufacturing process. The effects of process and machine tooling are always present in the surface textures. The former is called the roughness and the latter the waviness. Also, in addition to roughness and waviness, an even longer wavelength can be introduced into the surface geometry by weight deflection or long-term thermal effects. Filtration techniques are the means by which roughness, waviness, and form components of the surface texture are extracted from the measured data for further characterization. By separating surface profiles into

* For more examples and practical implementation, check the website: www.digitalsurf.com/freeform.

Advanced Metrology
https://doi.org/10.1016/B978-0-12-821815-0.00007-1

various bands, it is possible to map the frequency spectrum of each band to the manufacturing process that generated it [10].

The 1950s saw two attempts to separate the waviness from the profile so that the roughness could be characterized. One was graphical, simulating electrical filters in the meter circuit. The raw profile was divided into segments of equal length, and in each segment a mean line was drawn that captured the slope of the profile in that segment. The roughness profile was obtained by considering the deviation from the mean line. Thus it designated the mean line system (M-system). The other attempt was mechanically simulating the contact of a converse surface, for example, a shaft, with the face of the anvil of a micrometer gauge. It appeared as a large disk rolling over the profile from above and was entitled the envelope system (E-system); see Fig. 7.1.

The E-system based the reference line upon an envelope line that was generated from the center point of a rolling disk and then adjusted to the average height of the overall profile. The difficulty appeared in building practical instruments as two elements are needed: a spherical skid to approximate the "enveloping disk" and a needle-shaped stylus moving in a diametral hole of the skid to measure the roughness as deviation with respect to the "generated envelope." The supposed advantage of the E-system was that it was more physically significant as many engineering properties of a surface are determined by its peaks [12]. The standing objection was that the choice of the rolling disk radius is as arbitrary as the choice of cutoff in the M-system and no practical instruments using mechanical filtering could be made at that time. The evaluative discussion about the reference systems lasted at least a decade from 1955 to 1966 [5]. With the advent of digital processing techniques in 1960, the M-system became the preeminent technique and was improved by the 2RC digital filter and phase-corrected digital filter. Later the Gaussian filter, with better performance, was chosen as the standardized filter for separating differing wavelengths.

The Gaussian filter, although a good general filter, is not applicable for all functional aspects of a surface, for example, in contact phenomena; in instances such as this, the

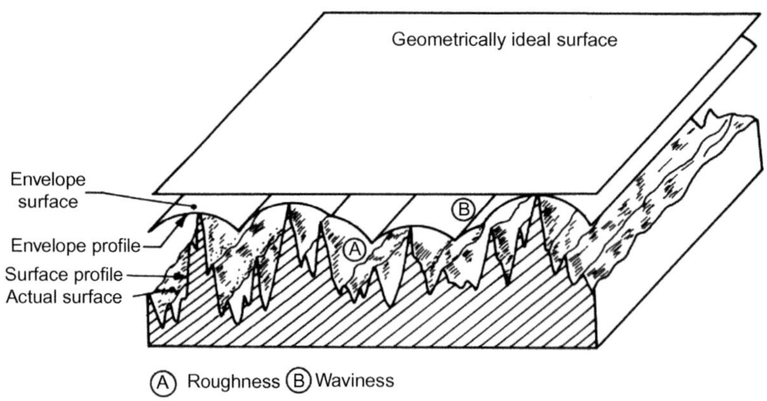

Fig. 7.1 The profile and surface envelope [11].

E-system method is more relevant. The advent of fast practical computers, which can be used in association with measurement instruments, had virtually eliminated the need for any hardware implementation for the E-system [13]. Furthermore, there was growing evidence showing that the E-system method gave better results in the functional prediction of surface finish in the analysis of mating surfaces, such as contact, friction, wear, lubrication, and failure mechanism [14]. With the M-system, there is little correlation between the standardized surface roughness parameters and functional requirements, while the E-system that depends on geometrical characteristics of the workpiece is more relevant [15]. In this respect the logic of the E-system was sounder against the M-system although both the M-system and the E-system approaches have their benefits and limitations. Arguing that one is better than the other without any concrete proof from the application area is not convincing [16]. In fact, rather than competing with each other, the M-system and the E-system are complementary to each other and contribute a better solution to surface evaluation.

In the last two decades, more advanced filtration techniques have emerged as a result of an urgent need for the analysis of surfaces produced by modern manufacturing technologies with complex geometry and high precision. The M-system was greatly enriched by incorporating advanced mathematical theories. The Gaussian regression filter overcame the problem of end distortion and poor performance of the Gaussian filter in the presence of significant form components and also solved the problem of outlier distortion [17, 18]. The spline filter is a pure digital filter, more suitable for form measurement [19]. Based on Lp norm the robust spline filter is insensitive with respect to outliers [19]. More recently a method of Gaussian filtering for free-form surfaces was developed by solving the diffusion equation that overcomes geometrical distortion in the presence of nonzero Gaussian curvature [20].

Meanwhile the E-system also experienced significant improvements. By introducing mathematical morphology, the superset of the early envelope filters, morphological filters, emerged but offered more tools and capabilities [21]. The basic variation of morphological filters includes the closing filter and the opening filter. They could be combined to achieve superimposed effects, referred to as alternating symmetrical filters. Scale-space techniques further developed morphological filtering techniques, which provided a multiresolution analysis to surface textures [22]. Even though morphological filters are generally accepted and regarded as the complement to mean line–based filters, they are not universally adopted due to limitations caused by their current implementation and lack of capabilities demanded by the analysis of modern functional surfaces, especially free-form surfaces.

Free-form surfaces are continuous surfaces dependent on global complex geometry and have no translational and rotational symmetry [20]. For free-form surfaces the description data might be specified by coordinates in two or three dimensions, rather than the regular height of the surface. Responding to all these requirements, novel methods based on the geometric computation have to be developed with an aim of supporting morphological filtering on free-form surfaces.

This chapter mainly discusses the origin of morphological operations and illustrates their applications in the field of surface metrology. It also presents traditional algorithms for morphological filtration and more advanced algorithms recently developed with better performance, aiming to support free-form surface filtration. Case studies are also provided to demonstrate the capability of the advanced methods.

7.2 Mathematical morphology and morphological operations

Mathematical morphology is a discipline established by two French researchers Jorge Matheron and Jean Serra in the early 1960s [23]. The central idea of mathematical morphology is to examine the geometrical structure of an image by matching it with small patterns at various locations in the image. By varying the size and the shape of the matching patterns, called structuring elements, one can exact useful information about the shape of the different parts of the image and their interrelation [24]. There are four basic morphological operations, namely, dilation, erosion, opening, and closing, which form the foundation of mathematical morphology.

7.2.1 Morphology operations on set

Dilation combines two sets using the vector addition of set elements. The dilation of A by B is

$$D(A, B) = A \oplus \check{B} \tag{7.1}$$

It is defined on the basis of vector addition, also known as the Minkowski addition, which was first introduced by Minkowski [25]. The Minkowski addition of two input sets is the following set:

$$A \oplus B = \{c | \, c = a + b, a \subseteq A \& b \subseteq B\} \tag{7.2}$$

Fig. 7.2 presents an example of dilating a square by a disk. The dilation of the light color square by a disk results in the dark color square with round corners.

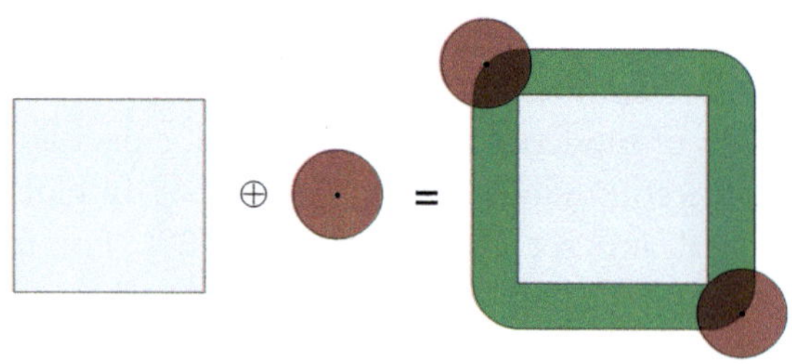

Fig. 7.2 Dilation of a square by a disk.

Erosion is the morphological dual to dilation. It combines two sets using the vector subtraction of set elements. The erosion of A by B is

$$E(A, B) = A \ominus \check{B} \tag{7.3}$$

where

$$A \ominus B = \overline{\overline{A} + B} \tag{7.4}$$

and $\overline{A}$ is the complementation of A.

An example of erosion is illustrated in Fig. 7.3. The erosion of the light color square by a disk generates the dark color square.

Opening and closing are dilation and erosion combined as pairs in sequence. The opening of A by B is obtained by applying the erosion followed by the dilation:

$$O(A, B) = D\left(E(A, B), \check{B}\right) \tag{7.5}$$

In Fig. 7.4 the opening of the light color square by a disk generates the dark color square with round corners.

Closing is the morphological dual to opening. The closing of A by B is given by applying the dilation followed by the erosion:

$$C(A, B) = E\left(D(A, B), \check{B}\right) \tag{7.6}$$

In Fig. 7.5 the closing of the light color shape (union of two squares) by a disk results in the union of the light color shape and the dark color areas.

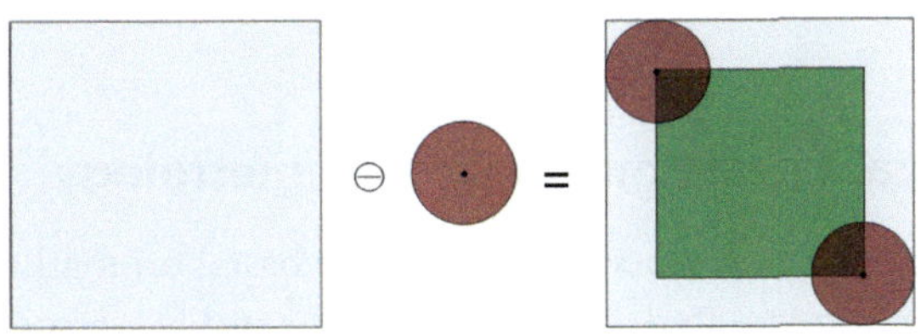

Fig. 7.3 Dilation of a square by a disk.

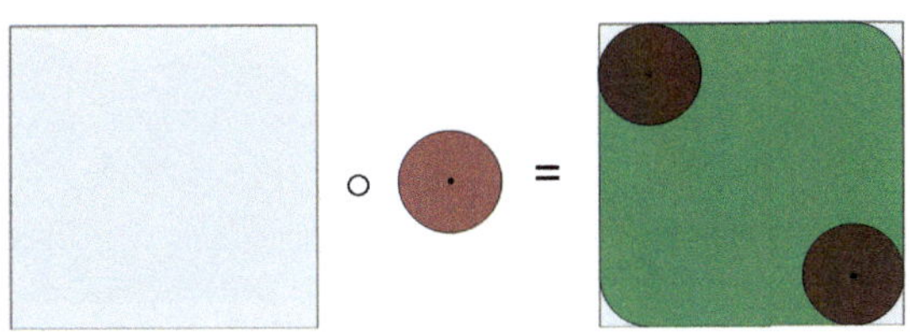

Fig. 7.4 Opening of a square by a disk.

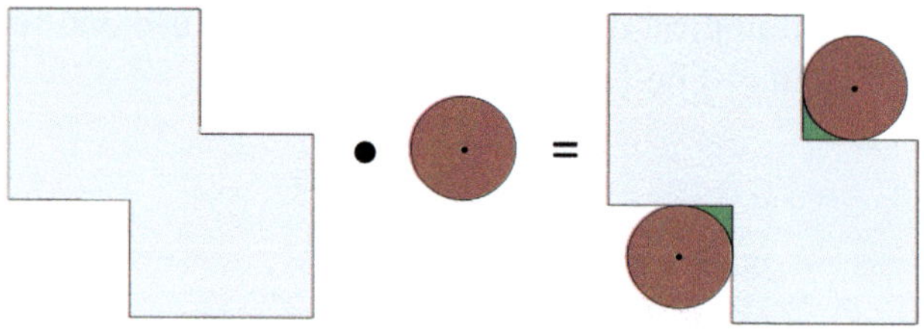

Fig. 7.5 Closing of a square by a disk.

7.2.2 Morphological operations on functions

In the literature of morphological image processing, both the input set and the structuring element of morphological operations are treated as sets. In binary morphology the sets are defined in $\mathbb{R}^2$. In grayscale morphology the morphological operations are invoked on the functions defined over a domain in $\mathbb{R}^2$ [26]. Morphological operations on functions can be mathematically linked to morphological operations on sets through "fill" transforms [27]. Fill transforms convert the curve defined by the function to a two-dimensional set and the surface defined by the function to a three-dimensional set [28]. If the function curve or surface is closed, then the fill transform produces the interior region of the closed curve or surface. If the function curve or surface is not closed, then a special fill transform called the "umbra transform" may be applicable. Fig. 7.6 demonstrates an example of the umbra transform. In this example a curve $f(x)$ is defined over a finite interval of x. Its umbra is the entire two-dimensional region under the curve of the function $f(x)$. Similarly the umbra of a surface $f(x, y)$ is the entire three-dimensional region under the surface of the function $f(x, y)$. In general, morphological operations of functions of n variables can be shown to be equivalent to the corresponding operations on sets in $\mathbb{R}^{n+1}$. Thus morphological operations of functions can be derived and justified from the set definitions.

7.3 Morphological operations in surface metrology

Morphological operations in surface metrology are based on mathematical morphology. The profile and surface are treated as functions of $\mathbb{R}^2$ and $\mathbb{R}^3$, respectively [21]. They are

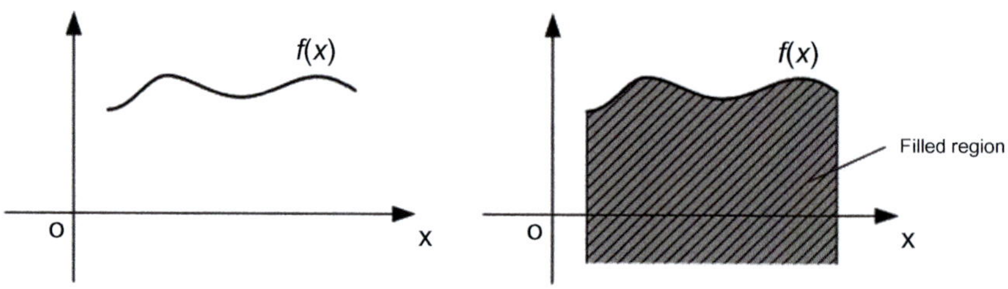

Fig. 7.6 Umbra transform of an open curve on the left to a two-dimensional set on the right.

carried out by performing morphological operations on the input profile or surface with circular or horizontal flat structuring elements [27]. In contrast to the widely used mean line–based evaluation techniques, for example, the Gaussian filter, morphological methods are not universally adopted in the field of geometrical metrology. Nonetheless, they are of great value even if they are not recognized in practice [28]. The initial application of morphological operations in geometrical metrology is the scanning of workpieces using tactile probes.

7.3.1 Surface scanning

The scanning of a workpiece surface using a tactile probe, for example, an analogue probe or a touch trigger probe, is a very common practice in geometrical measurement and a hardware implementation of morphological dilation operation [28]. The workpiece surface defined as the input set is dilated by the structuring element, the probe tip, to generate the morphological output, the measured surface, which is also called the traced surface. Fig. 7.7 illustrates the scanning process of a tactile probe. The scanning measurement is conducted by traversing the tip over the surface. The tip center data is recorded at each sampling position, and these sampled data form a discrete presentation of the measured surface. In [29] the traced surface profile is defined as "*locus of the centre of a stylus tip which features an ideal geometrical form (conical with spherical tip) and nominal dimensions with nominal tracing force, as it traverses the surface within the intersection plane.*"

In common practice the probe tips employed for scanning are small in size. However, the tip size still influences the precision in measurement of workpiece surfaces. Fig. 7.8 demonstrates the effect of the probe tip traversing over the workpiece surface. By comparing the traced profile with the real workpiece profile in the figure, it is evident

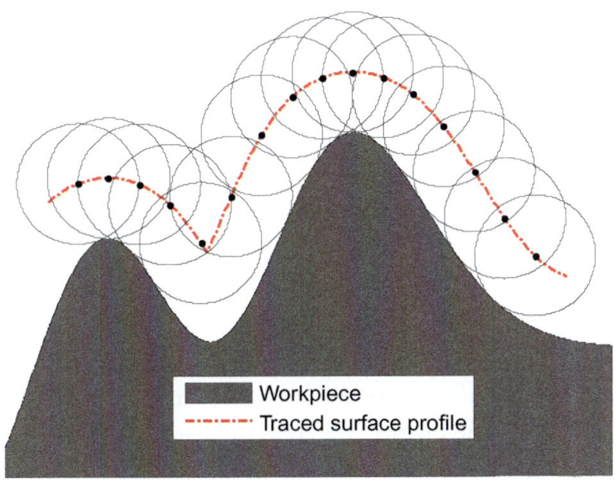

Fig. 7.7 Scanning the probe tip over the workpiece surface.

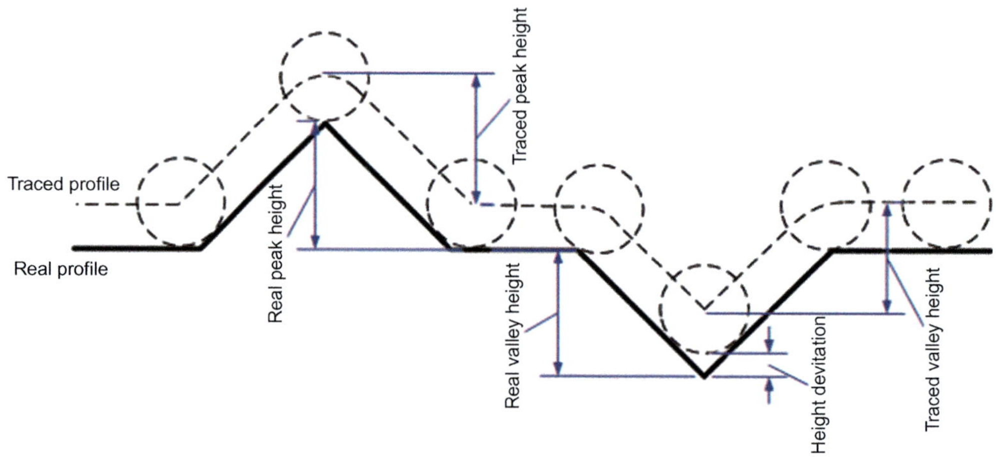

Fig. 7.8 Tip mechanical filtering effects.

that the probe tip tends to round off peaks on the profile making it broader; nevertheless, the peak height remains constant. The valleys on the profile are smoothed by the tip becoming narrow, meanwhile the valley depth is reduced as well [30]. This effect introduces distortion into measurement of workpiece surfaces and is called as the mechanical filtration effect of tips. For the measurement of workpiece surfaces, especially for the freeform shaped workpieces, the distortions caused by the tip mechanical filtration effect significantly influence the precision of measurement. Thus correction to the traced surface is desired to restore the real workpiece surface. However, the traced surface is unable to be perfectly reconstructed to the real surface, only to an approximate one, that is, the real mechanical surface.

7.3.2 Real mechanical surface reconstruction

ISO 14406 [31] presents the definition of mechanical surface as, "boundary of the erosion, by a sphere of radius r, of the locus of the center of an ideal tactile sphere, also with radius r, rolled over the real surface of a workpiece." Fig. 7.9 demonstrates the reconstruction process. Using an ideal sphere with the same size to the probe tip to roll over the traced profile, that is, the dilated profile by the probe tip (which is already presented in Fig. 7.7), the locus of the sphere center is treated as the real mechanical surface. Rolling the ball from below the traced surface is in essence an erosion operation.

It is clear from Fig. 7.9 that the morphological erosion operation is unable to perfectly reconstruct the original real surface of the workpiece. It was found that morphological operations can only reconstruct those portions of the surface where their local curvatures are larger than that of the probe tip [15, 32]. This indicates that the real mechanical surface differs from the real surface at the locations where the local surface curvature is small than the tips. Thus the reconstructed real mechanical surface varies with the probe tip size.

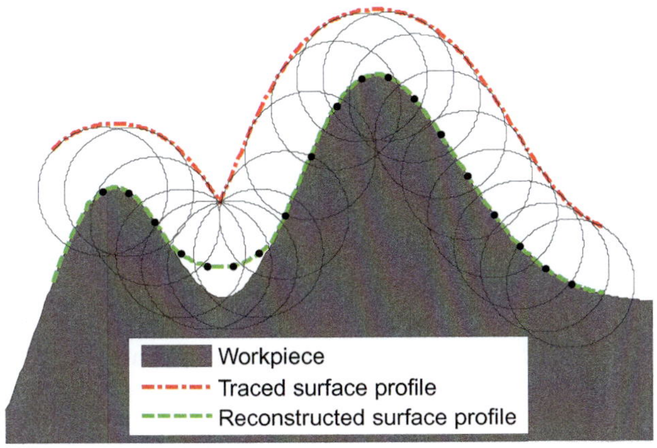

Fig. 7.9 Reconstruction of mechanical surface.

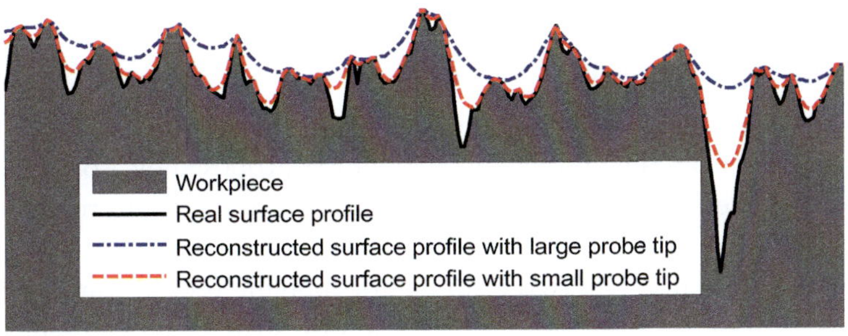

Fig. 7.10 Reconstructed real mechanical surfaces vary with the tip size.

Fig. 7.10 presents such an example. Large probe tips tend to reduce and smooth the surface irregularities, while small tips enable the reconstruction of a surface that is closer to the real surface. The smaller the tip is, the closer the mechanical surface can approximate to the real surface.

In industry the practical implementation of the reconstruction of the real mechanical surface varies from the application requirements. For surface texture instruments, for instance, the profilometer and the atomic force microscope, the reconstruction is usually performed by morphological image processing techniques [33], whereas in dimensional metrology, for example, coordinate measurement, the reconstruction is usually implemented by probe radius compensation. Compared with surface texture instruments, sampling of coordinate measurement machine (CMM) is usually less dense and the probe tip much bigger. As Fig. 7.11 illustrates a couple of sampling positions, the contact points of the probe tip to the workpiece surface are obtained by compensating the tip radius in the direction of the surface, normally at the contact point. The normal vectors for compensation are achieved either by estimating from the measured tip's center data, in the case

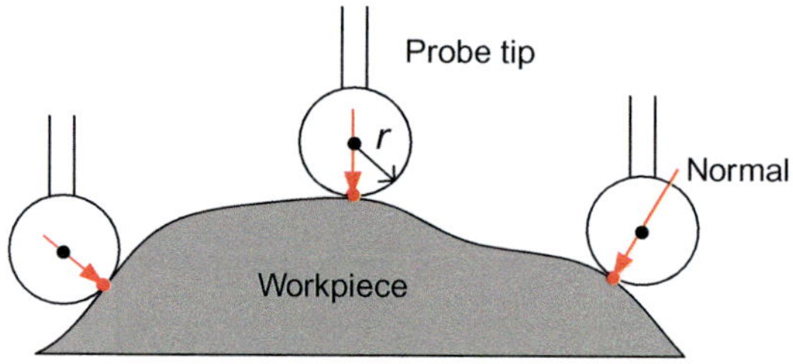

Fig. 7.11 The radius compensation of CMM measurement.

that the surface is densely scanned and nominal data are unavailable [34, 35], or by using the nominal vector at the matching point on the nominal surface model of the workpiece, for example, the CAD model [36, 37]. Although dimensional metrology and surface metrology employ different routes, both of them are essentially morphological reconstructions of the real surface.

7.3.3 Closing and opening filtering

Closing and opening filters are two envelope filters, whose output envelopes the input surface. They differ from the obsolete envelope filter proposed by Von Weingraber [38] in that the early envelope filter is essentially equivalent to the dilation operation, usually offset by ball radius.

Figs. 7.12 and 7.13 illustrate two examples of applying the closing operation and the opening operation on an open profile with the disk structuring element, respectively. The closing filter is obtained by placing an infinite number of identical disks in contact with the profile from above along the profile and taking the lower boundary of the disks [22]. In contrast the opening filter is achieved by placing an infinite number of identical disks in contact with the profile from below along all the profile and taking the upper boundary of the disks.

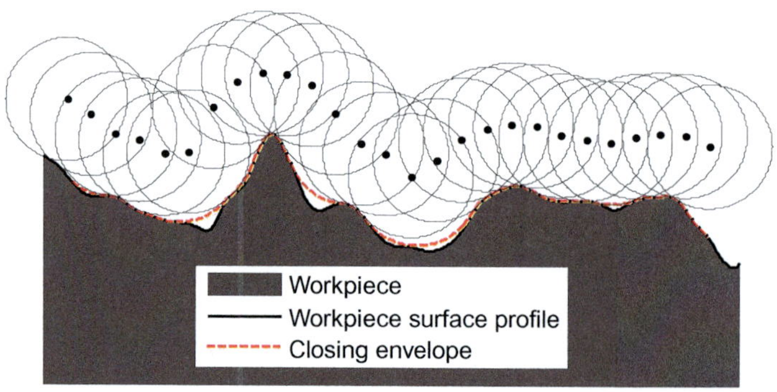

Fig. 7.12 The closing envelope of an open profile by a disk.

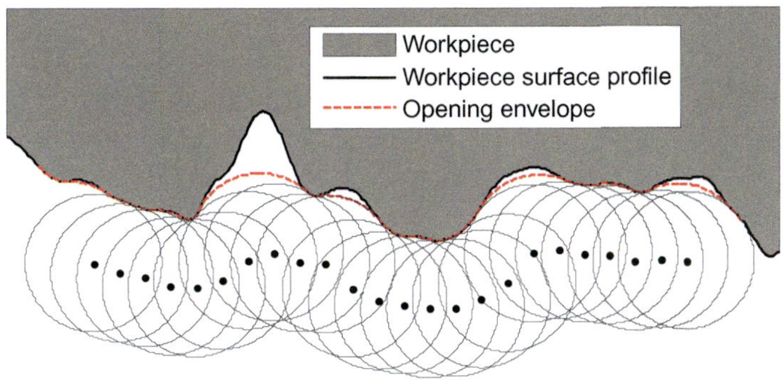

Fig. 7.13 The opening envelope of an open profile by a disk.

It can be clearly seen in the figures that the closing filter suppresses the valleys on the profile that are smaller than the disk radius in size; meanwhile the peaks remain unchanged. Conversely the opening filter suppresses the peaks on the profile that are smaller than the disk radius in size, while it retains the valleys. The selection of the disk radius depends on the size of physical features on the surface of the workpiece. Except for circular structuring elements, other most commonly used structuring elements recommended by ISO 16610 are flat structuring elements, for instance, the horizontal line segment for profile data.

7.3.4 Form approximation

It has been illustrated that morphological envelopes could be utilized to approximate the form of functional surfaces for conformable interfaces [39], for instance, a soft gasket in contact with a solid block to provide a sealing function. The long wavelength component of the block surface could be tolerated by the compatibility of the gasket material, while

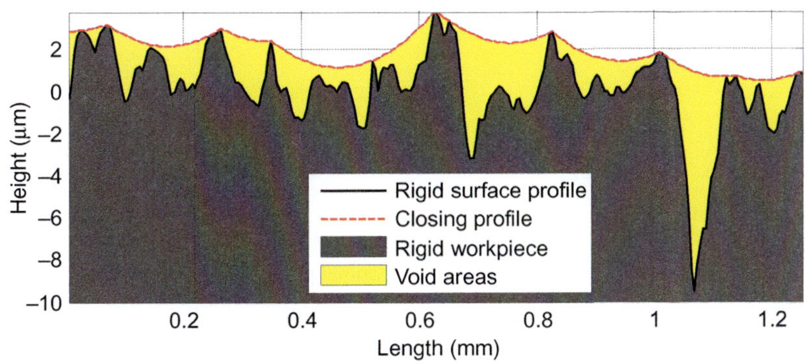

Fig. 7.14 Approximation of the form of a conformable surface profile.

the middle wavelength components result in highly localized contacts. See Fig. 7.14. The morphological closing envelope with the circular disk structuring element is used to approximate the conformable gasket surface so that the void areas between the conformable surface and the rigid surface can be obtained to characterize the sealing or load distribution. The radius of the circular structuring element should be chosen based on the compression and bending properties of the conformable component.

7.3.5 Contact points

In many engineering applications the contact between two surfaces is nonconforming; that is, the contact area is very small when compared with the geometry of the bodies in contact. Even in situations between conforming contacts, the contacts between the asperities that compose the surface topography are known to be nonconforming. To investigate the contact phenomenon, the closing operation can be employed, whereby the interaction is simulated by rolling a ball with a given radius, which is sized to simulate the largest reasonable radius at a contact, for example, peak curvature, upon the underlying surface. The contact points of the rolling ball against the rolled surface are then captured. This serves as an indication of the surface summits and surface portions that are in real contact.

In physics the contact points are those points on the surface that are in contact with the rolling ball. From a point of view of mathematical morphology, the contact points are those points on the surface that remain constant with the morphological closing operation. Therefore these points might be captured by computing the closing envelope using appropriate algorithms [40, 41]. Fig. 7.15 illustrates such an example with sampling length 8 mm and sampling interval 0.25 μm. For convenience of visualization the profile was translated and rotated. The contact points with disk radius 5 mm are then extracted from the profile texture. Finally the form error of the straightness of the profile was calculated by applying the minimum zone method to the contact point set. The obtained straightness is 1.434 m in the example.

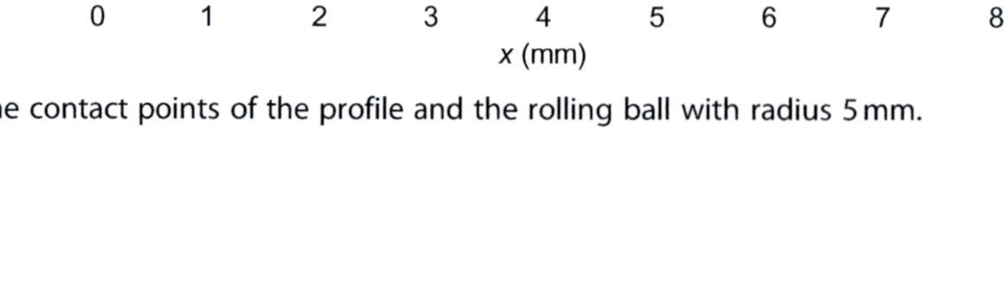

Fig. 7.15 The contact points of the profile and the rolling ball with radius 5 mm.

7.3.6 Uncertainty zone for continuous surface reconstruction

The workpiece surface is the set of features that physically exist and separate the entire workpiece from the surrounding medium. The inspection of the geometrical information of the workpiece surface is conducted by measuring the surface at certain sampling interval, using either the contact measurement instruments (e.g., CMM) or noncontact ones (e.g., interferometer). Either of these instruments generates a series of sampled points, which form a discrete representation of the original surface. It should be noted that the sampled points in this scenario differ from those presented in the preceding cases due to the fact they are supposed to be the contact points on the real workpiece surface, instead of the tip center points for tactile measurement. For noncontact measurement the sample data are all "contact points." It may be useful to reconstruct the original continuous workpiece surface from the discrete samples. In the theory of signal processing, the Nyquist theorem indicates that an infinitely long band-limited signal could be perfectly reconstructed without loss of information from the discrete data sampled at regularly spaced intervals if that interval is smaller than half of the minimal wavelength comprised by the original signal. In mathematical morphology, there is no theorem equivalent to the Nyquist theorem in that a universal equidistant sampling scheme can be found without loss of information; however, there are a number of morphological sampling theorems to limit the amount of information lost [42].

Fig. 7.16 illustrates an example of determining the uncertainty zone for the reconstruction of the original surface from a sequence of sampled points taken by a circular disk structuring element. The morphological sampling theorem makes the assumption that the surface profile Z under the examination remains unchanged after applying the opening and closing operation by a particular structuring element SE (e.g., a disk) of a given size (e.g., the disk radius), that is, $C(Z, SE) = Z = O(Z, SE)$. If the original surface Z is sampled with a sampling interval strictly less than the size of SE, yielding a sampled surface Z_s, the original profile is supposed to lie in the region constructed

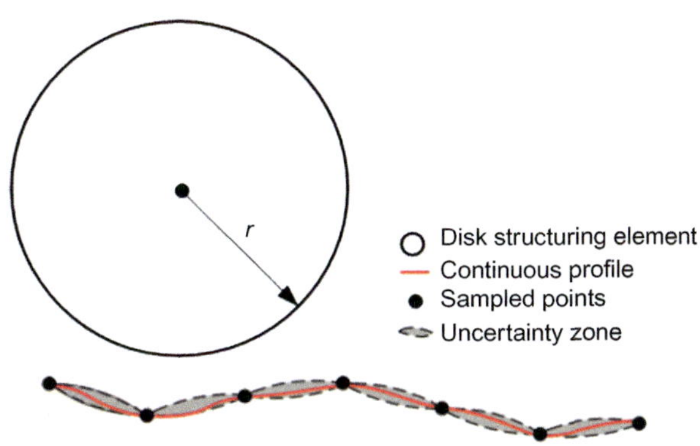

Fig. 7.16 Uncertainty zone for continuous surface profile reconstruction with the disk.

by the opening envelope $O(Z_s)$ and the closing envelope $C(Z_s)$. This region defines the uncertainty zone in which the original profile lies, that is, $C(Z_s, SE) \leq Z \leq O(Z_s, SE)$ [31].

7.3.7 Scale-space analysis

Scale-space techniques are a type of sieving technique, which could date back to the morphological granulometry [43]. It is similar to the process of sieving small solid particles through a series of sieves with mesh openings of increasing size, where the sieve with the smallest opening is used first. The grains that are bigger than the mesh opening are kept and counted, and the remaining grains are then sifted by the bigger sieve, and this process continues until all the sieves are used. In this way, grains are classified according to the size of mesh openings.

Scale-space techniques could decompose a signal (profile/surface) into objects of different scale. It uses alternating symmetrical filters of increasing scales to construct a ladder structure as shown in Fig. 7.17. The first rung S_0 is the original signal. At each rung in the ladder, the signal is filtered by an alternating symmetrical filter at the scale order $i+1(M_{i+1})$ to obtain the next space-scale representation of the signal S_{i+1} that becomes the next rung and a component d_{i+1} that is the difference between the two rungs. In this manner, signals at different scales are separated from each other.

Scale-space techniques provide a multiresolution method to decompose signals with wavelet-based filters being another type of multiscale approach in surface metrology [44, 45]. The scale of the alternating symmetrical filter at each rung works like a cutoff value λs. Therefore a "transmission bandwidth" can be defined by calculating the height difference between two rungs: $S_i - S_j$. The scale i is equivalent to the cutoff value λs, and the scale j is equivalent to the cutoff value λc.

In comparison with the famous Nyquist theorem used to sample and reconstruct a signal in the frequency domain [46], for morphological operations and filters, no universal equidistant sampling can be found without loss of information. However, there are a number of morphological sample theorems that limit the amount of information that is lost, scale-space techniques being one of them. The original signal could be sampled with various scales and can be reconstructed by reversing the ladder structure mentioned in the preceding text:

$$S_0 = S_n + \sum_{i=1}^{n} d_i. \tag{7.7}$$

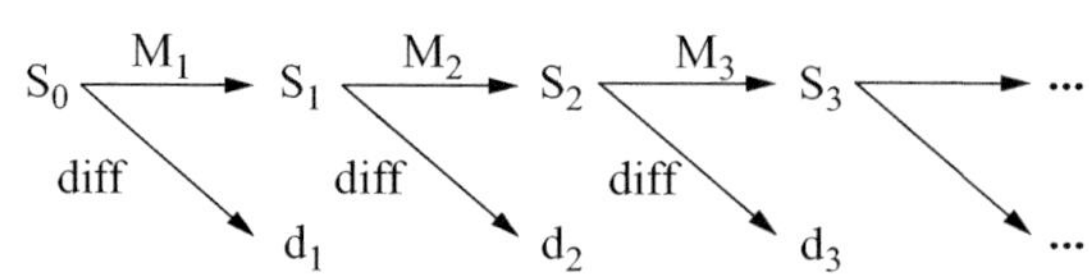

Fig. 7.17 The ladder structure of scale space.

The cutoff wavelengths of the multiscale analysis are always in a constant ratio of 2 to each other. This value is yielded by the experiences in dealing with multiscale analysis. This ratio is nearly optimal since this value is, on one hand, large enough to clearly differentiate the details of different levels; on the other hand, it is not so large that significant details are lost [47]. Based on this recognition, morphological scale-space techniques also choose a ratio of the scales of approximately 2: 1 μm, 2:2 μm, 2:5 μm, 2:10 μm, 2:20 μm, 2:50 μm, 2:100 μm, 2:200 μm, 2:500 μm, 2:1 mm, 2:2 mm, 2:5 mm, and 2:10 mm, ⋯. This series has an additional advantage that it is consistent with the recommended stylus tip radii of surface texture [48]. The smallest value of this series is limited by the morphological sampling theorem and therefore cannot be smaller than the value of the sampling interval in length. It is sensible to let the series start with the value of stylus tip radius used for the measurement. In principle, there is no upper limit to values for the scale series. Fig. 7.18 shows a profile that is from a milled surface. The series of scale values (1, 2, 5, 10, 20, 50, 100, 200, 500, and 1000 μm) was used starting with the first value (1 μm). Fig. 7.19 shows the differences between successively smoothing.

Muralikrishnan and Raja [47] employed scale-space techniques to analyze the cylinder liner whose inner is a plateau-honed surface. It is similar to the case presented by Decenciere and Jeulin [49], but the notable difference between the two cases is that the size of the features on the surface was not known in advance, while in the former case the size of features was estimated by physical comments.

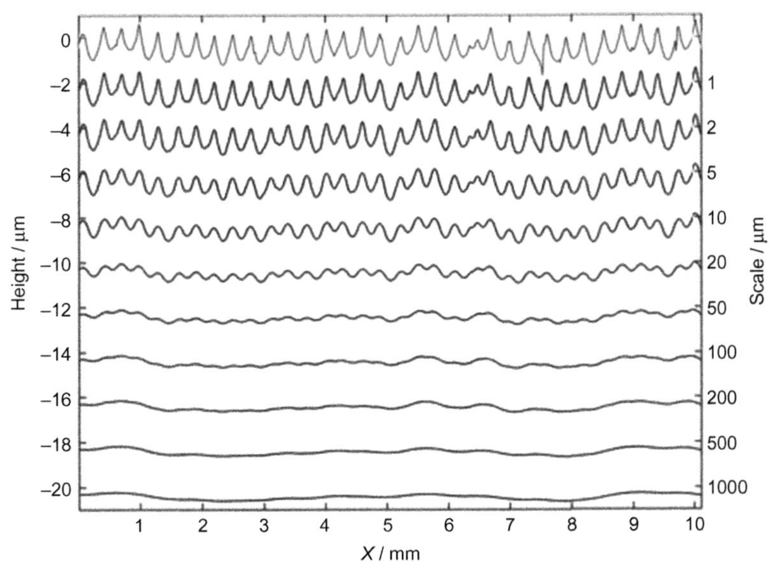

Fig. 7.18 Successively smoothed profiles from a milled surface using a circular disk.

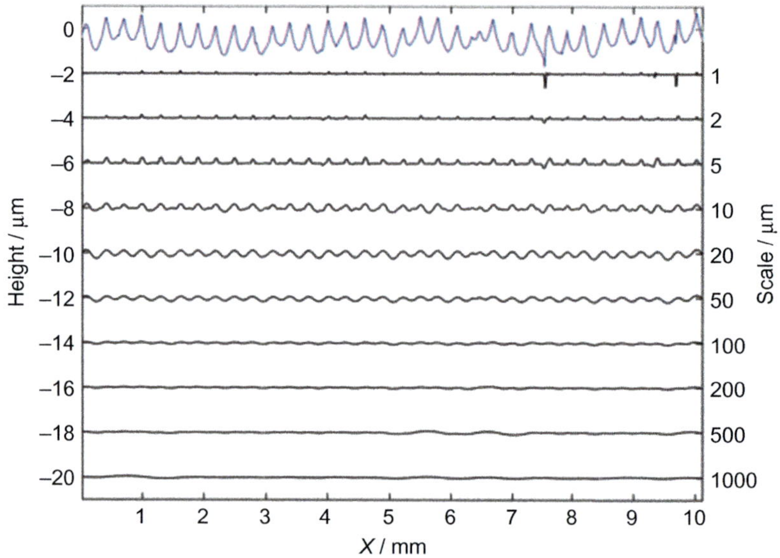

Fig. 7.19 Differences on a profile from a milled surface using a circular disk.

7.4 Alpha shape algorithms for morphological filtering

To overcome the limitations of the traditional computation method, a totally different method has been developed, where measured surfaces are no longer treated as grayscale images but 3D point clouds [50, 51]. Geometric computation techniques are adopted to solve the morphological envelope of the point cloud. The alpha shape, an ubiquitous geometrical structure in the field of surface reconstruction, is closely related to morphological envelopes and can be employed for their computation.

7.4.1 Alpha shape theory

The alpha shape was introduced by Edelsbrunner in the 1980s aiming to describe the specific "shape" of a finite point set using a real parameter to control the desired level of details [52]. Given a planar point set, a circular disk of radius α is rolled around both inside and outside (see Fig. 7.20) that will generate an object with arcs and points. The boundary of the resulted object is called the alpha hull. If the round faces of the object are straightened by line segments, it generates another geometrical structure, the alpha shape.

In the context of the alpha shape, the disk used in the aforementioned example is called the alpha ball. It is formally defined as an open ball of radius α. Given a point set $X \subseteq \mathbb{R}^d$, a certain alpha ball b is empty if $b \cap X = \varnothing$. With this a k-simplex σ_T is said

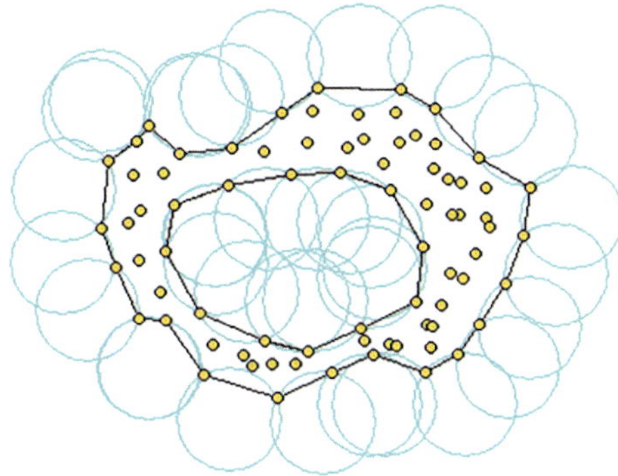

Fig. 7.20 The alpha shape of a planar point set.

to be α-exposed if there exists an empty alpha ball b with $T=\partial b \cap X$ ($|T|=k+1$) where ∂b is the surface of the sphere (for $d=3$) or the circle (for $d=2$) bounding b, respectively.

Definition 7.1 For $0 \leq \alpha \leq \infty$ the alpha hull of X, denoted by $H_\alpha(X)$, is defined as the complement of the union of all empty α-balls. $\partial S_\alpha(X)$, the boundary of the alpha shape of the point set X, consists of all k-simplices of X for $0 \leq k < d$ that are α-exposed:

$$\partial S_\alpha(X) = \{\sigma_T | \ T \subset X, |T| = k+1, \sigma_T \ \alpha - \text{exposed}\}. \tag{7.8}$$

The computation of the alpha shape is based on the Delaunay triangulation. Given a point set $X \subseteq \mathbb{R}^d$, the Delaunay triangulation is a triangulation $DT(X)$ such that no point in X is inside the circumsphere of any d-simplices σ_T with $T \subset X$. The relationship between the Delaunay triangulation and the alpha shape is that the boundary of the alpha shape ∂S_α is a subset of the Delaunay triangulation of X, that is,

$$\partial S_\alpha(X) \subset DT(X). \tag{7.9}$$

To further extract the simplices of $\partial S_\alpha(X)$ from $DT(X)$, another concept, the alpha complex $C_\alpha(X)$, was developed. Set ρ_T is the radius of the smallest circumsphere b_T of σ_T. For $k=3$, b_T is the circumsphere; for $k=2$, b_T is the great circle; and for $k=1$ the two points in T are antipodal on b_T. For a given point set $X \subseteq \mathbb{R}^d$, the alpha complex $C_\alpha(X)$ is the following simplicial subcomplex of $DT(X)$.

A simplex $\sigma_T \in DT(X)$ ($|T|=k+1$, $0 \leq k \leq d$) is in $C_\alpha(X)$ if

(1) $\rho_T < \alpha$ and ρ_T-ball is empty.

(2) σ_T is a face of other simplex in $C_\alpha(X)$.

The relationship between the alpha complex and the alpha shape consists of the boundary of the alpha complex making up the boundary of the alpha shape, that is,

$$\partial C_\alpha(X) = \partial S_\alpha(X) \subset DT(X). \tag{7.10}$$

7.4.2 Link between alpha hull and morphological envelopes

The boundary of the alpha hull is obtained by rolling the alpha ball over the point set. By intuition the alpha hull seems very similar to the secondary morphological operations, opening and closing, as the alpha ball acts as a spherical structuring element and the input set as the points set. In fact a theoretical link exists between the alpha hull and morphological opening and closing, as proved by Worring and Smelders [53]. They extended Edelsbrunner's work, taking the alpha graph and utilizing it to describe the boundary of the point set. They also found the relationship between the alpha graph and the opening scale space from mathematical morphology. Based on that, it was proved that the alpha hull is equivalent to the closing of X with a generalized ball of radius $-1/\alpha$. Hence, from the duality of the closing and the opening, the alpha hull is the complement of the opening of X^c with the same ball as the structuring element.

7.4.3 Morphological envelope computation based on the alpha shape

Based on the established relationship between the alpha shape and morphological envelopes, alpha shape theory is applied to the computation of morphological envelopes. The computation starts with the Delaunay triangulation of the point set that comprises the measured surface.

The Delaunay triangulation results in a series of k-simplices σ ($k=2$ for profiles, which are triangles; $k=3$ for surfaces, which are tetrahedrons). These k-simplices are categorized into two groups: k-simplices σ_p whose circumsphere radius is larger than the radius of the rolling ball α and k-simplices σ_{np} whose circumsphere radius is no larger than the radius of the rolling ball α.

σ_p consists of two parts: the $(k-1)$-simplices σ_{int} interior to σ_p and the $(k-1)$-simplices σ_{reg} that bounds its super k-simplices σ_p. We called σ_{reg} the regular facets. σ_{np} is composed of three components: the $(k-1)$-simplices σ_{ext} out to C_α, part of the regular facets σ_{reg}' shared by both σ_p and σ_{np}, and the $(k-1)$-simplices σ_{sing} that are the other part of ∂C_α. We call σ_{sing} the singular facets. σ_{sing} differs from σ_{reg} in that it does not bound any super k-simplices belonging to $C_\alpha.\sigma_{sing}$ satisfies two conditions as follows:

(1) The radius of its smallest circumsphere is smaller than α.

(2) The smallest circumsphere is empty.

An explanatory graph is presented by Fig. 7.21. The regular facets σ_{reg} and the singular facets σ_{sing} form the whole boundary of the alpha complex, that is, the boundary of the alpha shape, as Eq. (7.11) presents:

$$\partial S_\alpha = \partial C_\alpha = \sigma_{reg} + \sigma_{sing}. \tag{7.11}$$

Fig. 7.22 illustrates an example of the alpha shape facets extracted from the Delaunay triangulation of an experimental profile data. With the boundary alpha shape facets, the morphological envelope can be solved. For each sample point, there is a one-to-one

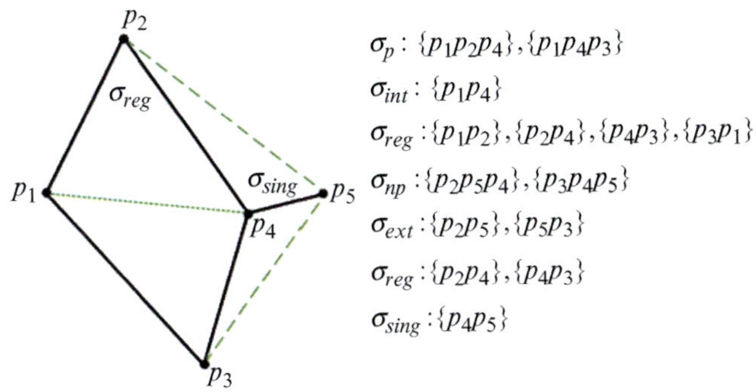

Fig. 7.21 Regular and singular facets.

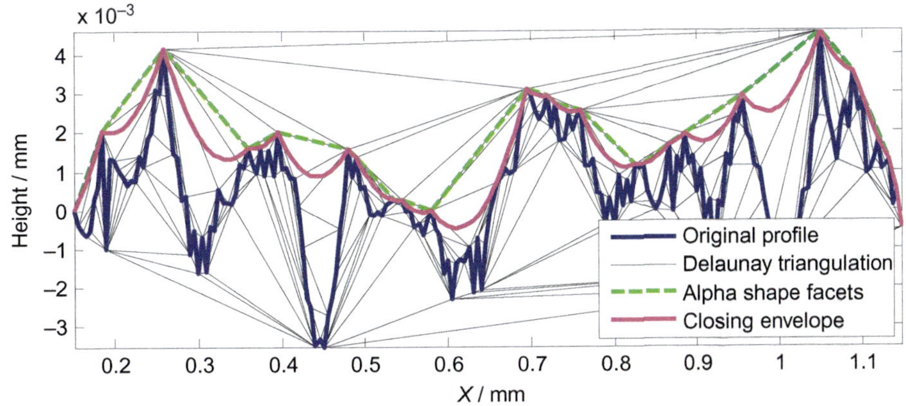

Fig. 7.22 Alpha shape facets extracted from the Delaunay triangulation of a profile data.

corresponding point on the envelope. These points form a discrete representation of the morphological envelope. Each boundary facet of the alpha shape determines its counterpart on the alpha hull. Due to the fact that the target envelope is contained within the alpha hull, the sample points are projected onto the alpha hull along the local gradient vector to obtain the envelope coordinates.

The alpha shape method calculates morphological filters geometrically and has a number of advantages over traditional methods. Firstly, by viewing the measured surface points as the 2D/3D point set, it breaks the constraints of the traditional method that only applies to planar profiles/surfaces. Secondly, alpha shape theory enables an arbitrary size of the disk or ball radius. Moreover the alpha shape method is better suited to nonuniform sampled data. The alpha shape method depends on the Delaunay triangulation, bringing in an additional merit that the triangulation could be reused for multiple attempts of various disk radii, saving a great deal of computing time because in practice a multitude of trials may be made for choosing an appropriate disk radius.

7.4.4 Divide and conquer optimization

The alpha shape method is more competent than traditional algorithms. However, its drawback is that the 3D Delaunay triangulation is costly in both computation time and memory for large areal data sets and it was reported that the data structure of the Delaunay triangulation was not suitable for data sets of millions of points [54]. In practice, measured engineering surfaces usually contain a large quantity of data, especially using fast optical measurement instruments. As a result, divide and conquer optimization is employed to speed up the computation of morphological envelopes and avoid the risk of running out of memory.

In the context of the alpha shape method, the vertices of boundary facets of the alpha shape are of tremendous importance because they are the surface points in contact with the rolling ball. The morphological envelope of a surface is only determined by these vertices while not affected by other points.

The basic scheme of the divide and conquer approach is to break a problem into several subproblems that are similar to the original problem but smaller in size, solve the subproblems iteratively, and then combine these solutions to create a solution to the original problem [55]. By applying the divide and conquer method, the surface under evaluation is divided into a series of small subsurfaces, and the ball rolls over each subsurface to generate a set of alpha shape vertices. Afterward the resulted vertices from each subsurface are merged to reconstruct a super set of vertices, which will be treated as the point set for the next iteration. It should be noted that some of the vertices around the joint sections in the previous iteration are removed. By repeating the iterations the final alpha shape of the original surface can be found, and the morphological envelope of the surface is then determined. A demonstration of the divide and conquer procedures is given by Fig. 7.23.

7.5 Contact points and their searching procedures

Using the alpha shape vertices, the divide and conquer optimization reduces the amount of points in the iteration processed by the Delaunay triangulation. However, it is not a fundamental change to the alpha shape method because the Delaunay triangulation is still required to extract the alpha shape boundary facets. It was revealed that the Delaunay triangulation is not necessary for searching the boundary facets and an alternative computational method is explored by searching the contact points [41].

7.5.1 Redundancy of the Delaunay triangulation

In Edelsbrunner's theory the alpha shape is extracted from the Delaunay triangulation. In fact the whole family of alpha shapes can be generated from the Delaunay triangulation, from the point set itself ($\alpha \to 0$) to the convex hull of the point set ($\alpha \to \infty$). Therefore the Delaunay triangulation data could be reused for multiple attempts of ball radii for the same data set. This could however be a drawback because the Delaunay triangulation

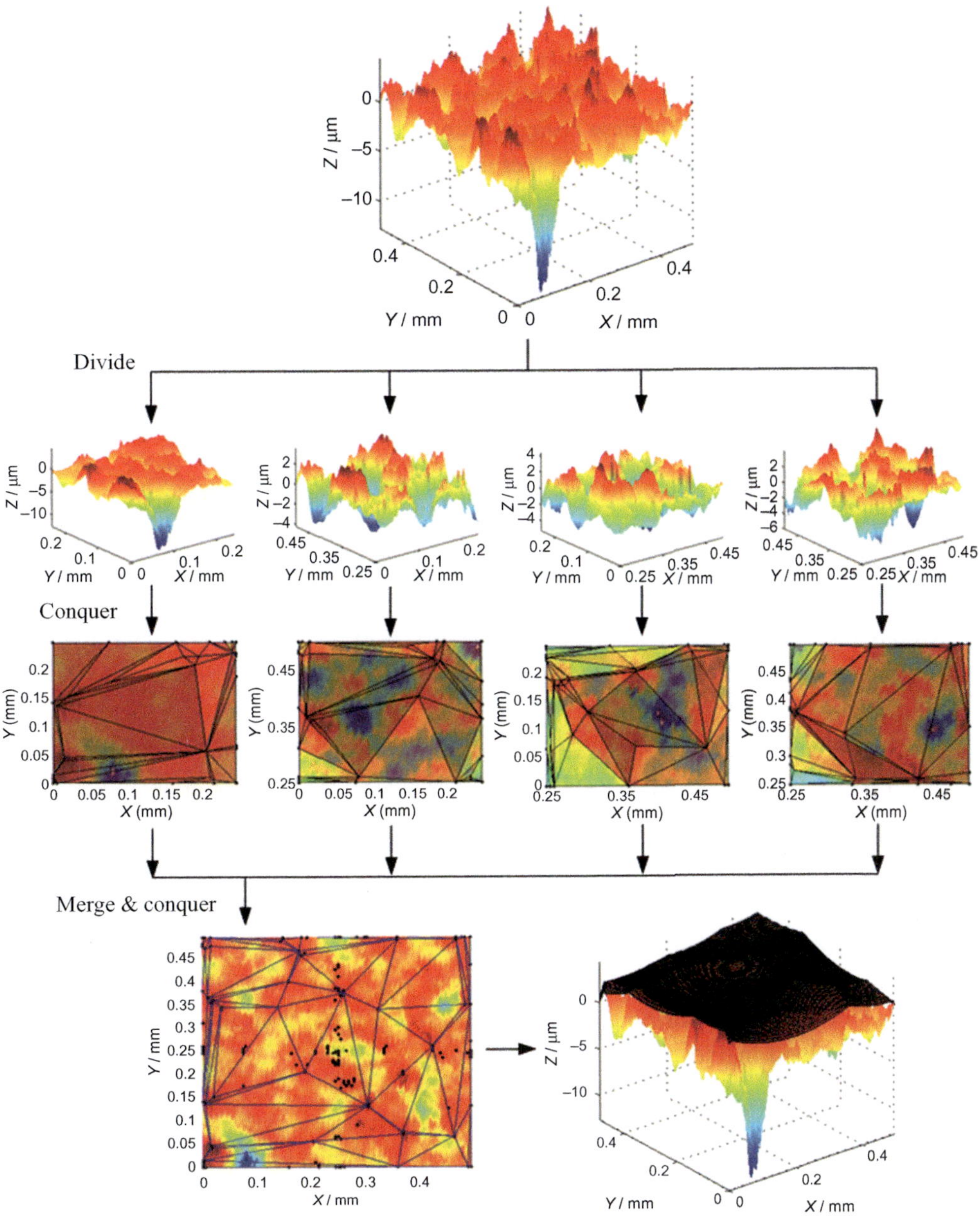

Fig. 7.23 Divide and conquer optimization for the alpha shape method.

is costly for large areal data sets. Given a single radius the Delaunay triangulation contains much more information than is necessary to generate the corresponding alpha shape for that radius. Thus, in this sense, it is a waste of time and memory to achieve the desired alpha shape with redundant computation.

7.5.2 Definition of contact points

In physics the contact points are those points on the surface that are in contact with the moving structuring element. Thus these points give an indication of which surface portions in the neighborhood of these contact points are most likely to be active in the contact phenomenon. From a point of view of mathematical morphology, the contact points are those points on the surface that remain constant before and after morphological closing or opening operations. Based on the mapping between the alpha hull and morphological opening and closing envelopes, the formal mathematical definition of the contact point is given by Definition 7.2 using alpha shape theory:

Definition 7.2 Given a sampled point set $X \subseteq \mathbb{R}^d$ $(d=2,3)$ and $\delta \le \alpha \le \infty$($\delta$: sampling interval), the contact points $P(\alpha)$ are those sampled points $\{p_i | p_i \in X\}$ that are on the boundary of the alpha hull $\partial H_\alpha(X)$:

$$P(\alpha) = \{p_i | \, p_i \in X, p_i \in \partial H_\alpha(X)\}.$$

7.5.3 Propositions for searching contact points

Proposition 7.1 *Given a point set* $X \subseteq \mathbb{R}^d$ $(d=2,3)$ *and* $\delta \le \alpha_1 \le \infty$, $\delta \le \alpha_2 \le \infty$, *if* $\alpha_1 \le \alpha_2$, *then* $P(\alpha_2) \subseteq P(\alpha_1)$.

Proof. $\alpha_1 \le \alpha_2 \Rightarrow H_{\alpha_1}(X) \subseteq H_{\alpha_2}(X)$ [56]. By Definition 7.2, $P(\alpha_1) = \{p_i | p_i \in X, p_i \in \partial H_{\alpha_1}(X)\}$, $P(\alpha_2) = \{p_i | p_i \in X, p_i \in \partial H_{\alpha_2}(X)\}$. Hence, $H_{\alpha_1}(X) \subseteq H_{\alpha_2}(X)$ implies $P(\alpha_2) \subseteq P(\alpha_1)$.

Proposition 7.2 *Given the point set* $X \subseteq \mathbb{R}^d$ $(d=2,3)$ *and* $\delta \le \alpha \le \infty$. *The convex hull points must all be contact points.*

Proof. Let $\alpha' \to \infty$, hence $\lim_{\alpha' \to \infty} H_{\alpha'}(X) = Conv(X)$. By Definition 7.2, $P(\alpha') = \{p_i | p_i \in X, p_i \in \partial H_{\alpha'}(X)\}$, then $P(\alpha') = \{p_i | p_i \in X, p_i \in \partial Conv(X)\}$, namely, $P(\alpha')$ is the convex point set. By Proposition 7.1, $\alpha \le \alpha' \Rightarrow P(\alpha') \subseteq P(\alpha)$. Thus the convex hull points must be contained in $P(\alpha)$.

Fig. 7.24 presents an example illustrating the boundary facets of the alpha shape of a planar point set relating to four different disk radii, respectively. The Delaunay triangulation of the point set is presented by a triangular mesh, and the boundary facets of the alpha shape are graphed as bold dotted lines. It can be easily verified that the results presented by four subfigures are consistent with Propositions 7.1 and 7.2. For instance the contact points of Fig. 7.24A with radius 1 mm are contained in Fig. 7.24B with radius 0.5 mm and so forth.

Following the definition of contact points and two associated propositions, another four theories/methods are proposed with proofs attached for searching contact points. For convenience of explanation the morphological closing profile filter with the disk structuring element is taken as the objective for demonstration. These propositions

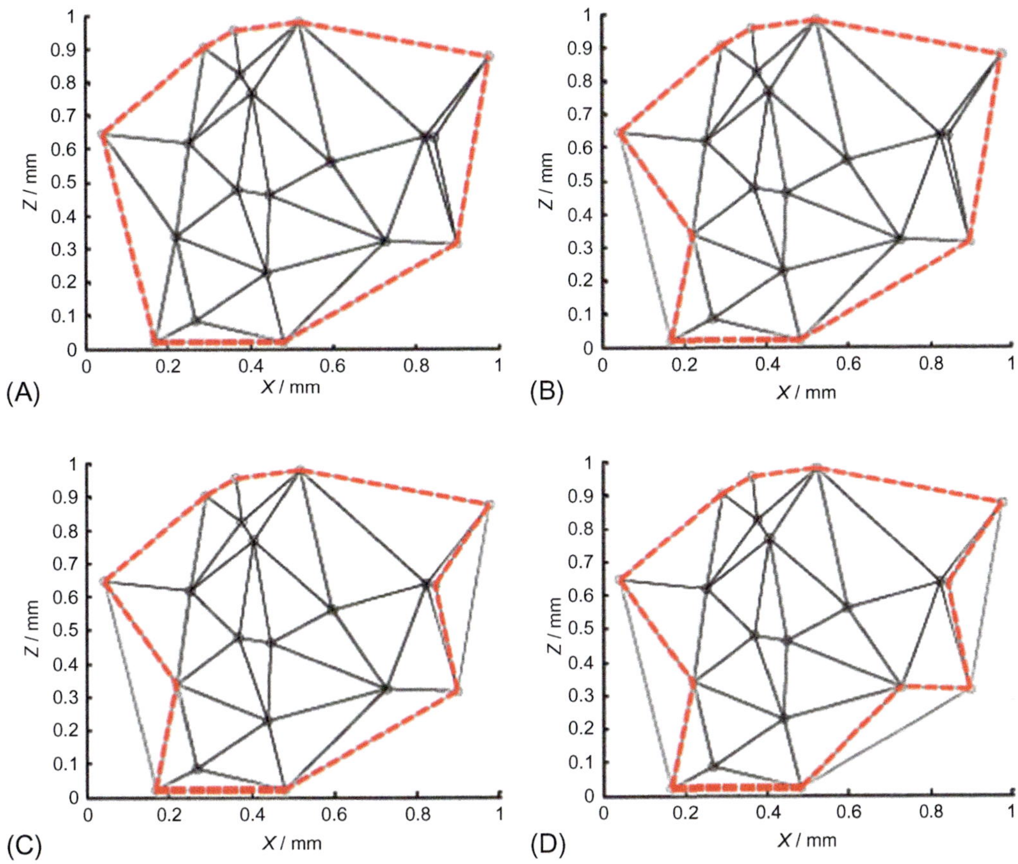

Fig. 7.24 The Delaunay triangulation of a planar point set and the boundary facets of the alpha shapes of four disk radii: (A) $\alpha = 1$, (B) $\alpha = 0.5$, (C) $\alpha = 0.4$, and (D) $\alpha = 0.3$ mm.

however can be easily extended to the opening filter, horizontal flat structuring elements, and areal data. In the context of the statement in the succeeding text, a and b are two known contact points, and r is the given radius of the ball (disk).

Proposition 7.3 *If there are points lying above σ_{ab} (left/positive side of $\overrightarrow{ab}$), then the contact point is the furthest point orthogonal to $\overrightarrow{ab}$.*

Proof. Suppose there exist some points above σ_{ab}. See Fig. 7.25. The furthest point c is the convex point for the point set $\{a, b, c, p_i\}$ [54]. By Proposition 7.2 the convex point must be the contact point. Thus c must be the contact point.

Proposition 7.4 *If there are no points lying above σ_{ab} and there exist points $\{p_i\}$ in the circular section $\widehat{ab}$ of the ball with radius $\alpha = \max\{r, ½|ab|\}$, then the contact point c is the one among the points $\{p_i\}$ in $\widehat{ab}$, which satisfies the condition: the circumscribed circle of σ_{abc} have the largest radius among the circumscribed circles of $\{\sigma_{abp_i}\}$.*

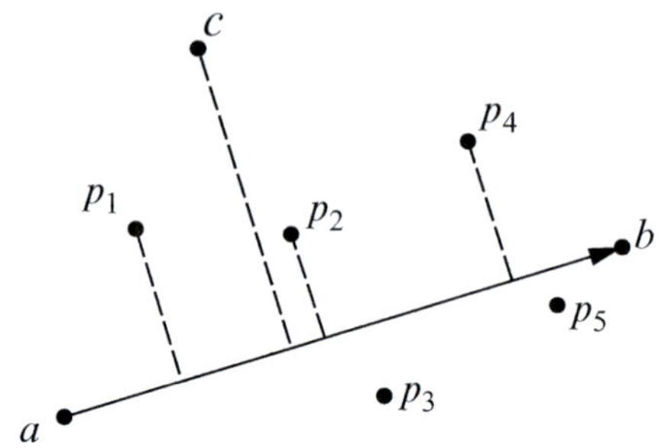

Fig. 7.25 Search the furthest point on the positive side of $\overrightarrow{ab}$ in orthogonal direction.

Proof. First, consider the case $|ab| \leq 2r$. See Fig. 7.26A. a and b could determine a unique alpha ball B with radius r. Since there exist points in the circular section $\widehat{ab}$ (the shadowed portion in the figure), $B \cap X = \{p_i\} \neq \emptyset$; thus σ_{ab} is not α-expose. By Definition 7.1, $\sigma_{ab} \notin \partial H_r(X)$. Let $\{\rho_i\}$ be the radii of the circumscribed circles of $\{\sigma_{abp_i}\}$ and c the point with $\max(\rho_i)$. The circumcircle of σ_{abc} must be empty; thus $c \in \partial H_{\max(\rho_i)}(X)$. By Proposition 7.1, $\max(\rho_i) > r \Rightarrow P(\max(\rho_i)) \subseteq P(r)$. By Definition 1, $c \in P(\max(\rho_i))$. Thus $c \in P(r)$, where c is the contact point.

Then, consider the other case $|ab| > 2r$. See Fig. 7.26B. Since $|ab| > 2r$, fit a great circle with radius $\alpha = ½|ab|$ passing through the points a and b with the center at the middle of σ_{ab}. Similar to the previous case, we could prove $c \in P(½|ab|)$. Then, by Proposition 7.2, $½|ab| > r \Rightarrow P(½|ab|) \subseteq P(r)$; thus $c \in P(r)$, where c is the contact point.

Definition 7.3 $\{p_i\}$ is a point lying below σ_{ab} (right/negative side of $\overrightarrow{ab}$). σ_{abp_i} has a unique circumscribed circle with radius α. If the center of the circumscribed circle is on the positive side of σ_{ab}, the circle has the positive radius $+\alpha$, otherwise the negative radius $-\alpha$.

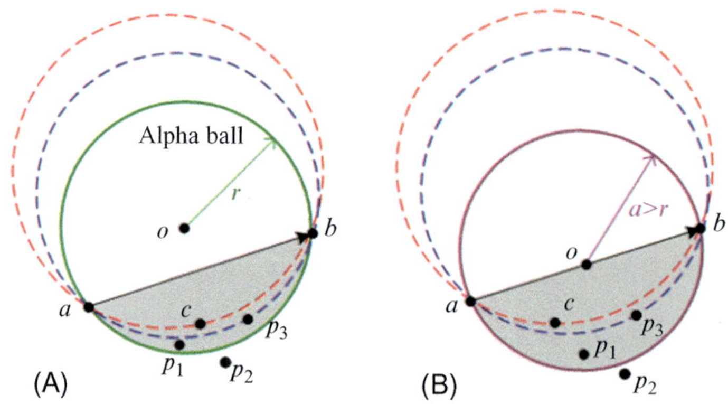

Fig. 7.26 Search the contact point with (A) $|ab| \leq 2r$ and (B) $|ab| > 2r$.

See Fig. 7.27. $\{p_1, p, p_2\}$ are three points below σ_{ab}. σ_{abp_1} has its circumcircle center o_1 above σ_{ab}; thus it has a positive radius. Conversely the center of the circumcircle of σ_{abp_2} lies below σ_{ab}; therefore the radius is negative. The critical case is that of σ_{abp} that has its circumcircle center o at the center point of σ_{ab}. In this case the radius is taken as positive.

Proposition 7.5 *If* $|ab| > 2r$ *and there are no points lying above* σ_{ab} *and also no points in the circular section* $\widehat{ab}$ *of the alpha ball with radius* $\alpha = \frac{1}{2}|ab|$*, then the contact point is the one* c *that satisfies the condition: the circumscribed circle of* σ_{abc} *has the largest radius among the circumscribed circles of* $\{\sigma_{abp_i}\}$.

Proof. See Fig. 7.28. There is no point in the circular section $\widehat{ab}$; thus the center of circumscribed circles of $\{\sigma_{abp_i}\}$ locates at the negative side of the chord $\overline{ab}$. Thus their radii are negative. The circumscribed circle with the largest radius (smallest in

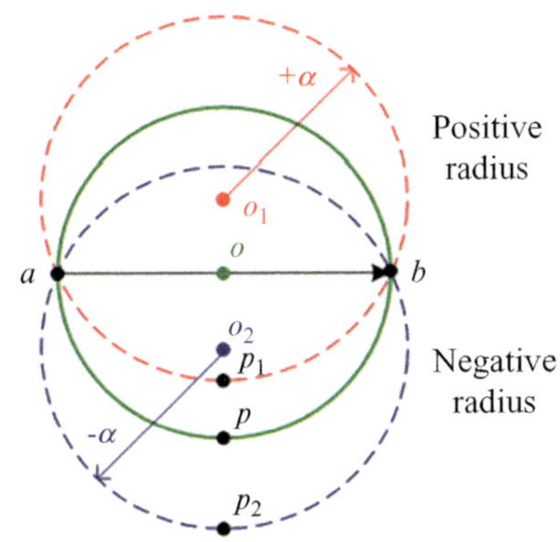

Fig. 7.27 The signed circumscribed circle radius.

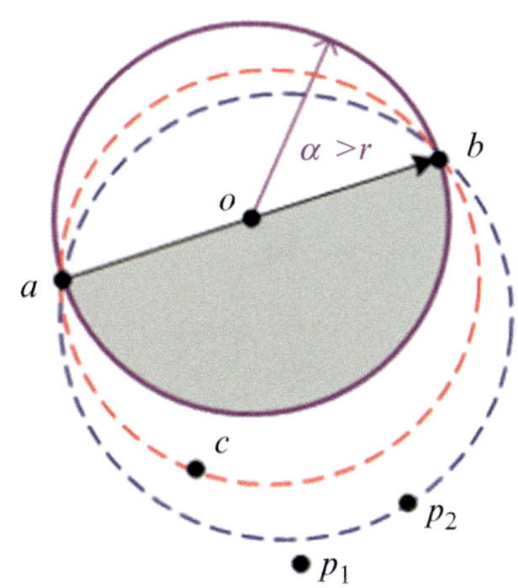

Fig. 7.28 Search the contact point with $|ab| > 2r$ and $\hat{\boldsymbol{ab}}$ empty.

absolute value) must be empty; thus $c \in \partial H_\alpha(S)$. $\alpha > r \Rightarrow c \in \partial H_r(S)$. By Proposition 7.1, we have $c \in P(r)$.

Proposition 7.6 *If* $|ab| \leq 2r$ *and there are no points lying above* σ_{ab} *and also no points in the circular section* $\widehat{ab}$ *of the alpha ball with radius r, then* $\sigma_{ab} \in \partial H_r(X)$.

Proof. See Fig. 7.29. a and b could determine an alpha ball B with radius r. If there is no point lying above σ_{ab} and no point in the circular section $\widehat{ab}$, then $\partial B \cap X = \{a, b\}$. Thus σ_{ab} is α-expose. By Definition 7.1, $\sigma_{ab} \in \partial H_r(X)$.

Propositions 7.1–7.6 establish the searching order of contact points, which first target convex hull points between two known contact points and which correspond to rolling a disk with an infinite large radius over the profile. If no convex hull point lies above the evaluating simplex, the contact point is found by computing the signed circumcircle radii. The contact point is the one that has the largest circumcircle radius. This is equivalent to rolling a disk with a radius smaller than the infinite big but larger than the given radius r. Finally, if the simplex in evaluation could hold an empty disk with radius r by its two ends, namely, the disk is in no contact with other sample points except two end points, then the simplex is a boundary facet of the alpha shape. In summary the algorithm is searching the contact points using disks with a radius ranging from the infinite big down to r. It should be notified that the aforementioned propositions also hold for areal data by replacing the disk with a ball and the circumcircle with a circumsphere.

Although the presented algorithm is specific to circular structuring elements, it is even easier to apply the basic scheme to horizontal flat structuring elements. In that case the contact point is examined by checking the highest point (say p_2 in $\{p_1, p_2, p_3\}$) between two known contact points (say a and b). If that point is lower in height than two given contact points (say p_2 is lower than a and b) and the horizontal distance between the two

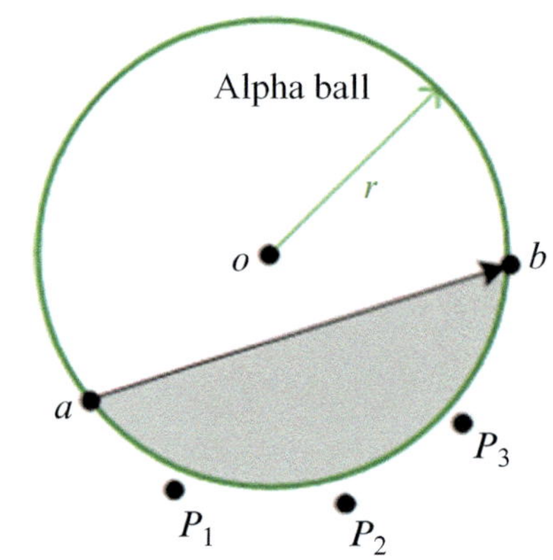

Fig. 7.29 σ_{ab} determines a facet of the alpha shape.

contact points is smaller than the length of the given horizontal line segment (say $|ab| < L$), the searching procedure exits, and the envelope height is determined by the lower height of the two contact points.

7.6 Case studies

Currently, state-of-the-art micromachining technologies enable the fabrication of high value-added optic components. The F-theta scanning lenses are aspheric optic components, commonly used in laser marking, engraving and cutting systems, image transfer, and material processing. F-theta lens has different curvatures on the front and back of the lens surface to maintain constant scanning speeds at both the lens's center and periphery. The quality of the F-theta lens is playing a decisive role in its optical performance. The lens surfaces are smooth, having ultraprecision accuracy with submicrometer shape error, and nanometer roughness. The mass product of precise F-theta-like lenses is usually enabled by advanced mold machining and injection molding. The key to the production of these optic components by injection molding is ultraprecision machining of an aspheric lens core and the development of geometrical measurement technology [57]. Fig. 7.30A shows a surface measured from a metal core for F-theta lenses. The ultraprecision surface of the metal core was processed by the diamond turning machine. While the raw measured surface is smooth and continuous, linear turning marks and small-scale material lumps are visible. The morphological filters are applied to extract these topographical features. The opening and closing with ball radius 0.5 mm is first applied, which allows small-scale material lump features to be extracted; see Fig. 7.30B and C. On the resulted opening surface, the closing filter with ball radius 2.5 mm is then applied; see Fig. 7.30D. The subtraction of the closing surface from the raw measured surface yields a residual surface where both turning marks and material lumps are clearly present.

A similar example is given in Fig. 7.31, where the presented surface is measured from an aspherical lens. The surface profile of an aspheric lens is not a portion of a sphere or cylinder, but often axially symmetric with quadric surface geometry. The precision aspheric lenses used in telescopes, high-precision scientific instruments, and nuclear energy devices are normally processed by grinding, polishing, and diamond turning. Fig. 7.31A presents the raw measured aspheric lens surface. The process marks and material defects are clearly visible and have almost groove- and pit-like features. The closing filter with ball radius 0.5 mm is applied to generate an upper envelope to estimate the form of the aspheric lens; see Fig. 7.31B. Subtracting the closing envelope from the raw measured surface allows process markings and defective materials to be displayed on a planar basis; see Fig. 7.31B, which facilitates the follow-on surface texture characterization.

Additive manufacturing (AM) is another technology used to build complex geometries by selectively adding materials in layers that are impossible to make using traditional machining processes. As a powerful technology, AM impacts a number of key industrial

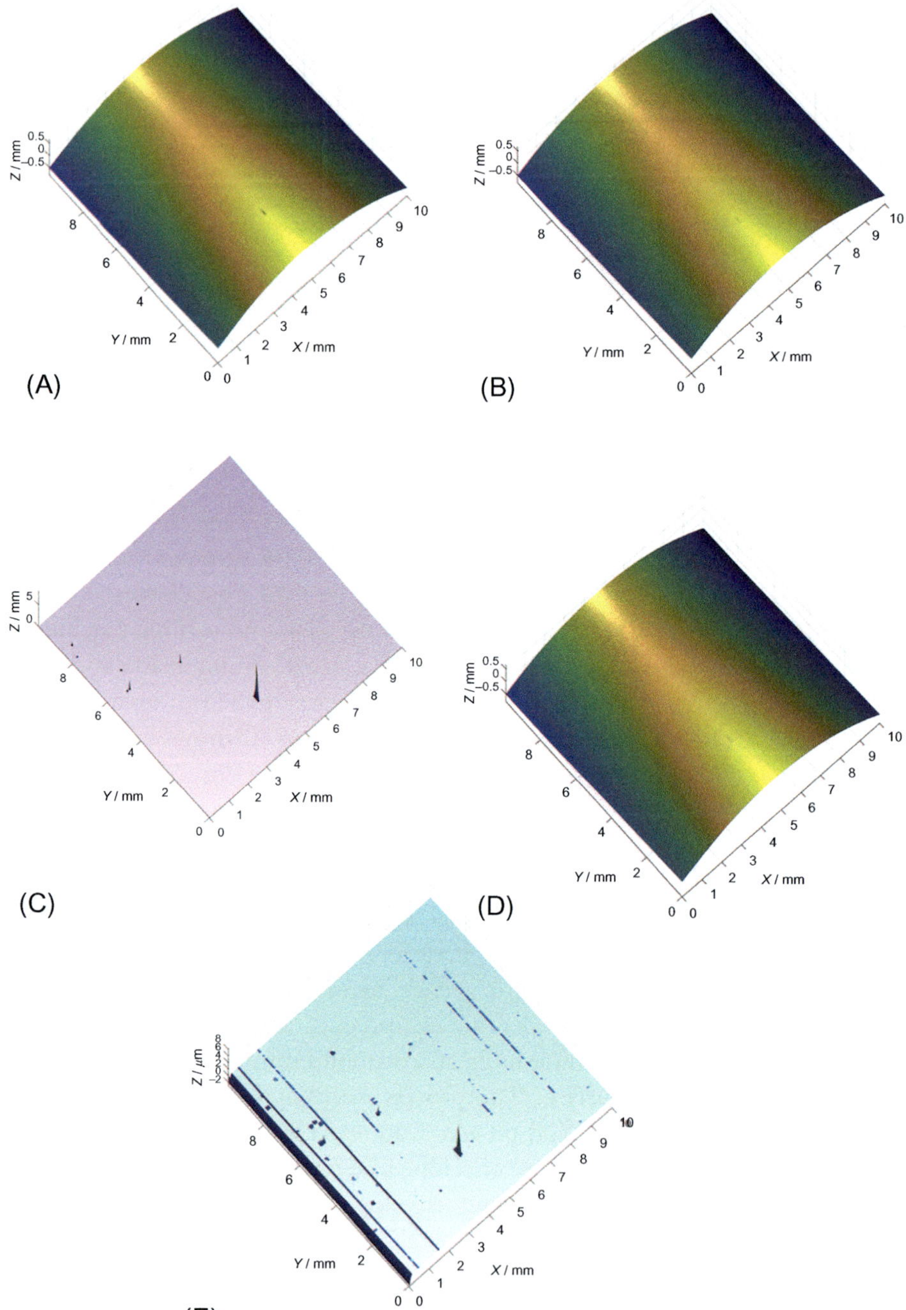

Fig. 7.30 Morphological filtering of a surface measured from a metal F-theta lens core: (A) raw measured surface, (B) opening envelope with ball radius 0.5 mm, (C) residual surface obtained by subtracting the opening envelope from the raw measured surface, (D) closing envelope with ball radius 2.5 mm, and (E) residual surface obtained by subtracting the closing envelope from the raw measured surface.

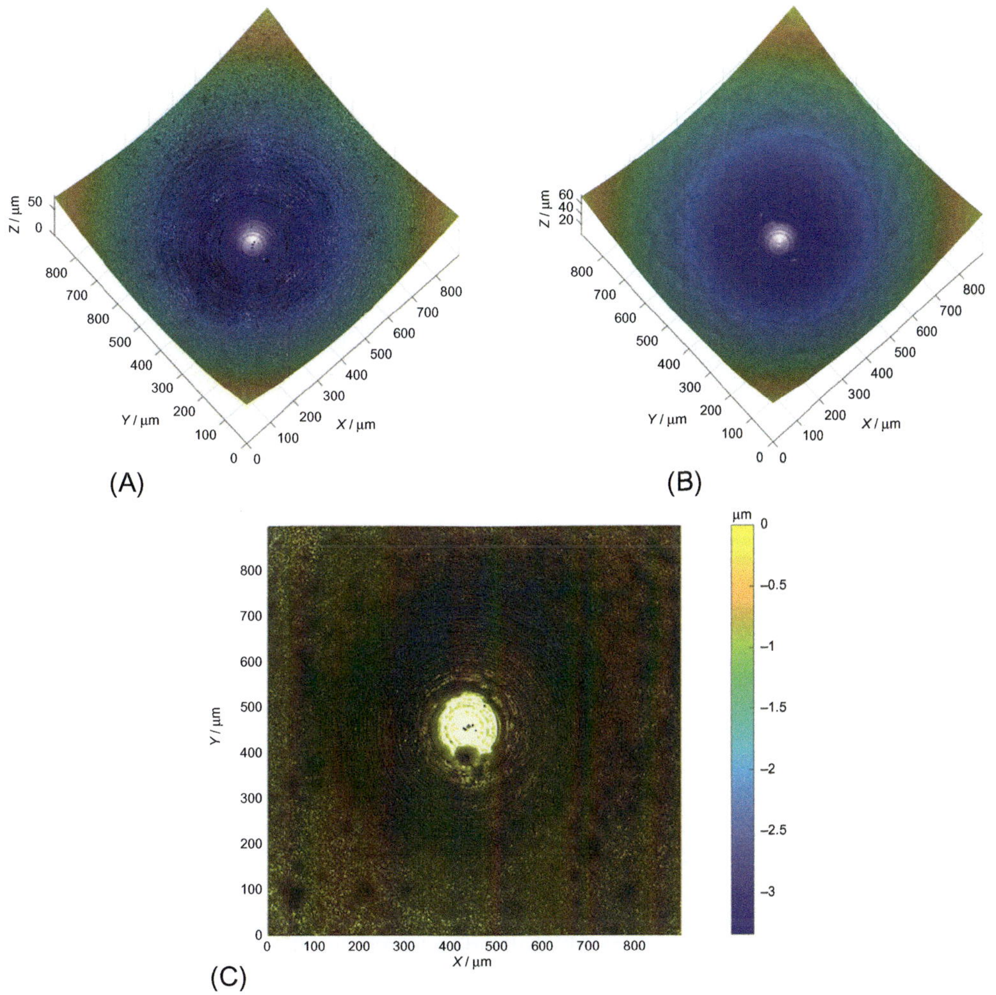

Fig. 7.31 Morphological filtering of a surface measured from an aspherical optic component: (A) raw measured surface, (B) closing envelope with ball radius 0.5 mm, and (C) residual surface obtained by subtracting the closing envelope from the raw measured surface.

sectors, such as aerospace, automotive, healthcare, defense, and electronics. New industrial trends like bionics, lightweight construction, or "mass customization" are all potential fields of application for AM. Various AM processes are available for printing components with diverse sizes and materials. Fig. 7.32A presents a surface measured from the metal cylindrical ring. The metal ring was built using wire laser AM (WLAM), a special AM process that is suitable for building median- and large-size AM components. WLAM process continuously feeds a wire into a melt pool generated by a laser beam. As the laser beam is moving along the component's contour, the deposited materials are

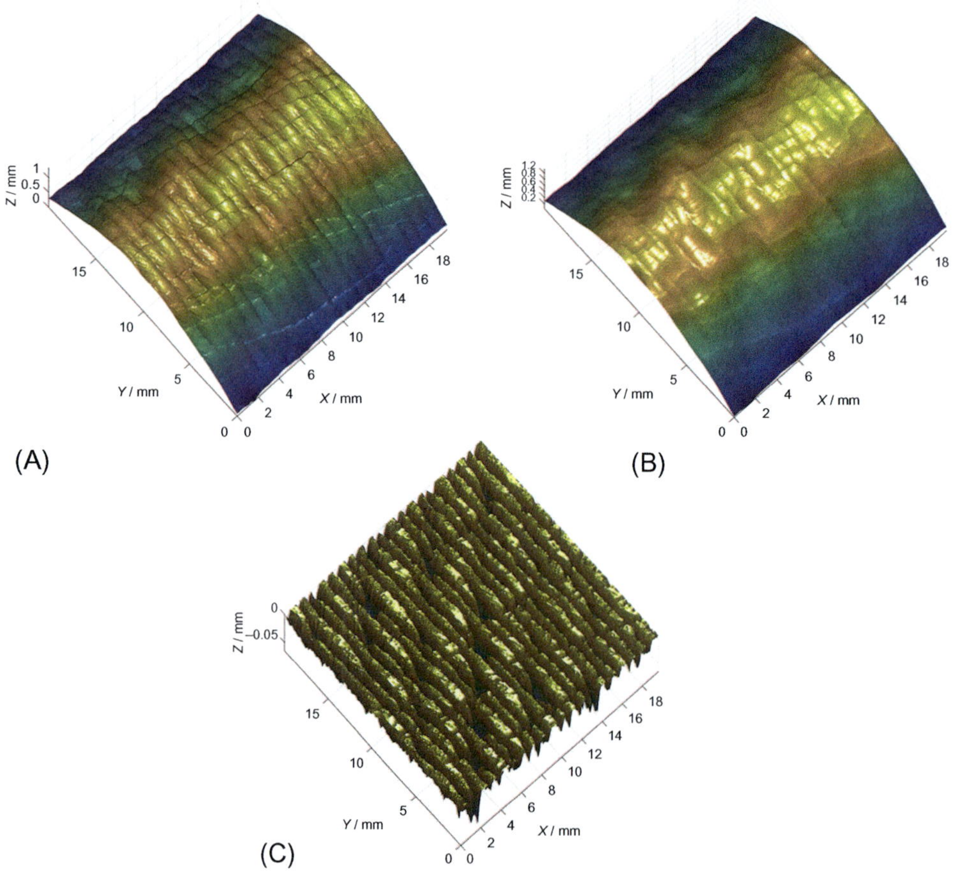

Fig. 7.32 Morphological filtering of a surface measured from an aspherical optic component: (A) raw measured surface, (B) closing envelope with ball radius 10 mm, and (C) residual surface obtained by subtracting the closing envelope from the raw measured surface.

fused to generate the shape of the surface. The WLAM processed surface usually presents a significant wave pattern resulted from layered material feeding, as is clearly visible in Fig. 7.32A. To extract the wavelike staircase pattern, the closing filter with ball radius 10 mm is applied; see Fig. 7.32B. A residual surface can be generated by subtracting the closing envelope from the original measured surface, where periodical pattern is exhibited and could be used to indicate interlayer bonding.

Fig. 7.33A illustrates a surface measured from a paraboloid-like AM polymer artifact built by fuse deposition modeling (FDM). FDM is a common material extrusion process where heated viscous material is drawn through a nozzle and then deposited layer by layer. FDM is a widespread and inexpensive process suitable for printing polymers and plastics. However, the accuracy and speed are limited to the nature of the process.

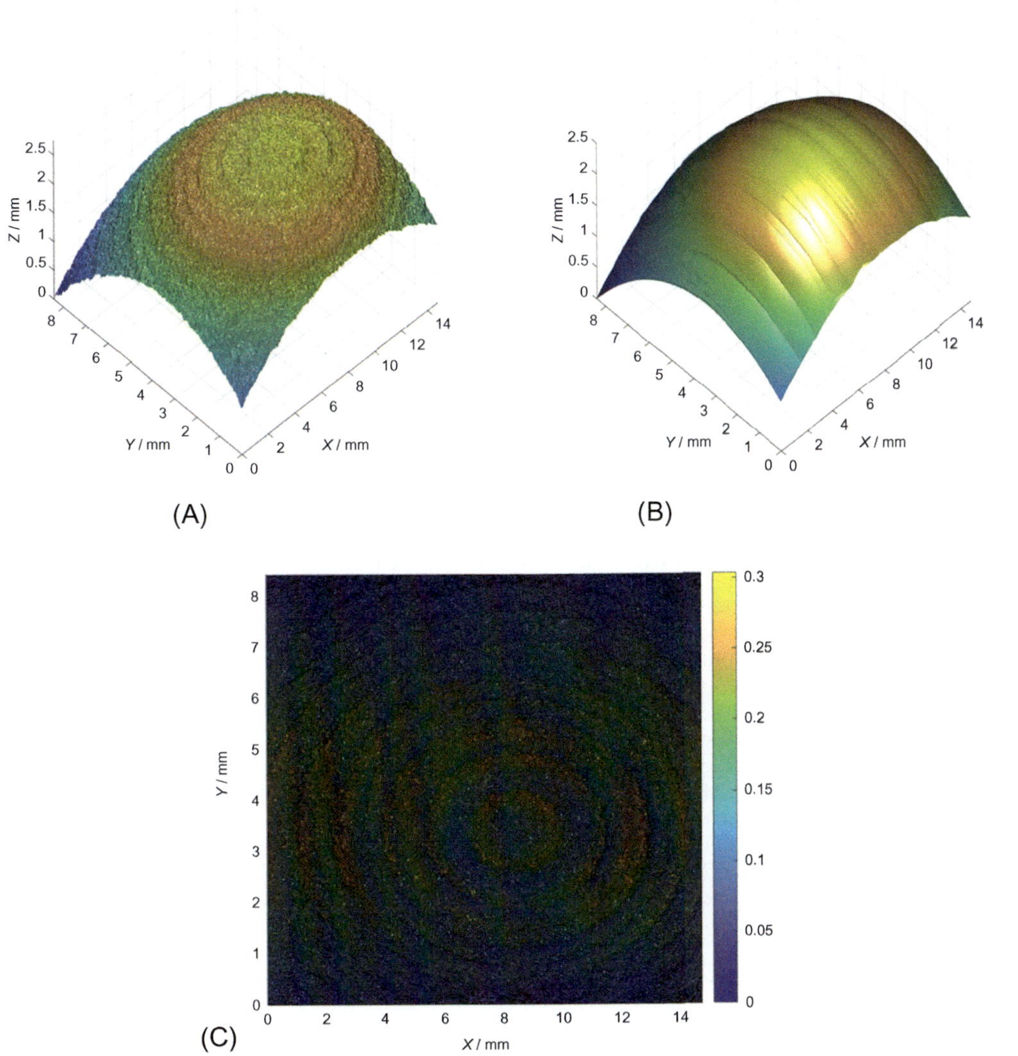

Fig. 7.33 Morphological filtering of a surface measured from an aspherical optic component: (A) raw measured surface, (B) opening envelope with ball radius 10 mm, and (C) residual surface obtained by subtracting the opening envelope from the raw measured surface.

Similar to the previously presented WLAM surface, the FDM surface in Fig. 7.33A also displays a dominant staircase pattern. In comparison with the closing filter that rolls the ball from the above of the surface, the opening filter rolling the ball from the below of the surface also facilitates the extraction of staircase features. Fig. 7.33B illustrates the opening enveloped result from a 20-mm ball. The circular interlayer joint pattern is observed on the residual surface; see Fig. 7.33C.

7.7 Summary

Morphological filters are useful tools in the surface analysis toolbox. Regarded as a complementary technique to the mean line–based filtration techniques, morphological filters are relevant to the functional performance of surfaces, especially contact phenomenon of two mating surfaces. Morphological filters are based on the mathematical morphology theory, which has a wider application in geometrical metrology aside from surface filtration. This chapter illustrates two advanced algorithms based on computational geometry, which enables a morphological filter to be applied to free-form surfaces. The alpha shape algorithm has many advantages, such as it suits nonuniform sampled surfaces and can work with an arbitrary disk/ball radius, but it suffers from relatively expensive computational costs in cases of large measurement data sets. The divide and conquer optimization can improve its performance and reduce the memory usage. The recursive algorithm that directly searches the contact points of the surface and the structure elements removes the dependence of the alpha shape method on the Delaunay triangulation. Case studies are presented, including example surfaces measured from ultraprecision optical components and AM-processed components that have free-form shapes. Morphological filters can be useful for estimating reference surfaces based on which topographical features can be extracted.

References

[1] Jiang X. The evolution of surfaces and their measurement In: The 9th International Symposium on Measurement Technology and Intelligent Instruments, pp. 54–60; 2009.
[2] Bruzzone AAG, Costa HL, Lonardo PM, Lucca DA. Advances in engineered surfaces for functional performance. CIRP Ann Manuf Technol 2008;57(2):750–69.
[3] Jiang X, Scott PJ, Whitehouse DJ, Blunt L. Paradigm shifts in surface metrology, Part II The current shift. Proc R Soc A 2007;463:2071–99.
[4] Whitehouse DJ. Surfaces- a link between manufacture and function. P I Mech Eng B-J Eng 1978;192:179–88.
[5] Peters J, Bryan JB, Eslter WT, Evans C, Kunzmann H, Lucca DA, et al. Contribution of CIRP to the development of metrology and surface quality evaluation during the last fifty years. CIRP Ann Manuf Technol 2001;50(2):471–88.
[6] Trumpold H. Process related assessment and supervision of surface textures. Int J Mach Tool Manuf 2001;41:1981–93.
[7] Unswoth A. Recent developments in the tribology of artificial joints. Tribol Int 1995;28:485–95.
[8] Sayles RS. How two- and three-dimensional surface metrology data are being used to improve the tribological performance and life of some common machine elements. Tribol Int 2001;34:299–355.
[9] Whitehouse DJ. Function maps and the role surfaces. Int J Mach Tool Manuf 2001;41:1847–61.
[10] Raja J, Muralikrishnan B, Fu S. Recent advances in separation of roughness, waviness and form. Precis Eng 2002;26:222–35.
[11] Haesing J. Bestimmung der Glaettungstiefe rauher Flaechen. PTB-Mittelungen 1964;4:339–40.
[12] Peklenik J. Envelope characteristics of machined surfaces and their functional importance. In: Proceedings of the Conference on Surface Technology, pp. 74–95; 1973.
[13] Tholath J, Radhakrishnan V. Three-dimensional filtering of engineering surfaces using envelope system. Precis Eng 1999;23:221–8.
[14] Westberg J. Opportunities and problems when standardising and implementing surface structure parameters in industry. Int J Mach Tool Manuf 1997;38:413–6.

[15] Dietzsch M, Gerlach M, Groger S. Back to the envelope system with morphological operations for the evaluation of surfaces. Wear 2008;264:411–5.
[16] Radhakrishnan V, Weckenmann A. A close look at the rough terrain of the surface finish assessment. P I Mech Eng B-J Eng 1998;212:411–20.
[17] Seewig J. Linear and robust Gaussian regression filters. J Phys: Conf Ser 2006;13:254–7.
[18] Zeng W, Jiang X, Scott PJ. Fast algorithm of the robust Gaussian regression filter for areal surface analysis. Meas Sci Technol 2010;21:055108.
[19] Goto T, Miyakura J, Umeda K, Kadowaki S, Yanagi K. A robust spline filter on the basis of L2-norm. Precis Eng 2005;29:157–61.
[20] Jiang X, Copper P, Scott PJ. Freeform surface filtering using the diffusion equation. Proc R Soc A 2011;467:841–59.
[21] Srinivasan V. Discrete morphological filters for metrology. In: Proceedings of the 6th ISMQC Symposium on Metrology for Quality Control in Production, pp. 623–628; 1998.
[22] Scott PJ. Scale-space techniques. In: Proceedings of the X International Colloquium on Surfaces, pp. 153–161; 2000.
[23] Serra J. Image analysis and mathematical morphology. New York: Academic Press; 1982.
[24] Heijmans HJAM. Mathematical morphology: a modern approach in image processing based on algebra and geometry. SIAM Rev 1995;37(1):1–36.
[25] Minkowski H. Volumen and oberflache. Math Ann 1903;57:447–95.
[26] Dougherty ER. An introduction to morphological image processing. Bellingham: SPIE Optical Engineering Press; 1992.
[27] ISO 16610-40, 2010. Geometrical product specification (GPS)-filtration, part 40: Morphological profile filters basic concepts, Switzerland.
[28] Krystek M. Morphological filters in surface texture analysis. In: XIth international colloquium on surfaces, pp. 43–55; 2004.
[29] ISO 3274, 1996. Geometrical Product Specifications (GPS)—Surface texture: Profile method—Nominal characteristics of contact (stylus) instruments.
[30] Dagnall H. Exploring surface texture. England: Rank Taylor Habson Limited; 1998.
[31] ISO 14406, 2003. Geometrical Product Specifications (GPS)—Extraction.
[32] Roger S, Dietzsch M, Gerlach M, Jeβ S. "Real mechanical profile"—The new approach for nano-measurement. J Phys: Conf Ser 2005;13:13–9.
[33] David JK, Fransiska SF. Envelope reconstruction of probe microscope images. Surf Sci 1993;294(3): 409–19.
[34] Mayer J.R.R, Mir Y.A., Trochu F., Vafaeesefat A., Balazinski M., 1997. Touch probe radius compensation for coordinate measurement using kriging interpolation, Proc Inst Mech Eng B 211, 11–18.
[35] Wozniak A., Mayer J.R.R, Balazinski M., 2009. Stylus tip envelop method corrected measured point determination in high definition coordinate metrology, Int J Adv Manuf Technol 42, 505–514.
[36] Liang SR, Lin AC. Probe-radius compensation for 3D data points in reverse engineering. Comput Ind 2002;48:241–51.
[37] Yin Z, Zhang Y, Jiang S. Methodology of NURBS surface fitting based on off-line software compensation of errors of a CMM. Precis Eng 2003;27:299–303.
[38] Von Weingraber H. Über die Eignung des Hüllprofils als Bezugslinie für die Messung der Rauheit. CIRP Ann Manuf Technol 1956;5:116–28.
[39] Malburg CM. Surface profile analysis for conformable interfaces. J Manuf Sci Eng, Trans ASME 2003;125:624–7.
[40] Lou S, Jiang X, Scott PJ. Algorithms for morphological profile filters and their comparison. Precis Eng 2012;36(3):414–23.
[41] Lou S, Jiang X, Scott PJ. Geometric computation theory for morphological filtering on freeform surfaces. Proc R Soc London, Ser A 2013;469(2159). 20130150.
[42] Haralick RM, Zhuang X, Lin C, Lee J. The digital morphological sampling theorem. IEEE Trans Acoust Speech 1988;37:2067–690.
[43] Matheron G. Random sets and integral geometry. New York: John Wiley & Sons; 1989.
[44] Jiang X, Blunt L, Stout K. Application of the lifting wavelet to rough surfaces. Precis Eng 2001; 25(2):83–9.

[45] Xiao S, Jiang X, Blunt L, Scott PJ. Comparison study of the biorthogonal spline wavelet filtering for areal rough surfaces. Int J Mach Tool Manuf 2001;41(13–14):2103–11.
[46] Nyquist H. Certain topics in telegraph transmission theory. Trans Am Inst Electr Eng 1928;47:617–44.
[47] Muralikrishnan B, Raja J. Functional filtering and performance correlation of plateau honed surface profiles. J Manuf Sci Eng 2005;127:193–7.
[48] ISO 16610-49, 2010. Geometrical product specification (GPS)-filtration, part 49: Scale space techniques, Switzerland.
[49] Decenciere E, Jeulin D. Morphological decomposition of the surface topography of an internal combustion engine cylinder to characterize wear. Wear 2001;249(5–6):482–8.
[50] Lou S, Jiang X, Scott PJ. Fast algorithm for morphological filters. J Phys: Conf Ser 2011;311(1). 012001.
[51] Jiang X, Lou S, Scott PJ. Morphological method for surface metrology and dimensional metrology based on the alpha shape. Meas Sci Technol 2012;23(1). 015003.
[52] Edelsbrunner H, Mucke EP. Three-dimensional alpha shapes. ACM Trans Graph 1994;13(1):43–72.
[53] Worring M, Smelders WM. Shape of arbitrary finite point set in R2. J Math Image Vision 1994;4:151–70.
[54] Bernardini F, Mittleman J, Rushmeier H, Silva C, Taubin G. The ball-pivoting algorithm for surface reconstruction. IEEE Trans Vis Comput Graph 1999;5(4):349–59.
[55] Cormen TH, Leiserson CE, Rivest RL. Introduction to algorithms. New Delhi: The MIT Press; 1990.
[56] Fischer, K. 2000 Introduction to Alpha Shapes, http://www.inf.ethz.ch/personal/fischerk/pubs/as.pdf.
[57] Lee D-K, Yang Y-S, Kim S-S, Kim H-J, Kim J-H. Development of a F-theta Lens for a laser scanning unit. J Korean Phys Soc 2008;53(5):2527–30.

CHAPTER 8

Segmentation of topographical features on free-form surfaces

8.1 Introduction

Surface is the interface that limits the body of an object and separates it from the surrounding medium. It is functionally important as it governs mechanical, tribology, biological, chemical properties, and also force, all of which impact the way that the object is interacting with others. Prime examples exist both in natural and man-made products. The lotus effect is a well-known hydrophobic phenomenon. The epidermis of a lotus leaf's surface is overlapped by many tiny bumps covered by epicuticular wax. This special double surface structure is water repellent, resulting in a self-cleaning effect. Based on this mechanism, many products were developed to achieve the lotus effect, such as the self-clean roof tile and waterproof nanomaterial coats [1]. Another example is the sharkskin surface, which is presented with many miniature teethlike ridges. These ridges or denticles can improve swimming efficiency and enable silent movement by reducing water drag. This principle is applied to diving suits and wet suits to improve an athlete's performance [2].

The importance of surface to functional behaviors gives rise to the initiative of studying the geometrical structures on the surface. Surface analysis techniques are available to extract and evaluate these microtopographical features. The analysis toolbox comprises the Gaussian filter, the spline filter, wavelet analysis, morphological operations, and motif analysis methods [3]. The motif method was originally developed to extract the significant peaks on the surface profile, which can date back to the 1970s. A huge amount of work was produced to build up a "roughness data bank" with more than 27,000 profiles from a range of workpieces. An empirical method based on the profile bank was developed to describe roughness and waviness, including the creation of motifs, the construction of the upper envelope, and the calculation of dedicated parameters called R&W [4]. The empirical method first became a French industry standard Comite de Normalization des Moyens de production (CNOMO) [5], and later, its main concepts and most of its parameters were adopted by ISO 12085 [6]. However, even though the set of rules for motif combination was standardized, the tweaking of the rules has continued because the method is not stable [7]. As a result the concept of motif analysis was generalized, and the rules for combination were laid out on a stable mathematical basis. With the advancement

* For more examples and practical implementation, check the website: www.digitalsurf.com/freeform.

Advanced Metrology
https://doi.org/10.1016/B978-0-12-821815-0.00008-3

of surface measurement instruments, the profile motif method was extended to the areal method based on image segmentation techniques, for example, the watershed transform [8]. The Wolf pruning technique was further employed as the discrimination criterion in the combination of areal motifs, which guarantees the uniqueness and stability of the solution [9, 10]. Areal motif methods provide useful tools for the evaluation of surface geometry, on the basis of which novel feature parameters are employed to characterize surface texture [11].

As modern manufacturing technologies enable the production of surfaces with free-form surfaces and structured surfaces (or even their combination), surface characterization of these complex surfaces is much more challenging than traditional planar surfaces. Particularly, free-form surfaces impede the application of the watershed segmentation technique because surface shapes are no longer planar, and measurement data often requires surface description in the form of triangular meshes. This chapter mainly discusses the development of the watershed segmentation used in surface texture characterization. Based on that a strategy to extend the watershed segmentation to free-form surfaces is discussed. Case studies then follow for illustration.

8.2 Watershed segmentation and its application

8.2.1 Watershed segmentation in geography

Although a large number of segmentation techniques are presented in the literature, the watershed method is a popular and natural choice. The essence of watershed originates from geography. Suppose a landscape is flooded by raining, the rainwater naturally flows down the steepest paths and eventually falls into a number of domains of attraction. The watersheds, being the dividing lines of these domains (called catchment basins), are the natural segmenting contours of the landscape.

The concept of watershed started with Maxwell more than 100 years ago [12]. He presented the physical characteristics of a landscape by means of hills, dales, and saddles (also called passes). An area with all of its maximal uphill paths leading to one particular peak forms a Maxwell hill. In opposition an area with all of its maximal downhill leading to one particular pit gives a Maxwell dale. He also highlighted the relationships between these hills, dales, and saddles: the boundaries between dales are ridge lines (watershed lines), and the boundaries of hills are course lines; ridge and course lines are steepest uphill and downhill paths emanating from saddles and terminating at peaks and pits. Based on this theory, Pfaltz proposed to use the surface network to organize large spatial databases and to describe geographical landscapes [13]. Surface networks, also called Pfaltz graph, are tripartite graphs, composed of the vertices representing peaks, pits, saddles, and the edges indicating their interrelations. As a result, Pfaltz graph possesses the ability to describe landscapes by selecting much smaller sets, involving only those critical points. More importantly a Pfaltz graph can be condensed by two holomorphic contractions to reduce the number of edges and vertices in graphs while preserving topological structures of topographical surfaces.

8.2.2 Watershed segmentation in image processing

More critical applications of watershed segmentation were found in the field of image processing, where the watershed method appears as one of the basic foundations. Lantuejoul defined and studied the skeleton by influence zones (SKIZ) of binary collections of grains to model polystalline alloy. His idea of constructing geodesic SKIZs of the local minima led to the first algorithm to compute watersheds [14]. With Beucher, they applied the watershed transform to the gradient image of gas bubbles, yielding the first application to image segmentation [15]. Since then, more algorithms were developed to compute the watershed transform. Two representative algorithms appeared in the 1990s, one based on simulating the immersion process [16] and the other based on topographical distance functions [17]. In the first approach a landscape is immersed in a lake, with holes pierced in local minima. Basins will be filled up with water starting at these local minima, and dams are built at points where water coming from different basins would meet. The immersion process stops once the water reaches the highest peak in the landscape. As a result the landscape is naturally partitioned into a number of basins separated by the dams, called watersheds. The second approach can be viewed as the generalized Voronoi tessellation defined on surfaces. Each catchment basin associated with a specific minimum is defined as the set of points that are topographically closer to this minimum than any other regional minimum. The watershed is the set of points that does not belong to any catchment basin.

The watershed segmentation has been a hot spot throughout the years. A quantity of algorithms appeared with enhanced performances and advanced capabilities [18–23]. In comparison with other segmentation techniques, the watershed paradigm possesses many advantages. It is generic and is applicable to many segmentation problems [24]; it can produce closed contours that are helpful for follow-on processing operations, for instance, pattern recognition [25–27]. In addition, existing watershed algorithms are usually computationally efficient.

8.2.3 Watershed segmentation in engineering surface analysis

For surface texture analysis of engineering components, the watershed method also found its application. It appeared as the areal extension of the profile motif analysis that aims to identify and characterize significant peaks on a surface profile [6]. Scott extended Maxwell's definition of hill and dale, wherein an areal motif is equivalent to a dale consisting of a single dominant pit surrounded by a ring of ridge lines connecting peaks and saddle points [9, 28]. A whole set of segmentation methods were constructed on the basis of Maxwell's theory, Pfaltz graph, and change tree. These techniques were successfully employed to analyze topographical features on engineering surfaces, such as grinding wheels and car body panels [9]. Barré and Lopez proposed another variation of areal motif extension, based on Vincent's recursive watershed algorithm [29]. This approach was adopted to characterize typical machined surfaces, such as turned surfaces, ground surfaces, knurled surfaces, and sandblasted surfaces. Watershed segmentation was also applied to characterize structured surfaces with deterministic functional patterns [29–31]. Prior to watershed segmentation, surface data were processed by edge detection operators, such as the Sobel operator, the Laplacian of Gaussian operator, and the morphological gradient operator, to enhance the boundaries of structured features so that the segmentation can yield better and more accurate results. Owing to all these merits, the watershed segmentation was accepted by ISO 16610-85 [32] as a powerful data analysis tool in surface metrology.

8.2.4 Watershed segmentation of free-form surfaces

Surface measurements performed by optical microscopes usually generate height maps, that is, a scalar height functions defined over the regular lattice in $\mathbb{R}^2$. In contrast, mesh has more general structures, being a combination of connected polygons. Particularly the lattice data can be regarded as a special case of mesh with four neighbor connections, six neighbor connections, or eight neighbor connections. Watershed segmentation on mesh data is inspired by needs of mesh reduction, mesh partitioning, and feature detection. Mangan and Whitaker proposed a downhill approach to segment 3-D meshes [33]. By tracking the steepest descent path, each vertex is assigned with the label of a certain minimum, and therefore the set of resultant basins tessellates the mesh surface. In their work, watershed segmentation is applied on the curvatures of vertices rather than the original mesh. The norm of the covariance matrix was employed to calculate the surface mesh curvature. A primary demonstration of using this method for 3-D mesh partitioning was presented where a dart model is divided into different components, for example, point, barrel, shaft, and flight. The watershed-cut method proposed by Cousty et al. [23] is also qualified for mesh segmentation that is built on the basis of weighted graph. A case study of using the watershed cut to segment the surfaces of some museum artworks was presented [34]. Another algorithm of watershed segmentation on triangular meshes can be found in [35], which is based on discrete Morse theory. The triangular mesh is taken as a special case of the Morse-Smale function that can be used to generate a natural tessellation of space into cells induced by the gradient to the function. Wolf [36] showed that surface networks, change tree, and Morse-Smale complexes reveal the same kind of information, although in different ways.

Our application differs from the aforementioned mesh segmentation cases. The topographical features lying on the free-form surface are our concern and desired to be extracted for further characterization. Take the human skin surface as an example. The skin topography, containing a network of furrows, is directly related to the tension of elastic and collagen fibers in the superficial dermis. Thus it is questioned how the skin morphology will be influenced by human aging [37]. The watershed segmentation of surface curvatures as adopted by two prior examples is not suitable for this application.

8.2.5 General strategy of applying the watershed method on free-form surfaces

Although the methods of using gradients in image segmentation and curvatures in mesh segmentation do not work for the segmentation of topographical features on free-form surfaces, they reveal an important fact. The watershed transform is applied on a scalar function. The selection of the scalar function can be image gradient, surface curvature, or any value of interest, depending on specific applications. The principles of watershed segmentation algorithms, however, are independent of the scalar function.

The watershed-cut method [23], the top-down tracking methods [33], and the Morse gradient method [35] are all capable algorithms for triangular mesh watershed segmentation. Nonetheless, Scott's method based on Maxwell's theory and Pfaltz graph [9] is adopted in this work in that its principle is based on topology and is independent of data structure, either lattice or mesh. The topological connection of the vertices and the chosen scalar function determine the watershed segmentation. Another reason for choosing Scott's method is its stability: small changes to the input set will not cause a significant change in the analysis of results, with "relevant" features appearing or disappearing [9], which is particularly important for engineering applications.

A general strategy of performing watershed segmentation on free-form surfaces can be found in [38], and the procedures are detailed in the succeeding text:

(1) Apply proper mathematical operations to extract surface topography from free-form surfaces, by which each of the vertices of a triangular mesh surface is associated with a surface height.

(2) Perform the watershed segmentation on the scalar height function based on Maxwell's theory.

(3) Merge the resultant segments to suppress insignificant ones.

(4) Finalize ridge/course lines.

8.3 Watershed segmentation of topographical features on free-form surfaces

8.3.1 Extraction of surface topography

To extract out surface topography, surface form needs to be removed, which is usually done by subtracting the reference surface from the measured surface. The residuals after form removal are viewed as surface topography that contains surface texture and other topographical features, for example, material defects. In traditional surface metrology, surface measurements are often performed on planar or nearly planar areas, wherein polynomial fitting suffices to remove surface form. Free-form surface measurements can involve larger measurement areas if surface texture is working with surface form together to achieve a desired functionality. In this case, polynomial fitting is no longer qualified, and other methods shall be considered to obtain reference surfaces.

Taking the designed nominal surface is a quick way to obtain a reference surface. For instance, in AM, the digital model of the part in the form of a triangular mesh is requested and sliced into a sequence of sections so that the AM machine can build the part layer by layer. To verify the dimensional accuracy of the built part, x-ray computed tomography (XCT) can be employed to scan the part. The measured surface is then aligned and compared with the nominal surface, and a color map is presented on the CAD surface with the colors indicating geometrical deviations. These deviations can be taken as surface topography if form errors are negligible.

In the case where the nominal CAD is not available or form errors are significant, mathematical operations, such as fitting and filtration, shall be employed to generate a smooth reference surface. In surface fitting, more work is reported for lattice data and point clouds than triangular meshes [39, 40]. Filtration is an alternative method to obtain a smooth reference surface. Capable filters include the Laplacian filter [40], the morphological filter [41], the curvature flow filter [42], and the anisotropic diffusion filter [43]. Comparing the original measured surface with the reference surface can yield the surface topography. The signed distances between two surfaces are computed, and they are stored as the scalars associated with each vertex of the mesh surface. Fig. 8.1 illustrates an example of using the Laplacian smoothing and the curvature flow smoothing to generate reference surfaces for a Rosetta comet mesh surface developed by European Space Agency.

8.3.2 Construction of the Pfaltz graph

To follow Scott's watershed method [9], peaks, pits, and saddles are firstly identified as the critical points to construct the Pfaltz graph. From the perspective of cartography, a new contour appears at the peak, and an existing contour disappears at the pit. Saddles are special cases: at these points a contour is divided into two (or more), or two (or more) contours are merged into one. Ridge lines and course lines are the connections of the critical points, emanating from a saddle and follows the steepest ascending (descending) path to reach a peak (pit).

Discrete algorithms for extracting critical points and lines can be found in [44–46] for lattice data structure but can be extended to triangular meshes. The identification of critical points is based on the extracted surface topography, that is, the scalar function associated with the mesh surface. A peak/pit is a local maximum/minimum point of the surface topography (the signed distances), while a saddle point has at least two lower neighbors and two higher neighbors in the ring of its connected vertices. For open surfaces a virtual pit of infinite depth is assumed to lie at the boundary of the data set to secure the stability of the algorithm.

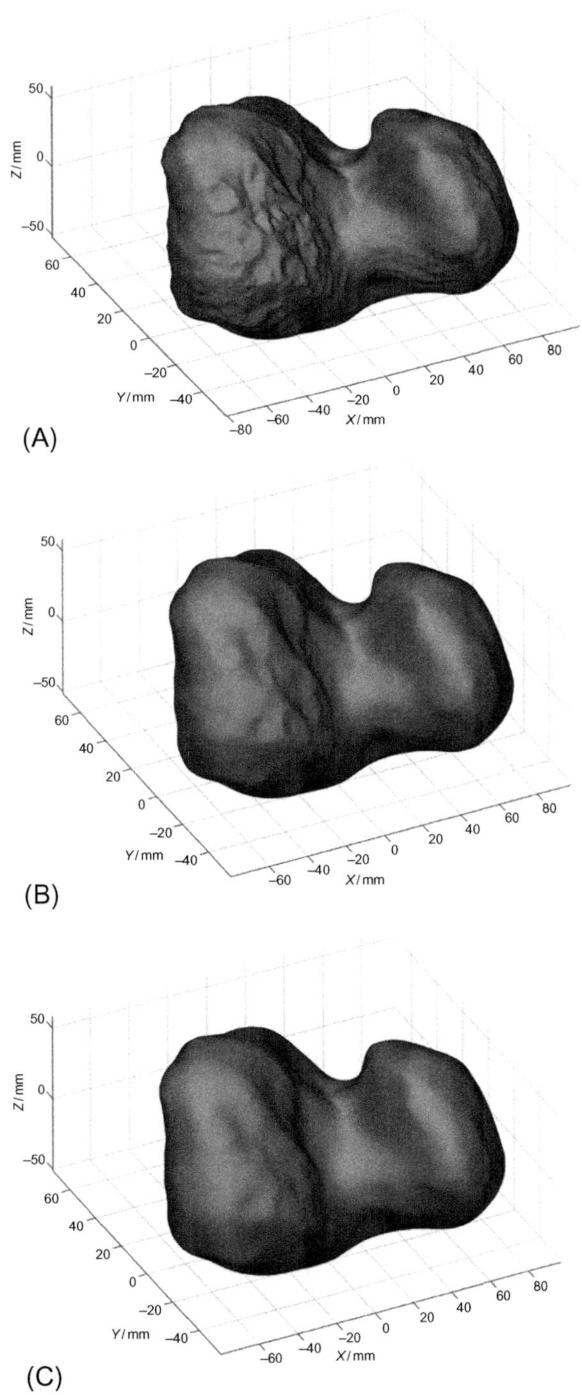

Fig. 8.1 Smoothing of a comet triangular mesh surface: (A) original mesh surface, (B) mesh surface filtered by the Laplacian smoothing, and (C) mesh surface filtered by the curvature flow smoothing.

Let p be a point on the surface mesh and n the number of the neighbor vertices of p, denoted by $\{p_i | i=1 \cdots n\}$. The height difference between p_i and p is indicated by Δ_i. Let Δ_+ be the sum of all positive height differences of Δ_i and Δ_- be the sum of all negative height differences of Δ_i. N_c indicates the number of sign changes in the sequence $\Delta_1, \Delta_2, \ldots, \Delta_n, \Delta_1$, which fulfills a whole ring of the neighbor of p. It can be found that for each peak, it has $|\Delta_+| = 0, |\Delta_-| > 0, N_c = 0$; for each pit, it has $|\Delta_-| = 0, |\Delta_+| > 0, N_c = 0$; and for each saddle, it has $|\Delta_+| + |\Delta_-| > 0$, $N_c = 2 + 2m$, $m = 1, 2, \ldots$ Particularly a saddle point is a common one if $N_c = 4$ ($m = 1$) and a super one if $N_c > 4$ ($m > 1$).

To track ridge/course lines, the searching route starts from each of saddle points and proceeds in the steepest slope paths. Let s be a saddle and then each ridge/course line emanating from s either (1) intersects the boundary of the surface mesh, (2) reaches a peak/pit, or (3) reaches a saddle other than s. A single tracking route terminates immediately when a peak or pit is encountered. Moreover the neighborhood of s falls into $2+2m$ regions with the positive Δ_i s and the negative Δ_i s alternating. There are then $1+m$ ascending ridge lines in positive regions and $1+m$ descending course lines in negative regions. With critical points and ridge/course lines, the Pfaltz graph can be constructed. Watersheds and catchment basins are then generated straightforwardly. The ridge lines are equivalent to the watersheds. Maxwell dales, each consisting of a pit and bounded by ridge lines, form catchment basins (also areal motifs). This gives the initial result of the watershed segmentation.

This methodology differs from Cousty's approach [23] due to the fact that it is more straightforward. Peaks and saddles must be on the watersheds. Any point on the upstream of a watershed point is itself of a watershed point [47]. Thus the steepest uphill paths from saddles to peaks yield the ridge lines (the watershed lines) directly. Similarly the steepest downhill paths from saddles to pits give the course lines. Having known the way that saddles are linked to peaks and pits, tracking of ridge/course lines can be clearly guided. The starting points of a track are limited to saddle points. A single track terminates immediately when a peak or pit is encountered. Saddles are playing critical roles in the whole computation. The method proposed by Mangan and Whitaker [33] behaves differently in this aspect. Their searching starts from each vertex in the mesh and follows the steepest path until it reaches a pit. All the vertices tracking to the same pit are labeled as a specific dale. Saddles in their algorithm are not identified separately.

For the convenience of visualizing the algorithm, a simple artificial triangular mesh surface is created; see Fig. 8.2. The underlying surface form of this simple mesh surface is intentionally made to flat. The z values of vertices are regarded as the scalar function to be segmented. Critical points, that is, four peaks, four pits (including one virtual pit), and six saddles, are extracted from the mesh surface first. The ridge/course lines are then tracked from each of the saddles. The corresponding Pfaltz graph is listed in Fig. 8.3. It is easy to find that all these saddles are common ones, each of them linked by two ridge lines and two course lines. In some cases, super saddles may appear depending on surface topography.

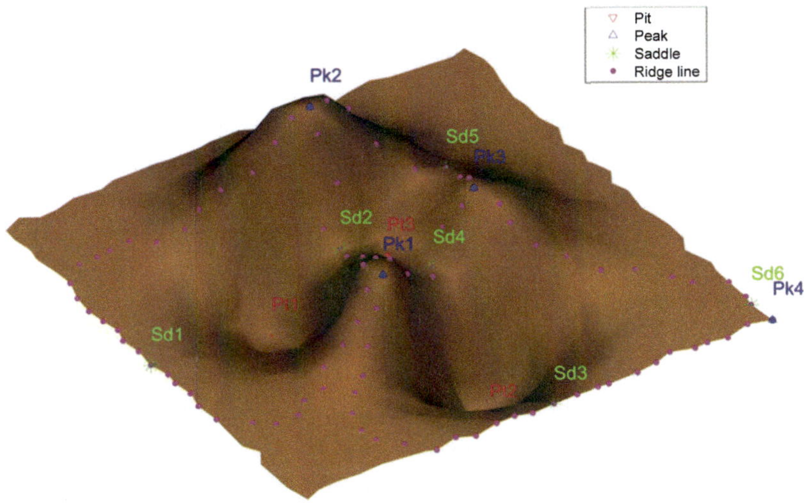

Fig. 8.2 Critical points and ridge lines of an arterial mesh surface.

8.3.3 Oversegmentation merging

Local minima/maxima, due to either measurement noises or real topographical features, can result in oversegmentation. Insignificant regions need to be combined into significant ones to enable a meaningful surface topographical analysis. Criteria should be placed to discriminate significance and insignificance regions, which can be based on heights, areas,

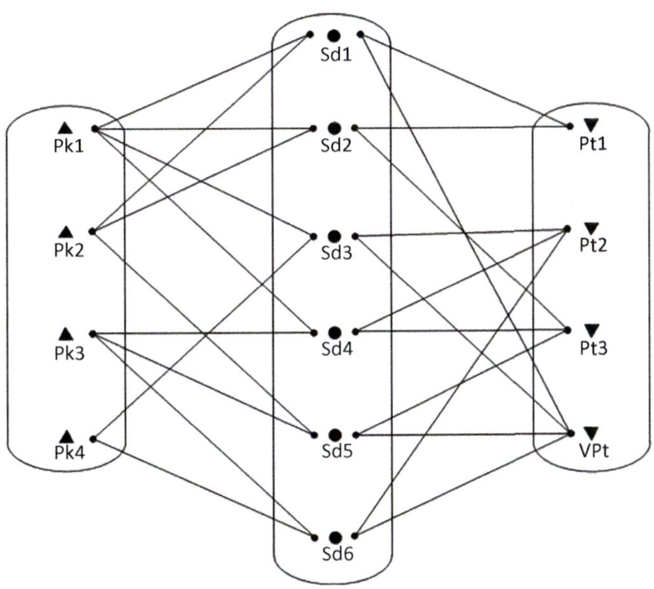

Fig. 8.3 Pfaltz graph associated with the artificial mesh surface.

volumes, or their combinations, depending on specific applications. A key criterion for analyzing surface topography is the height of peaks and pits specified in the scalar function.

The change tree [48] is adopted by Scott's method to clarify the height hierarchy and connectability of a surface [9]. The change tree is built from the contour map of the surface elevation data, representing the contour lines that continuously vary with the height. The vertical direction of the change tree represents the height elevation, while the tree nodes indicate critical points. Each node indicates a topographical change, either a contour line appears or disappears, or several contour lines are merged into one, or one contour line splits into more. The height of a peak/pit on the change tree is defined as the height difference between this peak/pit and its closest saddle. The change tree can be extended to mesh surfaces, where scalar functions can provide height hierarchy and edges of a mesh surface can give the connection information of critical points. Fig. 8.4A presents the change tree of the mesh surface illustrated in Fig. 8.2.

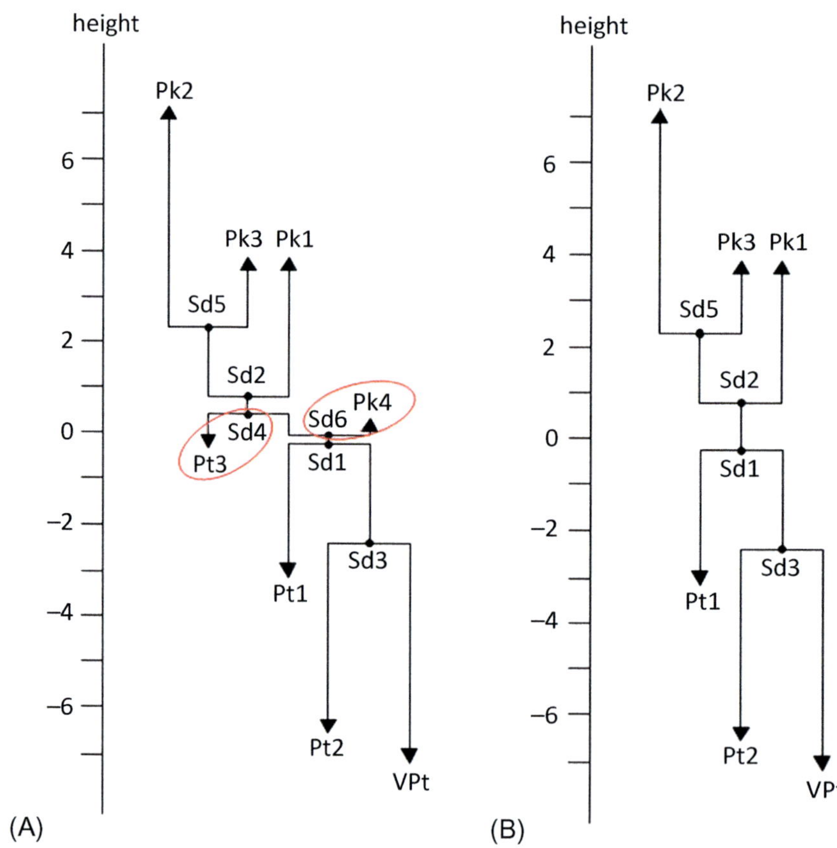

Fig. 8.4 Change tree of the example mesh surface: (A) before contraction and (B) after contraction.

To trim off the insignificant peaks/pits on the change tree, a threshold is applied to determine which of peaks or pits will be eliminated. This threshold can be given by a ratio of the peak/pit height against the maximum height of the change tree, known as Wolf pruning ratio [49]. For example, the threshold ratio 5% can be placed so that all peaks or pits with their heights below 5% of the maximum height difference will be contracted and combined. The contraction, however, will lead to local topological changes to the Pfaltz graph. Wolf pruning can be employed to prune the Pfaltz graph while preserving the general topological structure of the graph [50]. An example of applying 5% Wolf pruning to the example surface in Fig. 8.2 is given by Fig. 8.4. In this example the height difference between the peak *Pk*4 and the saddle *Sd*6 and also the pit *Pt*3 and the saddle *Sd*4 are smaller than the value corresponding to 5% Wolf pruning, and thus they are contracted. The steps of contracting *Pk*4 are detailed in the succeeding text:

(a) Identify the closest saddle point to *Pk*4, that is, *Sd*6.
(b) Remove the links between *Pk*4 and its connected saddles, that is, *Pk*4 and *Sd*6 and *Pk*4 and *Sd*3.
(c) Remove all the links between *Sd*6 and its connected peaks/pits, that is, *Pk*3 and *Sd*6, *Pt*2 and *Sd*6, and *VPt* and *Sd*6.
(d) Add a new link between *Pk*3 and *Sd*3.
(e) Remove *Pk*4 from the peak list.
(f) Remove *Sd*6 from the saddle list.

These steps are also depicted in Fig. 8.5, where the removed links are grayed out and the new created link is highlighted in bold. The contraction of *Pt*3 is similar. The contraction

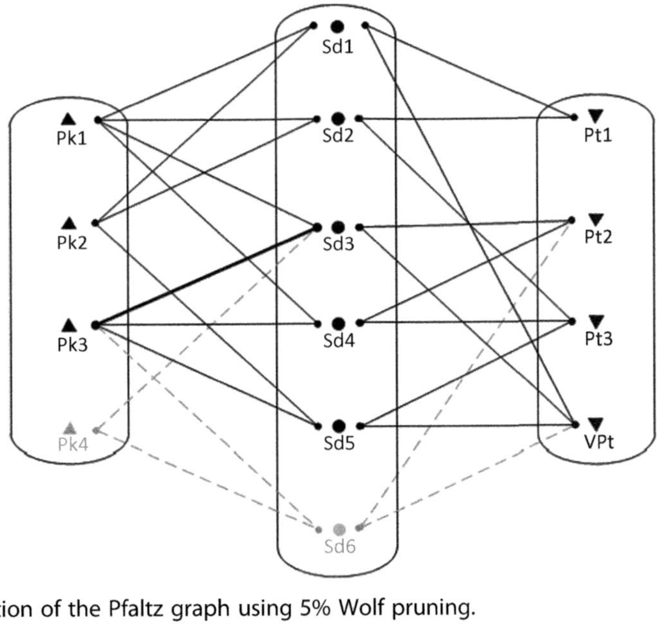

Fig. 8.5 Contraction of the Pfaltz graph using 5% Wolf pruning.

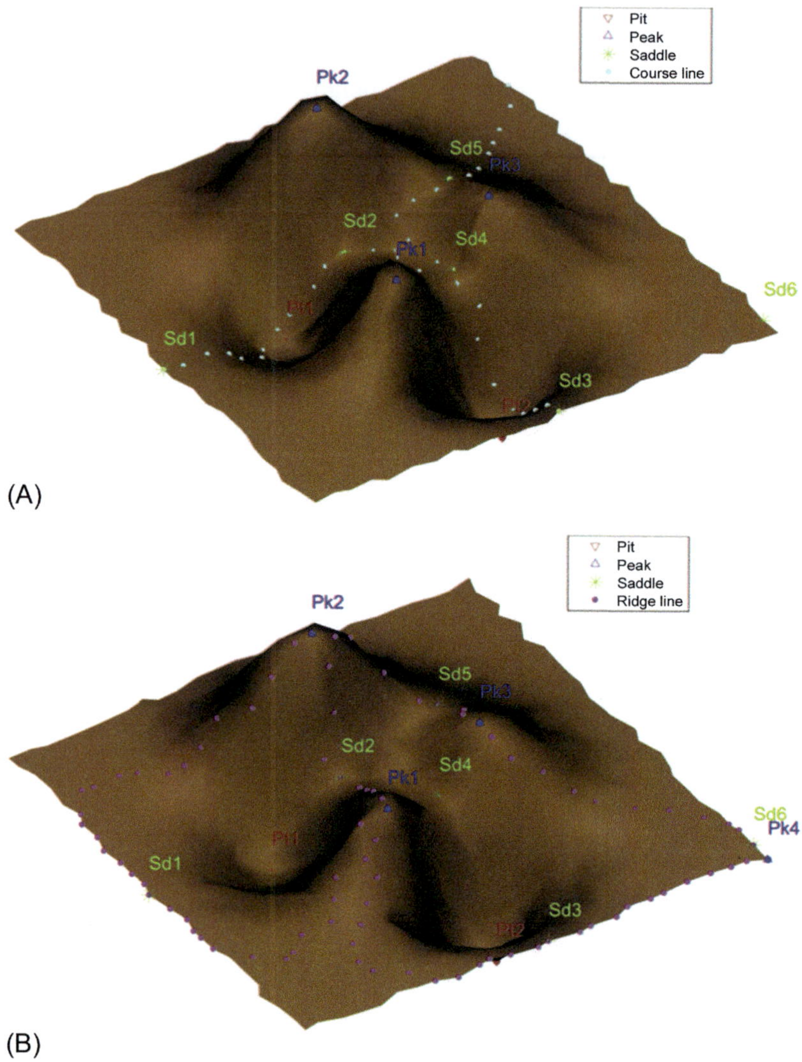

Fig. 8.6 Critical points and lines of the mesh after the contraction. (A) After the contraction of *Pk*4 and (B) after the contraction of *Pt*3.

of Pfaltz graph leads to the simplification of the change tree (see Fig. 8.5B). As a result, insignificant regions are merged. The contractions of *Pk*4 and *Pt*3 of the mesh surface are graphed in Fig. 8.6A and B, respectively.

8.4 Case studies

Human skin embraces bones and muscles, separating human body from the outside environment and protecting the body against pathogens and excessive water loss. Skin textures are closely related to the skin health. For instance, oily skin caused by overactive sebaceous glands usually results in sallow- and rough-textured skin that tends to have large, clearly visible pores. Aged skin often shows deep furrows and an increase of skin line textures. Thus the investigation of human skin surface textures is of necessity for the examination of healthy and aging skins. However, the bones and muscles that the skin encloses are usually in the free-form shape, and the skin textures are presented on these free-form surfaces. The segmentation of these textures leads to a prime example to demonstrate the effectiveness of the extended watershed method.

A measured human skin sample surface is given by Fig. 8.7A in the form of triangular mesh. Fig. 8.7B presents the initial result of the watershed segmentation. Since skin textures appear as furrows on the skin surface, the course lines instead of the ridge lines are of the interest for this segmentation. It is shown that the resulted course lines indicate the skin line network. However, it also includes some other lines that are obviously not part of it. These undesired course lines are caused by local trivial peaks. The application of Wolf pruning helps to trim off these insignificant peaks. Therefore the course lines associated with these

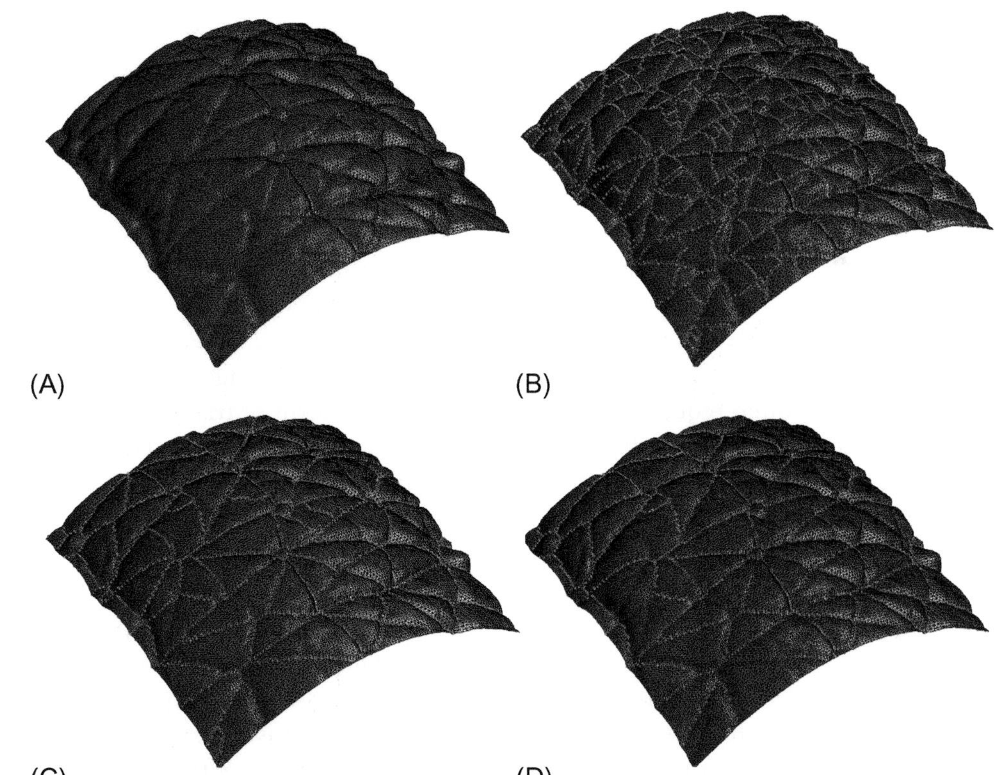

Fig. 8.7 Watershed segmentation of a human skin surface. (A) Original mesh surface, (B) without Wolf pruning, (C) with 5% Wolf pruning, and (D) with 10% Wolf pruning.

trivial peaks are eliminated. Fig. 8.7C presents the result of the watershed segmentation using 5% Wolf pruning. The better result is observed as the segmented regions basically reveal the skin texture cells. The result of 10% Wolf pruning is given by Fig. 8.7D. Some segment cells are merged, which does not conform to the truth. Therefore, based on these three trials, 5% Wolf pruning achieved the best result. Although the choice of the cutoff value is subjective, the value of Wolf pruning can be applied to generate a quantitative assessment of segmentation results from which a satisfied value can be found.

The complex nature of powder additive manufacturing (AM) processes tends to produce surfaces that are very rough, showing significant defect features, including large isolated globules due to partially melted particles attached to the surface, repeating steps generated by successively adding layers, open surface pores, and reentrant features [51]. The measurement and characterization of AM surface texture can be useful for both AM process control and AM product verification. Complementary to widely used filtration techniques, the watershed segmentation can provide a feature-based surface analysis.

Fig. 8.8A presents a part of an AM lattice structure measured by a Nikon XT H 225 industrial CT machine; the reconstructed triangular mesh has a total of 319,379 vertices and 634,889 faces. A total least squares cylinder has been chosen to estimate form surface. The descending watershed segmentations with no Wolf pruning and with 5% Wolf pruning are illustrated in Fig. 8.8B and C, respectively. The watershed segmentation can be useful to detect the regions with the biggest globules.

8.5 Summary

The analysis of surface topography is important for the evaluation of a product's performance. Nowadays, surfaces of modern high value-added components are changing from traditional simple geometries to free-form shapes. The traditional watershed segmentation cannot be applied to free-form surfaces because they are no longer planar, and also, measurement data often require surface description in the form of triangular meshes.

The watershed segmentation based on Maxwell's theory can be extended to triangular mesh surfaces. First, surface topography is extracted from the general shape of free-form surfaces using proper mathematical operations, that is, fitting and filtration. The signed distances between the measured surface and the reference surface are computed and taken as surface topography. Following that the watershed approach based on Maxwell's theory, Pfaltz graph, and change tree is extended to tackle triangular meshes. Critical points (i.e., peaks, pits, and saddles) are identified, and ridge/course lines are traced. Wolf pruning is applied to the change tree so that the oversegmented regions can be merged.

The watershed segmentation for free-form surfaces can be useful for the functional analysis of biosurfaces and the characterization of AM surfaces, where the surfaces under evaluation have complex geometry and significant surface texture.

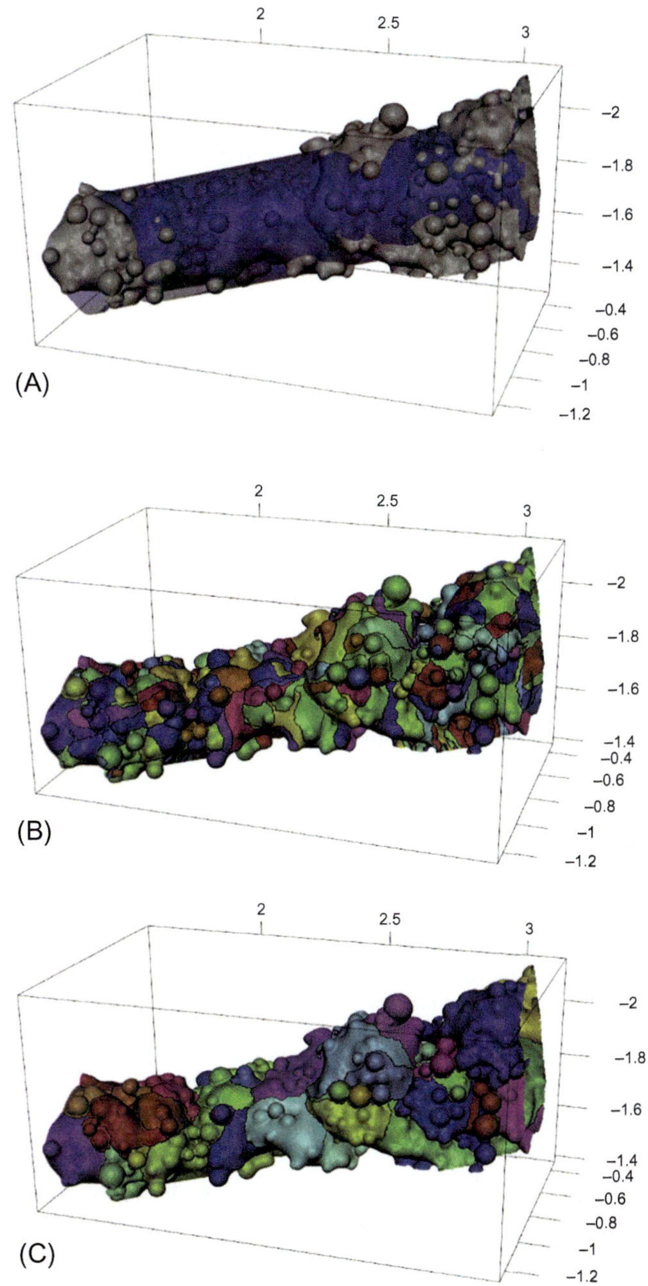

Fig. 8.8 3-D Watershed segmentation (values in mm): (A) XCT measured lattice model, (B) watershed segmentation without Wolf pruning, and (C) watershed segmentation with 5% Wolf pruning.

References

[1] Roach P, Shirtcliffe NJ, Newton MI. Progess in superhydrophobic surface development. Soft Matter 2008;4:224–40.
[2] Bechert DW, Bruse M, Hage W. Experiments with three-dimensional riblets as an idealized model of shark skin. Exp Fluids 2000;28:403–12.
[3] ISO 16610 series, Geometrical product specifications (GPS)—Filtration, 2013.
[4] Dietzsch M, Papenfuβ K, Hartmann T. The motif method (ISO 12085): a suitable description for functional manufactural and metrological requirements. Mach Tools Manufact 1998;38:625–32.
[5] CNOMO E00.14.015 N, Etats geometriques de surface, calcul des parametres de profil, 1983.
[6] ISO 12085, Geometrical Product Specification (GPS)—Surface texture: Profile method—motif parameters, 1996.
[7] Jiang X, Scott PJ, Whitehouse DJ, Blunt L. Paradigm shifts in surface metrology, Part I. Historical philosophy. Proc R Soc A 2007;463:2071–99.
[8] Barre F, Lopez J. Watershed lines and catchment basins: A new 3D-motif method. Int J Mach Tool Manuf 2000;40:1171–84.
[9] Scott PJ. Pattern analysis and metrology: the extraction of stable features from observable measurements. Proc R Soc A 2004;460:2845–64.
[10] Blanc J, Grime D, Blateyron F. Surface characterization based upon significant topographic features. J Phys Conf Ser 2011;311:012014.
[11] Barre F, Lopez J. On a 3D extension of the motif method (ISO 12085). Int J Mach Tools Manuf 2001;41:1873–80.
[12] Maxwell JC. On hills and dales. Phil Mag 1870;40:421–7.
[13] Pfaltz JL. Surface networks. Geogr Anal 1976;8:77–93.
[14] Lantuejoul C., La squelettisation et son application aux mesures topologiques des mosaques polycristallines, PhD thesis, Ecole des Mines, Paris, 1978.
[15] Beucher S, Lantuejoul C. Use of watersheds in contour detection. In: Proceedings of International Workshop Image Processing, Real-Time Edge and Motion Detection/Estimation; 1979.
[16] Vincent L, Soille P. Watersheds in digital spaces: an efficient algorithm based on immersion simulations. IEEE Trans Patt Anal Mach Intell 1991;13:583–98.
[17] Beucher S, Meyer F. The morphological approach to segmentation: the watershed transformation. In: Dougherty ER, editor. Mathematical Morphology in Image Processing; 1993. p. 433–81.
[18] Couprie M, Bertrand G. Topological grayscale watersheds transform. Proc SPIE 1997;3168:136–46.
[19] Bieniek A, Moga AN. An efficient watershed algorithm based on connected components. Pattern Recognit 2000;33:907–16.
[20] Lotufo RA, Falcão AX, Zampirolli FA. IFT-watershed from gray-scale marker. In: 15th Brazilian Symposium on Computer Graphics and Image Processing; 2002. p. 146–52.
[21] Osma-Ruiz V, Godino-Llorente JI, Saenz-Lechon N, Gomez-Vilda P. An improved watershed algorithm based on efficient computation of shortest paths. Pattern Recognit 2006;40:1078–90.
[22] Lin YC, Tsai YP, Hung YP, Shih ZC. Comparison between immersion-based and toboggan-based watershed image segmentation. IEEE Trans Image Process 2006;15:632–40.
[23] Cousty J, Bertrand G, Najman L, Couprie M. Watershed cuts: Minimum spanning forests and the drop of water principle. IEEE Trans Patt Anal Mach Intell 2009;311:362–1374.
[24] Beucher S. The watershed transformation applied to image segmentation. Scanning Microsc 1992;6:299–314.
[25] Shafarenko L, Petrou M, Kittler JV. Automatic watershed segmentation of randomly textured color images. IEEE Trans Image Process 1997;6:1530–44.
[26] Volkmann N. A novel three-dimensional variant of the watershed transform for segmentation of electron density maps. J Struct Biol 2002;138:123–9.
[27] Tarabalka Y, Chanussot J, Benediktsson JA. Segmentation and classification of hyperspectral images using watershed transformation. Pattern Recognit 2010;43:2367–79.
[28] Scott PJ. The mathematics of motif combination and their use for functional simulation. Int J Mach Tool Manuf 1992;32:69–73.

[29] Barré F, Lopez J. Watershed lines and catchment basins: a new 3D-motif method. Int J Mach Tool Manuf 2000;40:1171–84.
[30] Xiao S, Xie F, Blunt L, Scott PJ, Jiang X. Feature extraction for structured surface based on surface networks and edge detection. Mater Sci Semi Proc 2006;9:210–4.
[31] Blunt L, Xiao S. The use of surface segmentation methods to characterise laser zone surface structure on hard disc drives. Wear 2010;271:604–9.
[32] ISO 16610-85. Geometrical Product Specification (GPS)—Morphological areal filters: Segmentation, 2013.
[33] Mangan A, Whitaker R. Partitioning 3D surface meshes using watershed segmentation. IEEE Trans Visual Comput Graph 1999;5:308–21.
[34] Alcoverro M, Philipp-Foliguet S, Jordan M, Najman L, Cousty J. Region-based artwork indexing and classification. In: Proc. 2nd 3D TV Conference. IEEE; 2011. p. 393–6.
[35] Ćomić L, De Floriani L, Iuricich F, Magillo P. Computing a discrete Morse gradient from a watershed decomposition. Comput Graph 2016;58:43–52.
[36] Wolf GW. Scale independent surface characterisation: Geography meets precision surface metrology. Precis Eng 2017;49:456–80.
[37] Zahouani H, Vargiolu R, Piezanowski J. Surface morphology and elasticity of human skin during aging. In: 13th Met. and Props; 2011. p. 71–5.
[38] Lou S, Pagani L, Zeng W, Jiang X, Scott PJ. Watershed segmentation of topographical features on 3D freeform surfaces and its application to additively manufactured surfaces. Precis Eng 2020;63:177–86.
[39] Zhang LY, Zhou RR, Zhu JY, Wu X. Piecewise B-spline surfaces, fitting to arbitrary triangle meshes. CIRP Ann Manuf Technol 2002;51:131–4.
[40] Ma W, He P. B-spline surface local updating with unorganized points. Comput Aid Des 1998;30:853–62.
[41] Lou S, Jiang X, Scott PJ. Geometric computation theory for morphological filtering on freeform surfaces. Proc R Soc A 2013;469: 20130150.
[42] Desbrun M, Meyer M, Schröder P, Barr AH. Implicit fairing of irregular meshes using diffusion and curvature flow. In: SIG-GRAPH 99 Conference Proceedings; 1999. p. 317–24.
[43] Bajaj CL, Xu G. Anisotropic diffusion of surfaces and functions on surfaces. ACM Trans Graph 2003;22:4–32.
[44] Nackerman LR. Two-dimensional critical point configuration graphs. IEEE Trans Patt Anal Mach Intell 1984;6:442–50.
[45] Takahashi S, Ikeda T, Shinagawa Y, Kunii TL, Ueda M. Algorithms for extracting correct critical points and constructing topological graphs from discrete geographical elevation data. Comput Graph Forum 1995;14:181–92.
[46] Scott PJ. An algorithm to extract critical points from lattice height data. Int J Mach Tool Manuf 2001;41:1889–97.
[47] Roerdink JBTM, Meijster A. The watershed transform: Definitions, algorithms and parallelization strategies. Fundam Informat 2001;41:187–228.
[48] Kweon IS, Kanade T. Extracting topographic terrain features from elevation maps. CVGIP Image Understand 1994;59:171–82.
[49] ISO 25178-2, Geometrical product specifications (GPS)—Surface texture: Areal—Part 2: Terms, definitions and surface texture parameters, 2012.
[50] Wolf GW. A Fortran subroutine for cartographic generalization. Comput Geosci 1991;17:1359–81.
[51] Lou S, Jiang X, Sun W, Zeng W, Pagani L, Scott PJ. Characterisation methods for powder bed fusion processed surface topography. Precis Eng 2019;57:1–15.

Further reading

Vollmer J, Mencl R, Muller H. Improved Laplacian smoothing of noisy surface meshes. Comput Graph Forum 1999;131–8.

CHAPTER 9

Free-form surface filtering using wavelets and multiscale decomposition

9.1 Introduction

Wavelets and multiscale analysis have been applied to different types of engineering surfaces for decades, and they always produce very reliable results in analyzing the features of interest. However, engineering surfaces have undergone a significant development, and more complicated free-form surfaces are being produced. These new surfaces have more complicated geometries with non-Euclidean nature. The non-Euclidean surfaces require new wavelet tools that can analyze such surfaces.

In this chapter, two algorithms based on the lifting wavelet scheme and Laplacian mesh relaxation scheme are proposed to filter and decompose free-form surfaces represented by triangular meshes. This chapter begins by a brief review of wavelets and multiscale analysis including their implementations in Sections 9.2 and 9.3. Section 9.4 discusses the various generations of wavelets that are used to analyze the surface texture for ordinary surfaces. Then in Section 9.5 the new developments in multiscale analysis for surface represented by 3-D triangular meshes and its relevance to wavelets are discussed. Section 9.6 introduces an algorithm for filtering free-form surfaces based on lifting wavelets. Multiscale free-form surface decomposition using a Laplacian mesh relaxation is explained in Section 9.7. Computer-generated and bioengineering case studies are analyzed in Section 9.8, and finally the conclusions are drawn in Section 9.9.

9.2 Wavelet multiscale analysis

Wavelet multiscale analysis has been thoroughly investigated in the last three decades and has a variety of applications across different fields such as signal and image analysis, fault detection, and many other applications. In this section, we will briefly recall the basic theory of multiscale analysis, also known as multiresolution analysis. For more comprehensive revision of wavelet multiscale analysis and its applications, readers are advised to read dedicated wavelet books and research papers, such as [1–3].

The basic idea of a multiscale analysis of any given function defined in a dense space, $L^2(\mathbb{R})$, for example, is to decompose that function into a sequence of nested sparser, less

* For more examples and practical implementation, check the website: www.digitalsurf.com/freeform.

Advanced Metrology
https://doi.org/10.1016/B978-0-12-821815-0.00009-5

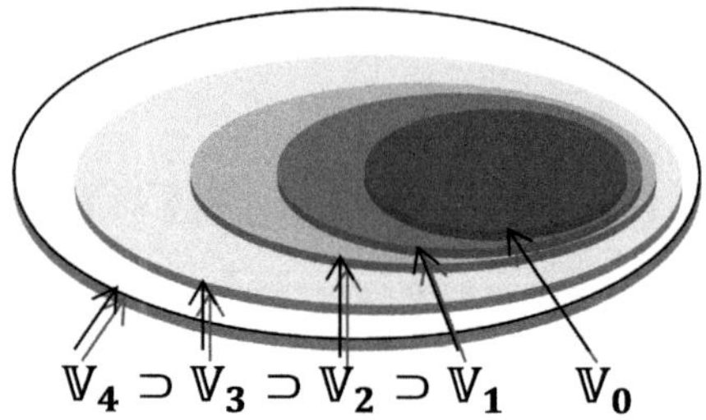

Fig. 9.1 Multiscale analysis nested spaces.

dense, subspaces. In other words a higher-resolution space can be represented using a sequence of lower-resolution spaces; also a higher-resolution space contains all lower-resolution spaces as shown in Fig. 9.1. Therefore, if we consider a function $f(x) \in L^2(\mathbb{R})$, the projection of this function onto different subspaces V_j allows us to view the function $f(x)$ under various scales with a certain level of details. In this chapter the larger the j, the denser the subspace.

Mathematically, these nested subspaces can be expressed by

$$\mathbb{V}_0 \subset \mathbb{V}_1 \subset \mathbb{V}_2 \subset \mathbb{V}_3 \subset \ldots \subset \mathbb{V}_{n-1} \subset \mathbb{V}_n \subset \ldots \subset L^2(\mathbb{R}) \tag{9.1}$$

Theoretically a set of vector spaces $\mathbb{V}_j$, which determines a multiscale approximation of $L^2(\mathbb{R})$, should satisfy the following properties:

$$\begin{aligned} &1. \mathbb{V}_j \subset \mathbb{V}_{j+1} \\ &2. \text{If} f(x) \in \mathbb{V}_j, \text{then} f(2x) \in \mathbb{V}_{j+1} \\ &3. \text{If} f(x) \in \mathbb{V}_j, \text{then} f(x+k) \in \mathbb{V}_j \\ &4. \cup_{j=0}^{\infty} \mathbb{V}_j = L^2(\mathbb{R}) \end{aligned} \tag{9.2}$$

Therefore a function $f(x) \in L^2(\mathbb{R})$ can be decomposed into different spaces using a set of basis function:

$$f(x) = \sum_j \alpha_j . \varphi_j(x) \tag{9.3}$$

where α_j is a set of coefficients needed to represent the function using a set of basis functions $\varphi_j(x)$.

The set of basis functions has to be carefully designed to the nested spaces, satisfying all the necessary conditions in Eq. (9.2). Stephane Mallat [3] showed that the nested spaces can be built by defining the basis functions $\varphi_j(x)$ to be scaled shifted, or dilated and translated, versions of a particular function $\varphi(x)$, that is,

$$\varphi_{j,k}(x) = \varphi(2^j x - k) \tag{9.4}$$

Therefore, using this set of basis functions, the original function $f(x)$ can now be represented as

$$f(x) = \sum_j \sum_k \alpha_{j,k}.\varphi\left(2^j x - k\right) \tag{9.5}$$

where $\varphi(x)$ is called the scaling function or mother wavelet function.

Using these basis functions the nested spaces can be defined as follows:

$$\mathbb{V}_0 = \text{span}\left(\varphi_{0,k}(x)\right), \mathbb{V}_1 = \text{span}\left(\varphi_{1,k}(x)\right), \ldots, \mathbb{V}_j = \text{span}\left(\varphi_{j,k}(x)\right).$$

where

$$\text{span}\left(\varphi_{i,k}(x)\right) = \sum_{k=0}^{\infty} \varphi_{i,k}(x).$$

One important observation for this set of basis functions is that a lower-scale basis functions can be built up using linear combinations of the higher-scale basis functions, or in other words the basis functions at any scale can be derived from the basis functions of the higher scale. This important relation can be mathematically formulated as

$$\varphi(x) = 2 \times \sum_k h_k \varphi(2x - k) \tag{9.6}$$

where h_k is the coefficient needed to produce a lower-scale basis function from its higher-resolution counterparts. h_k is also known to be the filter coefficient as we will see later in this chapter.

This equation is called the refinement equation, dilation equation, or the two-scale equation and is considered an important equation for designing basis functions for any multiscale analysis system.

Using the two-scale equation (9.6), Eq. (9.5) becomes

$$f(x) = \sum_k \alpha_{n,k}.\varphi(2^n x - k). \tag{9.7}$$

The examples in the succeeding text explain how to represent a function using a set of scaling functions and also explain the concept of the two-scale equation.

Example 9.1: A simple example of decomposing a function using the Haar scaling function

A Haar scaling function is a simple box function defined by

$$\varphi(x) = \varphi_0(x) = \begin{cases} 1; & 0 \leq x < 1 \\ 0; & \text{otherwise.} \end{cases}$$

A higher-scale function can be obtained by compressing the aforementioned function by the factor of 2, so

$$\varphi_1(x) = \varphi_0(2x) = \begin{cases} 1; & 0 \leq x < {}^1/_2 \\ 0; & \text{otherwise.} \end{cases}$$

The lower (zero-scale) resolution can be derived from the higher (first-scale), satisfying the refinement equation:

$$\varphi_0(x) = \varphi_1(x) + \varphi_1\left(x - \frac{1}{2}\right)$$

$$\varphi_0(x) = 2 \times \left(\frac{1}{2}\varphi_{1,0}(x) + \frac{1}{2}\varphi_{1,1}(x)\right)$$

Therefore h_k for Haar scaling function is equal to

$$h_0 = \begin{bmatrix} \frac{1}{2} & \frac{1}{2} \end{bmatrix}.$$

Now, let a function $f(x) \in \mathbb{V}_2$ be defined as

$$f(x) = \begin{cases} 1; & 0 \leq x < {}^1/_4 \\ 2; & {}^1/_4 \leq x < {}^1/_2 \\ -1; & {}^1/_2 \leq x < {}^3/_4 \\ 3; & {}^3/_4 \leq x < 1. \end{cases}$$

This function can be represented using the second scale Haar scaling functions ($j = 2$) by

$$f(x) = \varphi_{2,0}(x) + 2 \cdot \varphi_{2,1}(x) - \varphi_{2,2}(x) + 3 \cdot \varphi_{2,3}(x).$$

Alternatively the function can be represented at different scales, using Eq. (9.5) as

$$f(x) = \varphi_{0,0}(x) + 2 \cdot \varphi_{1,1}(x) + \varphi_{2,1}(x) - 4 \cdot \varphi_{2,2}(x).$$

Representing a function using different scales is in fact projecting that function onto different subspaces. Hence the approximation of a given function at the scales $j+1$ and j is equal to its projection, respectively, on the vector spaces $\mathbb{V}_j$ and $\mathbb{V}_{j+1}$. The difference between the two subspaces is called the wavelet space that contains all the details of the function that are not in $\mathbb{V}_j$. These details at the scale j are given by projecting the original function on the complement of $\mathbb{V}_j$ in $\mathbb{V}_{j+1}$, which is $\mathbb{W}_j$.

Let $\mathbb{W}_j$ be a complement subspace of $\mathbb{V}_j$ in $\mathbb{V}_{j+1}$, so

$$\begin{aligned} \mathbb{V}_{j+1} &= \mathbb{V}_j \oplus \mathbb{W}_j \\ \mathbb{W}_j &\subset \mathbb{V}_{j+1} \end{aligned} \tag{9.8}$$

where $\oplus$ denotes the inner sum of linear spaces and $\mathbb{W}_j$ are called wavelet subspaces. The complement wavelet subspaces are shown in the Fig. 9.2.

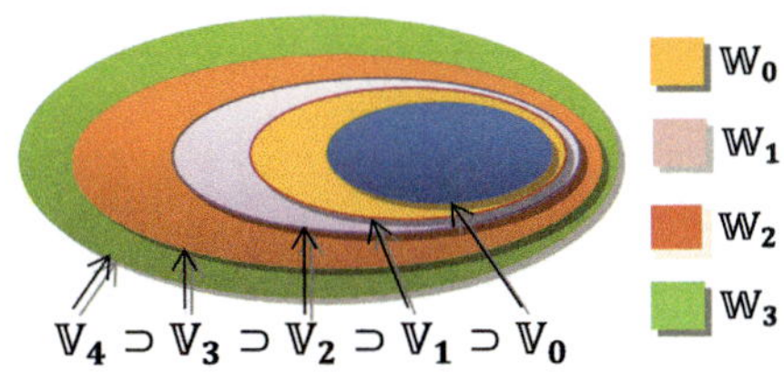

Fig. 9.2 Multiscale analysis nested spaces.

Using the wavelet subspaces a denser space $\mathbb{V}_n$ can then be written as a telescopic decomposition into a coarser-resolution scaling subspace and intermediate complement spaces as

$$\mathbb{V}_n = \mathbb{V}_0 \oplus \mathbb{W}_{n-1} \oplus \mathbb{W}_{n-2} \oplus \ldots \oplus \mathbb{W}_1 \oplus \mathbb{W}_0. \tag{9.9}$$

In the same manner that we previously defined the scaling functions $\varphi(x)$ for building the different vector subspace $\mathbb{V}_{\mathrm{j}}$, we need now to define the wavelet function $\psi(x)$ that by scaling and translating will allow us to build every complement subspace $\mathbb{W}_{\mathrm{j}}$, so

$$\psi_{j,k}(x) = \psi\left(2^j x - k\right). \tag{9.10}$$

As with the scaling functions, the wavelet functions should satisfy the two-scale equation, therefore

$$\psi(x) = 2 \times \sum_k h_k \varphi(2x - k). \tag{9.11}$$

Using the scaling and wavelet subspaces, the function $f(x)$ can now be viewed using many levels of details as

$$f(x) = \sum_k \alpha_k.\varphi(x-k) + \sum_j \sum_k \beta_{j,k}.\psi\left(2^j x - k\right). \tag{9.12}$$

This equation is very important, as it will be the main building block to implement the multiscale decomposition of any given function using the filter bank method. This is shown in the following section.

Using Eq. (9.12), Example 9.2 in the succeeding text explains how to represent a function using the Haar scaling and wavelet functions:

Example 9.2: A simple example of decomposing a function using the Haar scaling and wavelet functions

A Haar wavelet function is defined by

$$\psi(x) = \psi_0(x) = \begin{cases} +1; & 0 \leq x < {}^1/_2 \\ -1; & {}^1/_2 \leq x < 1. \end{cases}$$

The Haar wavelet function can now be expressed in terms of higher-resolution scaling functions as

$$\psi_0(x) = \varphi_1(x) - \varphi_1\left(x - \frac{1}{2}\right)$$

$$\psi_0(x) = 2 \times \left(\frac{1}{2}\varphi_{1,0}(x) - \frac{1}{2}\varphi_{1,1}(x)\right)$$

Therefore h_k, for the Haar wavelet function, is equal to

$$h_1 = \begin{bmatrix} \frac{1}{2} & -\frac{1}{2} \end{bmatrix}.$$

Now the function $f(x)$ defined in the previous example can be decomposed using Haar scaling and wavelet functions as

$$f(x) = 1.25 \cdot \varphi_{0,0}(x) + 0.25 \cdot \psi_{0,0}(x) - 0.5 \cdot \psi_{1,0}(x) - 2 \cdot \psi_{1,1}(x).$$

Note that we only needed two wavelet levels as $f(x) \in \mathbb{V}_2$, so we only need $\mathbb{V}_0$, $\mathbb{W}_0$, and $\mathbb{W}_1$ to represent this function, whereas we needed to use $\mathbb{V}_0$, $\mathbb{V}_1$, and $\mathbb{V}_2$ when we represented this function using only the scaling spaces.

9.3 Implementations methods of wavelet multiscale analysis

In this section the most important two methods for implementing wavelet multiscale analysis to decompose any given function or signal are briefly explained. The first method is carried out by filtering the input signal with two filters; this method is called the filter bank algorithm. The second method includes using the lifting scheme and is explained later in this section.

9.3.1 First-generation wavelet: The filter bank method

Using the multiscale analysis described in the previous section, let the function to be analyzed be denoted by $f(x) \in \mathbb{V}_n$. From the previous section the higher-resolution vector space $\mathbb{V}_n$ can be decomposed into coarser lower-resolution space $\mathbb{V}_{n-1}$ and detail space $\mathbb{W}_{n-1}$ according to the equation

$$\mathbb{V}_n = \mathbb{V}_{n-1} \oplus \mathbb{W}_{n-1}.$$

The scaling subspace $\mathbb{V}_{n-1}$ is called the approximation, and the wavelet subspace $\mathbb{W}_{n-1}$ is called the details as shown in Fig. 9.3.

From Eq. (9.12) the function $f(x) \in \mathbb{V}_n$ can be expressed using the wavelet and scaling functions as

$$f(x) = \sum_k \alpha_k \cdot \varphi(x-k) + \sum_{j=0}^{n-1} \sum_k \beta_{j,k} \cdot \psi\left(2^j x - k\right).$$

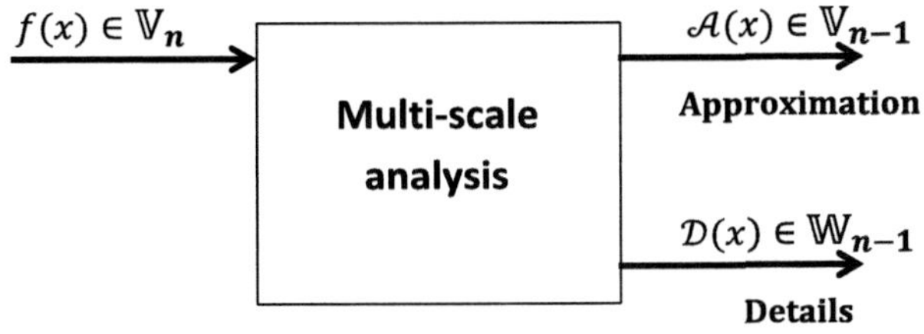

Fig. 9.3 Multiscale analysis nested spaces.

Therefore decomposing the function $f(x) \in \mathbb{V}_n$ using only one level is given by

$$f(x) = \sum_k \alpha_{n-1,k} \cdot \varphi\left(2^{n-1}x - k\right) + \sum_k \beta_{n-1,k} \cdot \psi\left(2^{n-1}x - k\right). \tag{9.13}$$

The approximation, $\mathcal{A}(\mathrm{x}) \in \mathbb{V}_{n-1}$, and the detail coefficients, $\mathcal{D}(\mathrm{x}) \in \mathbb{W}_{n-1}$, are calculated using the inner products with the scaling and wavelet functions at this level, that is,

$$\begin{aligned} \alpha_{n-1,k} &= \ < f(x), \varphi(2^{n-1}x - k) > \\ \alpha_{n-1,k} &= \ < \sum_k \alpha_{n,k} \cdot \varphi(2^n x - k), \varphi\left(2^{n-1}x - k\right) > \end{aligned} \tag{9.14}$$

$$\begin{aligned} \beta_{n-1,k} &= < f(x), \psi(2^{n-1}x - k) > \\ \beta_{n-1,k} &= < \sum_k \alpha_{n,k} \cdot \varphi(2^n x - k), \psi\left(2^{n-1}x - k\right) > \end{aligned} \tag{9.15}$$

By performing the aforementioned inner products and by using the two-scale equations for the scaling and wavelet functions, as shown in Eqs. (9.6) and (9.11), the approximation and details can be simplified to

$$\alpha_{n-1,k} = \sum_i h_0(i - 2k) \cdot \alpha_{n,i} \tag{9.16}$$

$$\beta_{n-1,k} = \sum_i h_1(i - 2k) \cdot \alpha_{n,i} \tag{9.17}$$

where h_0 and h_1 represent the coefficients of the two-scale equation for the scaling and wavelet functions, respectively.

Recall that the discrete convolution between two signals $x[n]$ and $g[n]$ is given by

$$y[i] = x[i] \star g[i] = \sum_n x[i - n] g[n].$$

Therefore Eqs. (9.16) and (9.17) show that the approximation and detail parameters are obtained by convolving or filtering the input coefficients $\alpha_{n,i}$ with time-reversed

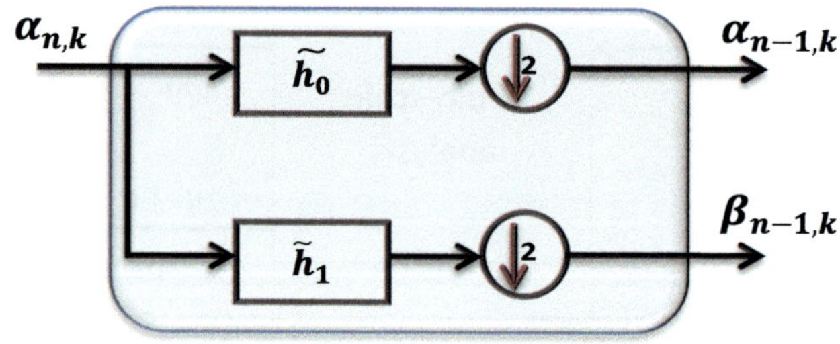

Fig. 9.4 Multiscale analysis nested spaces.

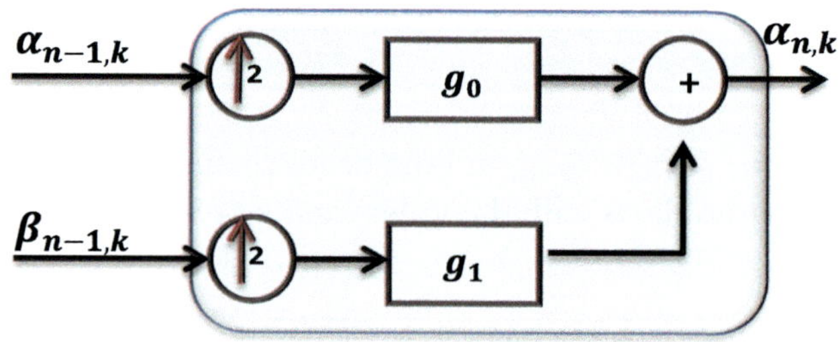

Fig. 9.5 Multiscale analysis nested spaces.

versions of the two filters h_0 and h_1 and $\widetilde{h}_0$ and $\widetilde{h}_1$, respectively, followed by a subsampling, or downsampling, operation as shown in Fig. 9.4. In practice, $\widetilde{h}_0$ and $\widetilde{h}_1$ represents a low-pass and high-pass filters, respectively:

$$\widetilde{h}_0[n] = h_0[-n]$$
$$\widetilde{h}_1[n] = h_1[-n].$$

Fig. 9.4 shows only one level of decomposition, where the input function is represented using the scaling and wavelet functions of next lower scale. However, representing the input function using lower and lower scales is possible by further decomposing the approximated functions $\alpha_{n-1,k}, \alpha_{n-2,k}, \alpha_{n-3,k}, \ldots \alpha_{1,k}$ using nested levels of the filter bank implementation.

Finally the original input function can be reconstructed by upsampling and filtering the approximated and detail signal with a new set of filters g_0 and g_1. These new filters are carefully designed to ensure a perfect reconstruction of the input signal. Fig. 9.5 shows how to reconstruct the input function from its approximation and details.

Example 9.3 shows how to decompose a given function using the filter bank algorithm.

Example 9.3: Haar decomposition using the filter bank method

In Examples 9.1 and 9.2, it was shown that the two-scale equation coefficients for the scaling and wavelet functions of the Haar are given by

$$h_0 = \left[+\frac{1}{2} \ \ +\frac{1}{2} \right] \text{ and } h_1 = \left[+\frac{1}{2} \ \ -\frac{1}{2} \right]$$

Therefore their filter coefficients will be

$$\tilde{h}_0 = \left[+\frac{1}{2} \ \ +\frac{1}{2} \right] \text{ and } \tilde{h}_1 = \left[-\frac{1}{2} \ \ +\frac{1}{2} \right]$$

These coefficients are used to build our filter bank decomposition algorithm to decompose the function $f(x)$ defined earlier in Example 9.1; $f(x)$ can be represented as

$$f(x) = [1 \ \ 2 \ -1 \ \ 3]$$

First level of decomposition:

The output of the filters will be

$$f(x) \star \tilde{h}_0 = [0.5 \ \ 1.5 \ \ 0.5 \ \ 1.0 \ \ 1.5]$$
$$f(x) \star \tilde{h}_1 = [-0.5 \ \ -0.5 \ \ 1.5 \ \ -2.0 \ \ 1.5]$$

The output of the downsampling operator will be

$$\alpha_{1,k} = [1.5 \ \ 1]$$
$$\beta_{1,k} = [-0.5 \ \ -2.0]$$

Therefore the signal can be represented as

$$f(x) = 1.5 \cdot \varphi_{1,0}(x) + 1 \cdot \varphi_{1,1}(x) - 0.5 \cdot \psi_{1,0}(x) - 2 \cdot \psi_{1,1}(x).$$

Second level of decomposition:

The output of the filters will be

$$\alpha_{1,k} \star \tilde{h}_0 = [0.75 \ \ 1.25 \ \ 0.5]$$
$$\alpha_{1,k} \star \tilde{h}_1 = [-0.75 \ \ 0.25 \ \ 0.5]$$

The output of the downsampling operator will be

$$\alpha_{0,k} = [1.25]$$
$$\beta_{0,k} = [0.25]$$

Therefore the input function can be represented as

$$f(x) = 1.25 \cdot \varphi_{0,0}(x) + 0.25 \cdot \psi_{0,0}(x) - 0.5 \cdot \psi_{1,0}(x) - 2 \cdot \psi_{1,1}(x).$$

9.3.2 Second-generation wavelet: The lifting scheme method

In the previous section, it was shown that a given function can be decomposed using the multiscale wavelet analysis by convolving the function with a series of filter banks. Convolution operations can be carried out in spatial domain or by using the Fourier transform, both of which require the input signal to be regularly sampled along the whole interval. Furthermore, constructing the wavelet analysis using the filter bank method is only valid when the scaling and wavelet functions are translations and dilations of particular functions.

Constructing a multiscale wavelet analysis without using the filter bank method is vital. Wavelets are not necessarily a translate and dilate of one function and can be constructed without the need of the convolution operation or the Fourier transform. This type of construction will extend multiscale wavelet analysis into many new applications such as wavelets on bounded domains, wavelets on irregular sampled signals, and also wavelets on curves and surfaces, which is our concern in this chapter.

In 1995 Swelden [4] proposed the lifting scheme as a new generation of wavelets, which he referred to as the second generation of wavelets. The lifting scheme generalizes the first generation and could be applied to both regular and irregular data sets [4–6].

The lifting scheme allows the construction of the filter banks entirely within the spatial domain and eliminates the need of Fourier or convolution operations, which limits the first generation to only regular data sets. Instead of explicitly designing and specifying the scaling and wavelet functions, the lifting scheme decomposes that data through three major operations: splitting, prediction, and updating as shown in Fig. 9.6. These

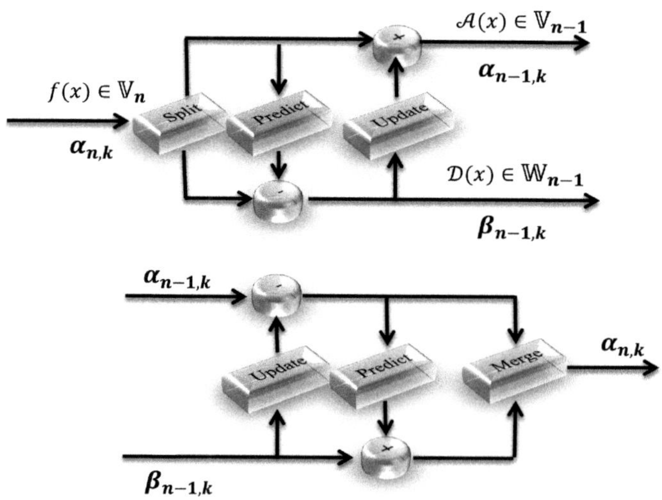

Fig. 9.6 The Lifting scheme decomposition and reconstruction process.

properties of the lifting scheme enable wavelet analysis to be extended to irregular data sets such as free-form surfaces and graphs [7, 8], as will be explained in the coming sections.

One major property of the lifting scheme is that it starts with a very simple wavelet called the lazy wavelet that splits the data into even and odd sets. This lazy wavelet is then lifted up to produce the desired wavelet and scaling functions by using the prediction and the update operations. A simple example of how the prediction and update operators can be used to lift up the lazy wavelet to the Haar wavelet is shown in [6].

As shown in Fig. 9.6, the three major building blocks for the lifting scheme are split, predict, and update.

The split operator splits the input data into two disjoint sets or groups. There is no restriction of how to split the data; however, splitting the data into even and odd is the most popular technique, which is called the lazy wavelet transform. The split operator can be expressed as

$$Split(\alpha_n) = \{even_n, odd_n\}. \tag{9.18}$$

The predict operator attempts to predict the values of the odd subset using the even subset. The input data will have some local correlation; therefore the even and odd subsets will be correlated. The prediction step attempts to propose a correlation between the two sets and to predict one set using the other based on that proposed correlation. For example, given the even set, it should be possible to predict the odd one with reasonable accuracy. The predicted odd values are then subtracted from the original odd values to represent a measure of the error between the prediction and original as shown in Fig. 9.6. The detail coefficients β_{n-1} are regarded as the output of the prediction difference and are given by

$$\beta_{n-1} = odd_n - P(even_n). \tag{9.19}$$

There are a number of ways to construct the prediction operator; most of these methods depend on a subdivision scheme and can be summarized in the following three categories:

1. *Interpolating subdivision*: Here, the even set is used to construct a polynomial function, and the odd set is predicted as an interpolated value between these even points.
2. *Average interpolating*: Here, the even set does not represent the values of a polynomial; instead, it represents the average values of a polynomial over the interval between two successive even points. The predicted odd values represent, in this case, the new average values over the new intervals when the odd set is interpolated between the even points.
3. *B-spline interpolating* and in particular cubic B-spline is a very common prediction tool in wavelet analysis. As seen in the previous interpolating methods, B-spline interpolating predicts the odd values by constructing a B-spline function based on the even set and then uses these functions to predict the odd.

Interested readers are advice to read [9] for more details.

The update operation is designed to preserve the average and some higher moments of all coarser signals so that they all have the same average value as the original signal. The update operator can be described mathematically by

$$\alpha_{n-1} = even_n + U(\beta_{n-1}). \tag{9.20}$$

These three blocks are the major building blocks to construct a wavelet transform using the lifting scheme; however, some wavelets require an additional block that contains a simple multiplication operation to build the appropriate wavelet and scaling functions, as will be explained later.

How the lifting scheme is used to calculate the discrete wavelet transform can be explained with the help of the following examples:

Example 9.4: Haar wavelet decomposition using the lifting scheme method

The Haar wavelet can be built using the lifting scheme. The prediction operation simply uses the even samples to predict the odd samples, and then the even samples are updated by adding half of the prediction outcome as shown in the diagram in the succeeding text; therefore

$$\begin{aligned}\beta_{n-1} &= odd_n - even_n \\ \alpha_{n-1} &= even_n + \frac{1}{2}(\beta_{n-1})\end{aligned}.$$

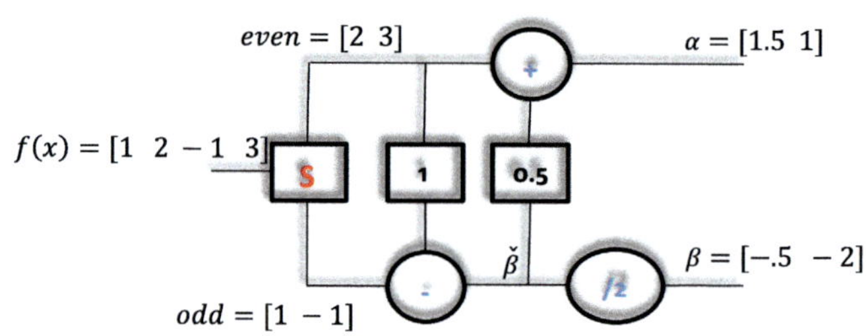

Given the function $f(x) \in \mathbb{V}_2$ defined in Example 9.1, the Haar wavelet decomposition can be calculated using the lifting diagram shown in the preceding text.

Firstly the function $f(x) = [1\ 2 -1\ 3]$ has to be split into two subsets: even and odd as shown in the figure. Then the odd values are predicted using the even values. In this example the first detail value is $(1-(2)=-1)$, and the second detail value is $(-1-(3)=-4)$. The even values are then updated by adding half of the difference in the prediction; therefore the first approximation value is $(2+0.5*(-1)=1.5)$, and the second is $(3+0.5*(-4)=1)$. Finally the detail coefficients are scaled by factor of 0.5 to produce the correct results; therefore the results of the first decomposition level are

$$\begin{aligned}\alpha_{1,k} &= [1.5\quad 1] \\ \beta_{1,k} &= [-0.5\quad -2.0]\end{aligned}$$

which is the same result obtained in Example 9.3 using the traditional filter bank method.

Example 9.5: Haar wavelet reconstruction using the lifting scheme method

The inverse wavelet transform, wavelet reconstruction, using the lifting scheme is simply carried out by reversing the order of the operations and flipping the signs. Therefore the inverse Haar wavelet transform for the previous example can be found using the following diagram:

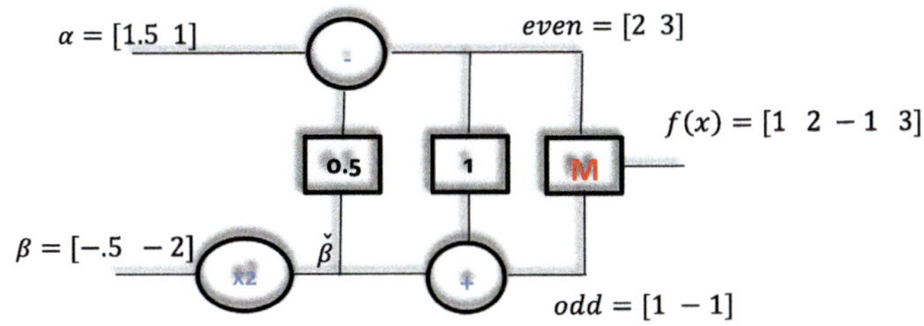

As seen in the diagram, the detail coefficients are multiplied by a factor of 2 to reverse the effect of dividing by 2 in the decomposition stage. After that, these details are used to update the approximated coefficients to find the even values. In this example the first even value is (1.5–0.5*(−1)=2), and the second value is (1−0.5*(−4)=3). Then the odd values are found using the even, and for this particular example, the first odd value is (2+−1=1), and the second is (3+−4=−1). Finally the two subsets are merged to gather to reconstruct the original function, which is $f(x)=[1\ 2-1\ 3]$ in our example.

Example 9.6: Haar scaling and wavelet functions using the lifting scheme

The scaling or wavelet function for a specific coefficient can easily be extracted using the lifting scheme by assigning one to that coefficient and setting all other coefficients to zero and then running the inverse transform. In our example the scaling and wavelet functions for each coefficient are calculated in the succeeding text.

The scaling function for the approximate coefficient $\alpha_{1,0}$ is given by

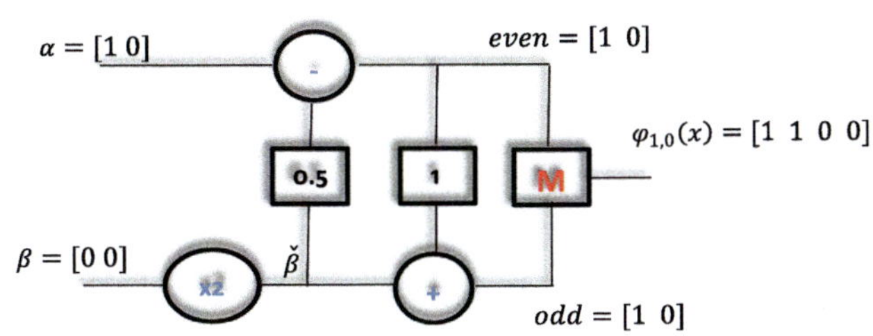

Running the inverse wavelet transform by setting only the approximate coefficient $\alpha_{1,0}$ to 1 and setting all others to zero will result in a scaling function given by $\varphi_{1,0}(x)=[1\ 1\ 0\ 0]$.

The scaling function for the approximate coefficient $\alpha_{1,1}$ is given by

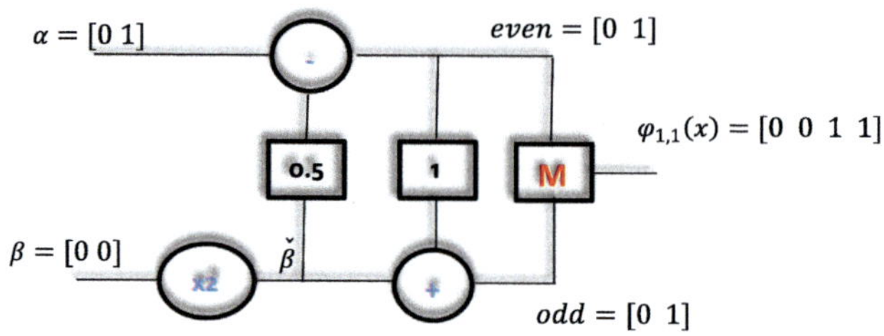

Running the inverse wavelet transform by setting only the coefficient $\alpha_{1,1}$ to 1 and setting all others to zero will result in a scaling function given by $\varphi_{1,\ 1}(x) = [0\ 0\ 1\ 1]$.

The wavelet function for the detail coefficient $\beta_{1,\ 0}$ is given by

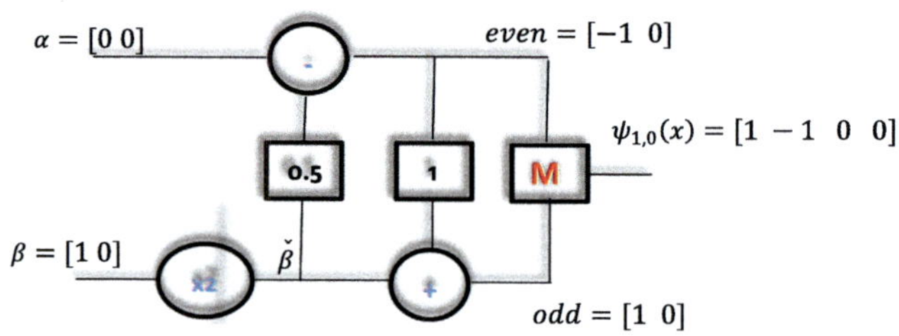

Running the inverse wavelet will give a wavelet function $\psi_{1,0}(x) = [1 - 1\ 0\ 0]$.

The wavelet function for the detail coefficient $\beta_{1,1}$ is $\psi_{1,1}(x) = [0\ 0\ 1 - 1]$. This can be found by setting $\alpha = [0\ 0]$ and $\beta = [0\ 1]$ and then run the inverse wavelet using the diagram in the preceding text.

Therefore the signal can be represented as

$$f(x) = 1.5\bullet\varphi_{1,0}(x) + 1\bullet\varphi_{1,1}(x) - 0.5\bullet\psi_{1,0}(x) - 2\bullet\psi_{1,1}(x).$$

This confirms the results obtained earlier in Example 9.3.

9.4 Wavelets generations for surface texture characterization

Surface characterization and parameterization can be defined as the process of describing the surface using a set of parameters that give an indication of the quality of the surface and also interpret functional properties such as optical quality, service life, and reliability of the surface. These parameters are usually defined as a statistical measurement of the surface texture; for example, areal parameters could be calculated based on the average, standard deviations, or specific features of the surface texture. Therefore the extraction of surface texture and in particular surface roughness is required as a prerequisite of surface parameterization and characterization [10].

Surface filtering is the key operation that can isolate the nominal surface from its texture. Novel filtering algorithms based on Fourier, Gaussian, robust Gaussian regression, and wavelet have been developed during the last two decades, for example [5, 11–15], and many of these algorithms have become industrial standards [16–21]. Among all of these different techniques, wavelet filtering technique enjoys particular interest in surface filtering and provides a mathematical microscope that is capable of decomposing a given surface into different scales and then analyzing each scale with different resolution. Different wavelet-based algorithms for surface filtering have been proposed and have proven their robustness and the reliability of filtering different kinds of surfaces and in many different applications [22, 23].

Three generations of surface filtering using wavelet transform have been developed [22, 24–26], with each generation advancing on its predecessor and overcoming some limitations and problems that could not be solved before.

The keystone of the first-generation model is that the wavelet analysis is generated using dilation and translation of one particular function called the mother wavelet; these translations and dilations are implemented by convolving the input surface data with different filters or so-called filter banks. Using filter banks means that the calculation of the wavelet transformation requires Fourier analysis and/or a convolution operation that not only is time consuming but also limits the application of such algorithms to only regular data sets. The design of filter banks in the first-generation models can be roughly classified into orthogonal and biorthogonal filters [27, 28]. The biorthogonal filters are more reliable for surface filtering as they have a linear phase property that leads to real outputs without frequency aliasing and phase distorting. First-generation wavelet was successfully applied to filter many surfaces; for example, it was successfully applied in denoising areal surface stylus instrumentation [29].

Swelden proposed the lifting scheme as a new wavelet construction tool, which he referred to as second-generation wavelets. The lifting scheme generalizes the first generation and could be applied to both regular and irregular data sets [4, 6, 30]. The lifting scheme allows the construction of the filter banks entirely in the spatial domain and eliminates the need for Fourier or convolution operations that limit the first generation to only regular data sets. Instead of explicitly designing and specifying the scaling and wavelet functions, the lifting scheme decomposes that data through three major operations: splitting, prediction, and updating. With the introduction of the lifting scheme, a new generation of wavelets has been adopted and applied to surface filtering. Models for the second-generation wavelet have helped analyzing different surfaces in different industrial applications. For example, in the steel industry, wavelets have helped to identify multiscalar surfaces such as roughness, waviness, and form [28, 29]. The second generation has also helped to diagnose pump failures in the pump industry [21]. Furthermore, in the bioengineering industry, it has helped to separate, extract, and reconstruct isolated morphological features and multiscalar surfaces [28, 29] as shown in Fig. 9.7. The lifting scheme models for surface filtering have now been transferred to ISO/TS 16610-29 for geometrical filtration [26].

Despite all of the advantages of the aforementioned generations, the first-/second-generation wavelet models were built using a real discrete wavelet transform that has no shift-invariant property. This means that small shifts in surface signal (step height, linear, or curved scratches) can cause large variations in the distribution of energy between

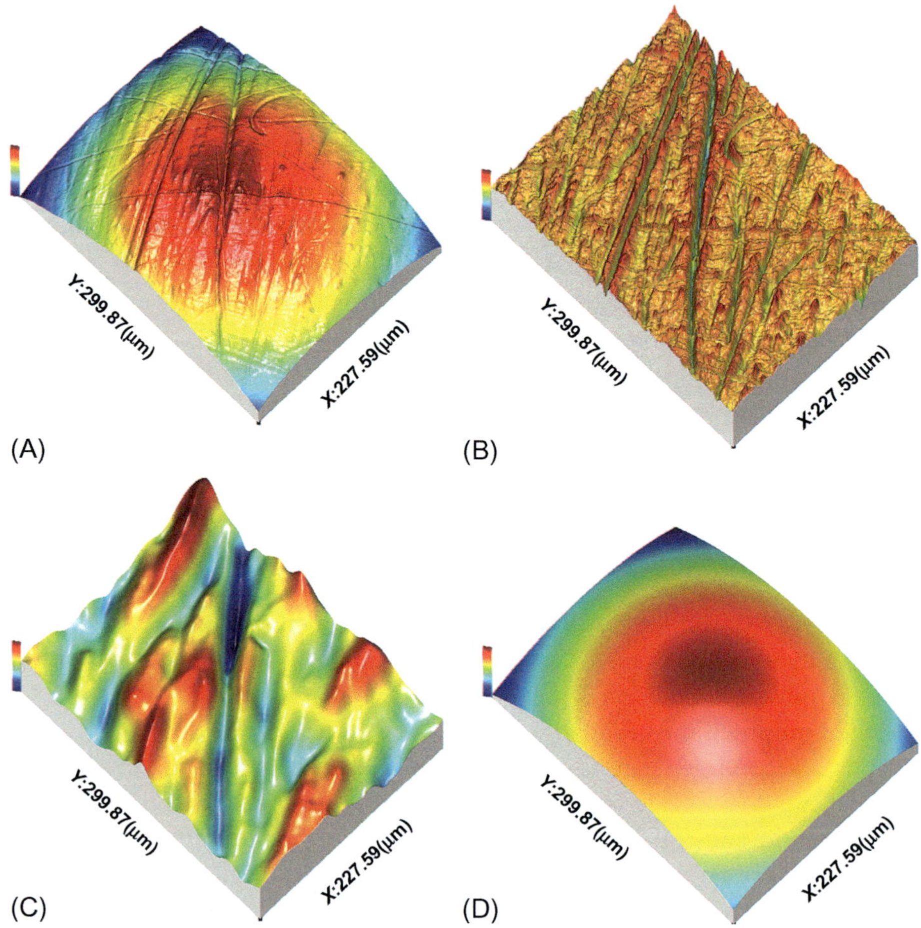

Fig. 9.7 Wavelet decomposition of a precision-lapped ceramic sphere used as a femoral bearing surface in a replacement hip joint. (A) The raw measured surface, (B) roughness surface, (C) wavy surface, and (D) form surface.

the wavelet coefficients at different scales. The lack of shift invariance makes these models less sensitive to detecting and extracting certain morphological features, such as surfaces with linear and curved scratches. In many cases, morphological features have directional properties; however, first- and second-generation methods for extraction of such features are less effective as they do not sense direction and anisotropy. Consequently the accurate extraction of directional morphological features is impossible under these wavelet models. To overcome these problems a new generation of wavelet models was required.

A third-generation wavelet model based on the complex wavelet transform using biorthogonal complex wavelet transforms was developed [22, 24]. Different case studies were conducted using a series of engineering and biomedical surfaces to demonstrate the application of the third-generation models. The results showed that the complex model

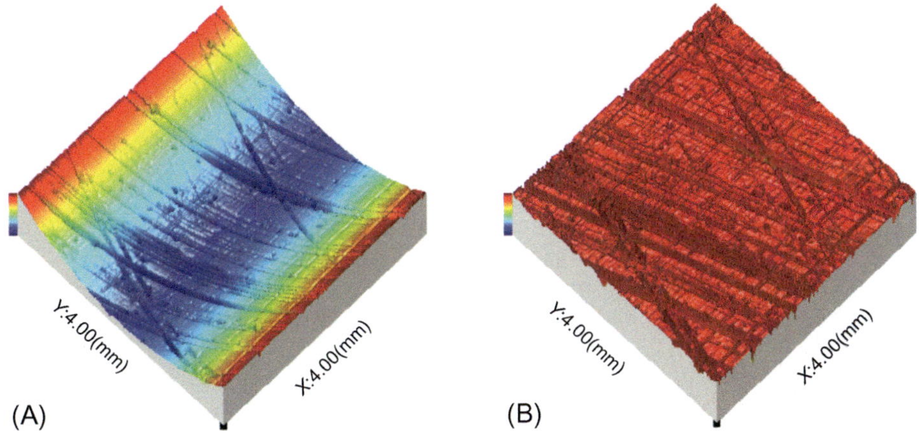

Fig. 9.8 Extraction of honing marks on a plateau-honed surface. (A) The raw measured surface and (B) the honing marked surface.

has extended the previous generations and successfully applied to the extraction of linear and curved features of different surfaces. An example of extraction of honing marks on a plateau-honed surface is shown in Fig. 9.8 [21, 31].

The three generations of wavelets have been successfully applied to the filtration and extraction of features on many different surfaces. Unfortunately the three generations were designed to filter, analyze, and decompose simple surfaces that have Euclidean geometries; they do not apply to non-Euclidean geometries. Therefore the previous wavelet generations cannot be applied to today's complex free-form surfaces without distortion of the results, and thus a new generation of wavelet models capable to filter, analyze, and extract features from such surfaces is required.

9.5 Wavelet multiscale decomposition for 3-D triangular meshes

Multiscale analysis for 3-D meshes has been an active research area for the past decade. The introduction of the second-generation wavelets and lifting Scheme [4, 6, 30] made the extension of wavelets and multiscale analysis possible for all types of 3-D meshes, and subsequent algorithms have been proposed [32–35]. The main idea behind multiscale analysis is to decompose a high-resolution mesh into a lower-resolution mesh (approximation) and extract details that are needed to recover the original mesh; this operation is repeated iteratively starting from the finest mesh and ending with the coarsest base mesh as shown in Fig. 9.9.

Lounsbery et al. [35] have proposed a biorthogonal filter bank to decompose regular and semiregular 3-D meshes into a lower-resolution counterpart and a series of wavelet coefficients as shown in Fig. 9.9. In their method, they made the connection between the nested spaces of scaling functions and 3-D mesh decomposition through the subdivision operation. They showed that the subdivision scheme can be used to create the nested linear spaces required to build the multiscale analysis [32]. The decomposition is

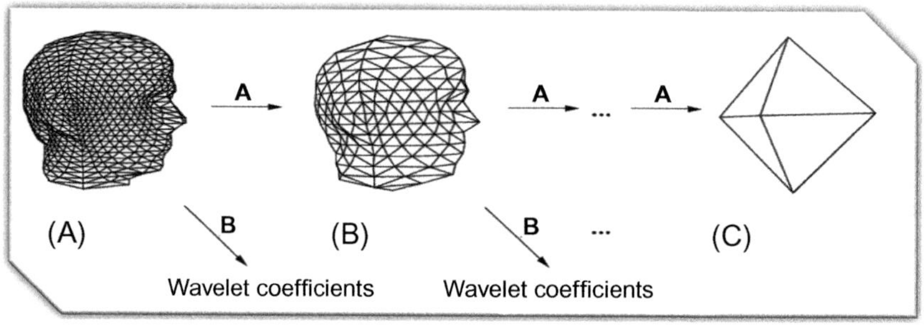

Fig. 9.9 Decomposition of 3-D mesh into approximation and details [35].

computed with two analysis filters, A_j and B_j, for each resolution level j. The reconstruction is done with two synthesis filters P_j and O_j. They showed that coarser mesh and its wavelet coefficients (V_j and W_j, respectively) can be calculated from a finer mesh V_{j+1} using the following equations:

$$V_j = A_j V_{j+1} \tag{9.21}$$

$$W_j = B_j V_{j+1} \tag{9.22}$$

The finer mesh V_{j+1} could be recovered from its coarser approximation and wavelet coefficients using a pair of synthesis filters P_j and Q_j:

$$V_{j+1} = P_j V_j + Q_j W_j \tag{9.23}$$

where the connection between the analysis and synthesis filters that ensures a perfect reconstruction is given by

$$\left[\frac{A^j}{B^j}\right] = \left[P^j \quad Q^j\right]^{-1} \tag{9.24}$$

This technique works only on regular and semiregular meshes with subdivision connectivity but fails to handle the irregular cases.

Daubechies et al. [33] proposed another technique that can handle irregular 3-D meshes. Their technique is based on mesh simplification and subdivision schemes; the authors use a Burt-Adelson pyramid Scheme [36] as shown in Fig. 9.10. The design of the subdivision scheme is carried out by inserting new values in such a manner that the second-order differences are minimized [33].

Schroder and Swelden [9] have proposed an extension of the lifting scheme to decompose spherical surfaces. In their technique, they divide the mesh into two sets of vertices; the first set contains the new vertices resulted from subdivision, and the second set contains the old vertices that are used to determine the new values. Then the new vertices are predicted using interpolation techniques.

Bonneau [35] was the first to introduce multiresolution analysis over nonnested spaces, which are generated by BLaC wavelets. BLaC wavelet is a combination of the

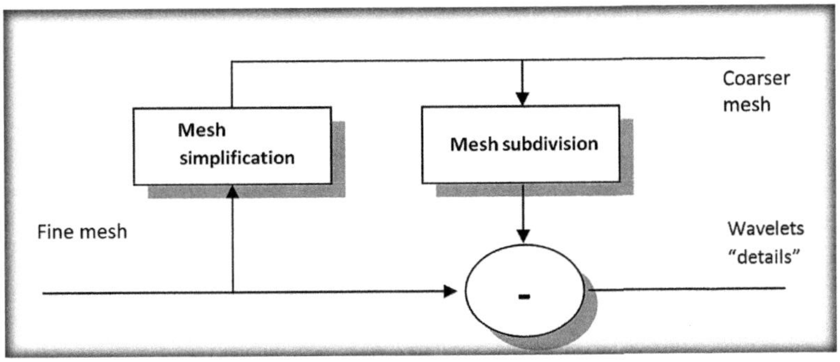

Fig. 9.10 Decomposition of irregular 3-D meshes using a Burt-Adelson pyramid-like scheme.

Haar function with the linear B-spline function. In his work, two major operators were proposed: the smoothing operator to compute the coarse mesh and an error operator to determine the difference between the approximation and the original meshes.

Roy et al. [33] proposed a multiscale analysis for irregular meshes based on split and predict operations. This algorithm consists of three main steps: split, predict, and downsampling. The split operator separates the odd and even vertices, and the odd vertices were defined as a set of independent vertices that are not directly connected by an edge. All the selected odd vertices are removed by a mesh simplification algorithm in the global downsampling stage and then predicted back using the prediction operator that relaxes the curvature based on the Meyer smoothing operator [9, 37].

Valette et al. [38] resented a wavelet-based multiresolution decomposition of irregular surface meshes. The method is essentially based on Lounsbery decompositions. However, the authors introduced a new irregular subdivision scheme. Their algorithm uses a complex simplification technique to define surface patches suitable for the irregular meshes.

Szczensa [39, 40] proposed a new multiresolution analysis for irregular meshes using the lifting scheme, in which she proposed a new prediction operator using Voronoi cells in a local neighborhood.

9.6 Free-form surface filtering using lifting wavelets

Extending the lifting scheme form profile and areal regular cases into 3-D irregular meshes is very challenging. In this section, all the different blocks required to build the lifting scheme on 3-D meshes are given in detail. The framework of a generalized lifting scheme for 3-D meshes is represented in Fig. 9.11. Fig. 9.11A shows the mesh decomposition stage; an input mesh is decomposed into a coarser mesh (wavelet approximation) and details (wavelet details) by splitting the mesh vertices into two groups: evens and odds. Odd vertices are used to update the even vertices. Even vertices are chosen to rebuild the coarser mesh that approximates the original mesh. The odd vertices are subsequently removed. The updated even vertices are used to predict the odd vertices, and then the detail coefficients are calculated as the difference between the prediction and the original odd vertices.

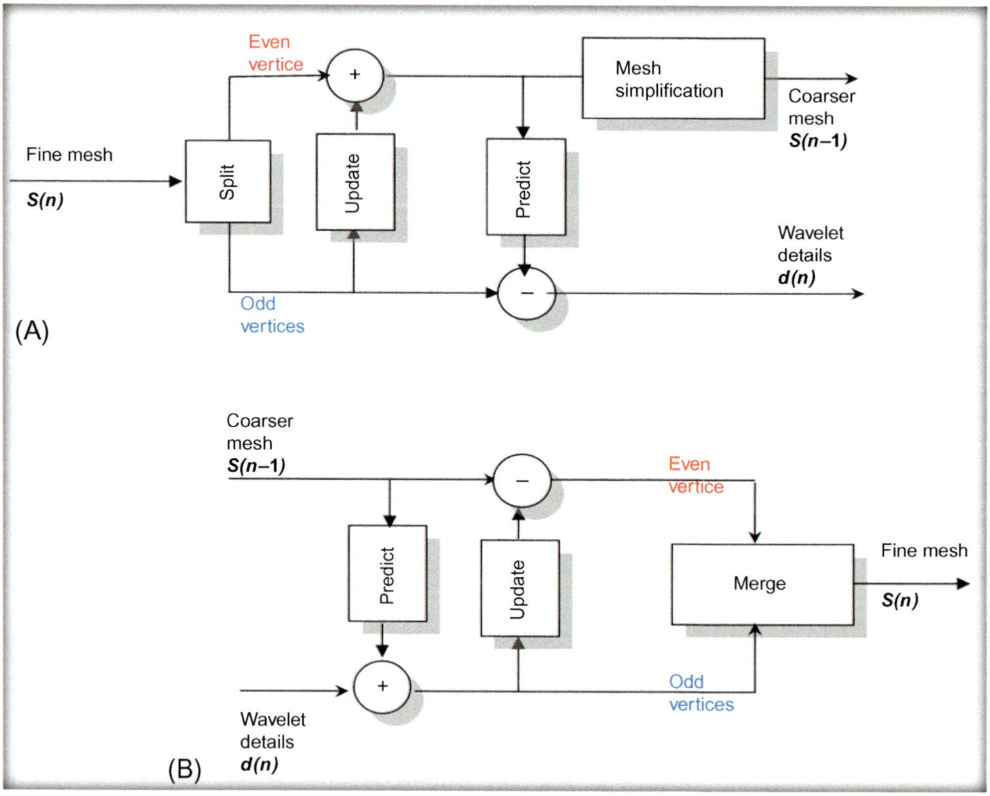

Fig. 9.11 The generalized lifting scheme on 3-D meshes. (A) Mesh decomposition and (B) mesh reconstruction.

Fig. 9.11B shows how to reconstruct the original mesh using its approximation and details. All of these different blocks are explained in the following subsections.

9.6.1 Split operator

The first stage in building the lifting scheme is to split the input data into even and odd. In the case of profile input, this task is trivial, but not for 3-D meshes.

One important point to note is that in the profile case, each odd index is surrounded by even indices. This observation is kept true in the proposed algorithm, so all odd vertices have to be surrounded by even vertices, and no two odd vertices can share an edge. Even vertices can be adjacent to each other and form edges in the mesh. The output of the split operator can be mathematically described as

$$S(M^j) = \begin{cases} {v_{\text{odd}}}^j \\ N({v_{\text{odd}}}^j) \end{cases} \tag{9.25}$$

where M^j represents an input mesh at level j. ${v_{\text{odd}}}^j$ is the set of odd vertices at level j, and $N({v_{\text{odd}}}^j)$ is the set of the even vertices that represent one-ring neighborhood of the odd vertices.

Different methods can be used to select the odd vertices, and the quality of the output coarser mesh depends entirely on the selected odd vertices. Better selection algorithms will produce better approximations. Three different split selection operators are implemented: random, shortest edge, and quadric error metric (QEM) split operators.

9.6.1.1 Random split operator

In random split an initial vertex is randomly selected to be odd, and then, all its neighbors are set to be even; then a new unprocessed vertex is selected to be odd, and all of the neighbors are even. This process ends when no more vertices can be selected.

9.6.1.2 Shortest-edge split operator

The second split operator is based on the shortest edges in the mesh. Initially the length of all edges is calculated and then sorted in ascending order in a list. One vertex of the shortest edge is randomly selected to be odd, and all the adjacent vertices are locked to be even. In this method, it does not matter which of the two vertices is odd and which is even. Then a second shortest edge is selected, and if one of its vertices are not processed yet, (neither even nor odd), then that vertex is selected to be odd, and all the adjacent vertices are set to be even and so on. The algorithm continues until all edges have been processed.

9.6.1.3 Quadric error metric (QEM) split operator

The third splitting algorithm is based on the quadric error metrics that were originally proposed by Garland and Hechbert [41] to simplify triangle meshes with high accuracy. The algorithm uses iterative vertex-pair contraction to simplify a surface and maintain a geometric error approximation of the triangular meshes using the quadric matrices. These vertex pairs are used to identify the odd vertices in our splitting module.

Vertex-pair contraction is carried out iteratively based on the cost of the contraction. Small cost contractions are performed first keeping higher costs to the end. We summarize the algorithm of calculating the cost of the contraction using the following steps. Readers are referred to the Garland paper [41] for more details.

1. For each vertex in the mesh, calculate the error quadric (Q) matrix using the following equation:

$$Q(v) = \sum_{p \epsilon faces(v)} K_p \tag{9.26}$$

where

$$K_p = \begin{bmatrix} a^2 & ab & ac & ad \\ ab & b^2 & bc & bd \\ ac & bc & c^2 & cd \\ ad & bd & cd & d^2 \end{bmatrix} \tag{9.27}$$

and

[a b c d] represents the plane defined by the equation $ax+by+cz+d=0$ where $a^2+b^2+c^2=1$.

faces(v) are the set of faces that share the vertex v. Each facet of these faces is part of the plane defined by the coefficients *[a b c d]*.

2. For each edge $e_{i,j}$ or $(v_i \rightarrow v_j)$ in the mesh, calculate the contraction cost by

$$E(e_{i,j}) = \min \begin{cases} v_i.(Q_i + Q_j).v_i^T \\ v_j.(Q_i + Q_j)v_j^T \end{cases} \tag{9.28}$$

where

$E(e_{i,j})$is the cost of contracting the edge $e_{i,j}$, v_i^T is the transpose of v_i, and $v_i=[x_i y_i z_i\ 1]$. Note that if the cost using the vertex v_i is less than the cost obtained by using v_j, then v_i is an even vertex, and v_j is an odd vertex that has to be removed.

3. Sort the edges according to their costs and start selecting the odd vertices based on the criteria described in step 2.

Similar to the shortest edge method, this method not only selects the odd vertices but also chooses the even partners that will be needed in mesh simplification algorithms as will be described later in this chapter.

9.6.2 Prediction operator

The design of the prediction operator plays a key role in surface filtration using the lifting scheme. It has to predict the properties of the odd vertices using the even vertices. In our application, these properties could be the vertex's position or the vertex texture, the residual normal distance between the nominal and the measured surfaces at that vertex.

In the split operator, odd vertices are chosen so that each odd vertex is surrounded by even neighbors. Therefore the prediction operator depends on the even-ring neighborhood of the odd vertex.

Traditionally the predicted odd value is a weighted summation of the values of its even neighbors. Many algorithms have been proposed to design these weights, and the cubic spline prediction operator is one of the common methods to calculate these weights in a traditional lifting scheme.

The predicted odd value is calculated using one-ring even neighbors by the equation

$$\mathcal{P}f(v_i) = \sum_{j \epsilon \mathcal{N}_1(v_i)} w_{i,j}.f(v_j) \tag{9.29}$$

where $f(v_i)$ could be any function or attribute defined over the vertex v_i; this function in our application is the surface texture.

Designing the weights is very important, and different weights can significantly improve the filtration process. These weights could be calculated using different methods depending on the application. The simplest algorithm to calculate these weights is by using equal weights for all the surrounding neighbors:

$$w_{i,j} = \frac{1}{\mathcal{NN}} \tag{9.30}$$

where $\mathcal{NN}$ is the number of even neighbors in the even-ring neighborhood.

This method is very primitive and will produce large errors between the predicted values and the original values, and therefore better prediction methods are required.

9.6.2.1 Laplacian prediction operator

Laplacian operator is widely used across many different disciplines, and researchers show that it is a very useful prediction operator for triangular meshes as proposed by Roy et al. [42, 43]. In the case of triangular meshes, the Laplacian operator is defined by the discrete Laplace-Beltrami operator, which is given by

$$w_{i,j} = \frac{\cot\left(\alpha_{i,j}\right) + \cot\left(\beta_{i,j}\right)}{\Sigma_{l \in \mathcal{N}_1(v_i)}\left\{\cot\left(\alpha_{i,l}\right) + \cot\left(\beta_{i,l}\right)\right\}} \tag{9.31}$$

where $\alpha_{i,j}$ and $\beta_{i,j}$ are the angles opposite to the edge $e_{i,j}$ as shown in the Fig. 9.12.

9.6.2.2 Gaussian prediction operator

Gaussian filters play an important role in filtering different kinds of surfaces. The simplicity of the algorithm, ease of implementation, and the robustness of the results make this type of filtration the first choice for filtration in many applications. The linear Gaussian filter is very popular in surface characterization, it has been widely used among

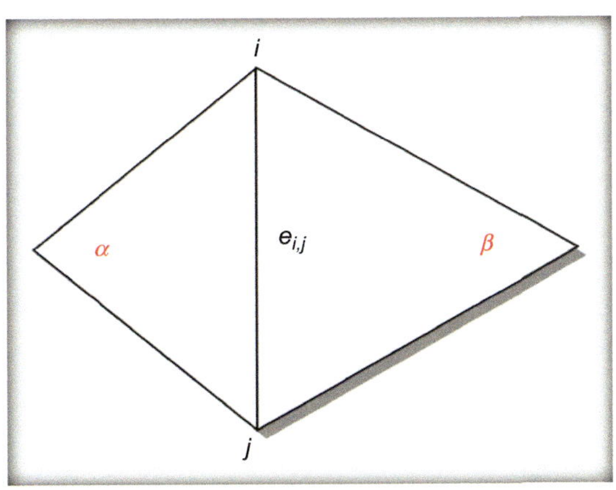

Fig. 9.12 The Laplace-Beltrami angles α and β associated with the edge $e_{i,j}$.

researchers, and it has become an industrial filtration standard. Gaussian filters can be applied to the input surface by convolving the measured surface with a Gaussian weighting function. The Gaussian weighting function has the form of a bell-shaped curve as defined by the equation

$$g(x) = \left(\frac{1}{\delta\lambda_c}\right) \exp\left[-\pi\left(\frac{x}{\delta\lambda_c}\right)2\right] \tag{9.32}$$

where δ is given by $\delta = \sqrt{(\ln(2/\pi))}$ and λ_c is the cutoff wavelength.

In case of the lifting wavelets, the weights of the prediction operator could be calculated using the Gaussian function by

$$w_{i,j} = \left(\frac{1}{\delta\lambda_c}\right) \exp\left[-\pi\left(\frac{x_{i,j}}{\delta\lambda_c}\right)2\right] \tag{9.33}$$

where $w(i,j)$ is the weight of the Gaussian kernel for the vertex v_j for a central vertex v_i. $x_{i,j}$ is the Euclidean distance between the two vertices v_i and v_j:

$$\delta = \sqrt{\ln\left(\frac{2}{\pi}\right)} \approx .4697$$

λ_c is the nesting index, corresponding to the cutoff wavelength in regular cases, which determines the smoothness of the output. λ_c is constant, and it is set to 0.01 in this chapter.

9.6.3 Update operator

Traditionally the update operator preserves some features from the input higher-resolution (fine) signal to the output lower-resolution (coarse) signal. For example, the update operator for designing a Haar wavelet using the lifting scheme ensures that the average of the fine input signal is equal to the average of the coarse output signal. Therefore the update operator in the triangular mesh case has also to preserve some important features in all approximated meshes at all different decomposition levels.

In this chapter the authors choose to preserve the average value of the vertex ring before and after removing the odd vertex; thus the update is given by

$$U[(f(v_{even})] = Ring_{Const} + f(v_{even}) \tag{9.34}$$

where

$$Ring_{Const} = \frac{\mathcal{NN} \cdot (f(v_{odd}) - \sum_{j \in N(v_{odd})} (f(v_j))}{\mathcal{NN} \cdot (\mathcal{NN} + 1)}. \tag{9.35}$$

$\mathcal{NN}$ is the number of even neighbors in the even-ring neighborhood. $\sum_{j \in N(v_{\text{odd}})} (f(v_j))$ is the summation of the function values of the even neighbors in the even-ring neighborhood.

9.6.4 Mesh simplification (the approximation)

Mesh simplification or downsampling is the process of reducing the number of faces, edges, and vertices while attempting to preserve the overall geometry, shape, and boundaries as much as possible. It is the step that produces the approximated coarser meshes (wavelet approximations), which can be used as an input for further decomposition levels.

Many algorithms have been proposed for mesh simplification as shown in Fig. 9.13. These algorithms can be roughly divided into three major groups: the first group simplifies the mesh by selecting an edge to be collapsed, the second defines a face to be removed, and the third type relies on selecting vertices to be removed as shown in the figure. The half-edge collapse does not introduce a new vertex position, but rather, it simplifies (subsample) the mesh using the same vertices locations; hence, half-edge collapse is adopted to perform the simplification step in implementation of the lifting scheme over surfaces represented by triangular meshes.

9.6.5 The details (wavelet coefficients)

The detail coefficients are defined as the Euclidean distant or a vector between the odd vertex properties and its prediction. These coefficients must have all the information needed to perform a perfect reconstruction of the original mesh.

The details are stored as a number of records; the number of detail records is equal to the number of odd vertices selected by the split operator. Each record preserves all the information that is required for the reconstruction; this information includes the detail vectors and all the edges and topological information before removing the odd vertex.

9.6.6 Merge operator

The merge operator, mesh upsampling, defines how to reinsert new vertices into the mesh. The insertion of a new vertex into the mesh is controlled by the detail records of that vertex. The mesh topology before removing the odd vertex must be preserved and perfectly reconstructed by the merge operator. In this chapter the merge operator was carried out using the vertex split algorithm adopted from the progressive mesh (PM) procedure [44].

9.6.7 Filtration algorithm

The next step is to setup a filtration framework based on the lifting. Fig. 9.14 shows the filtration framework that has been used. As shown in the figure, the input surface mesh is

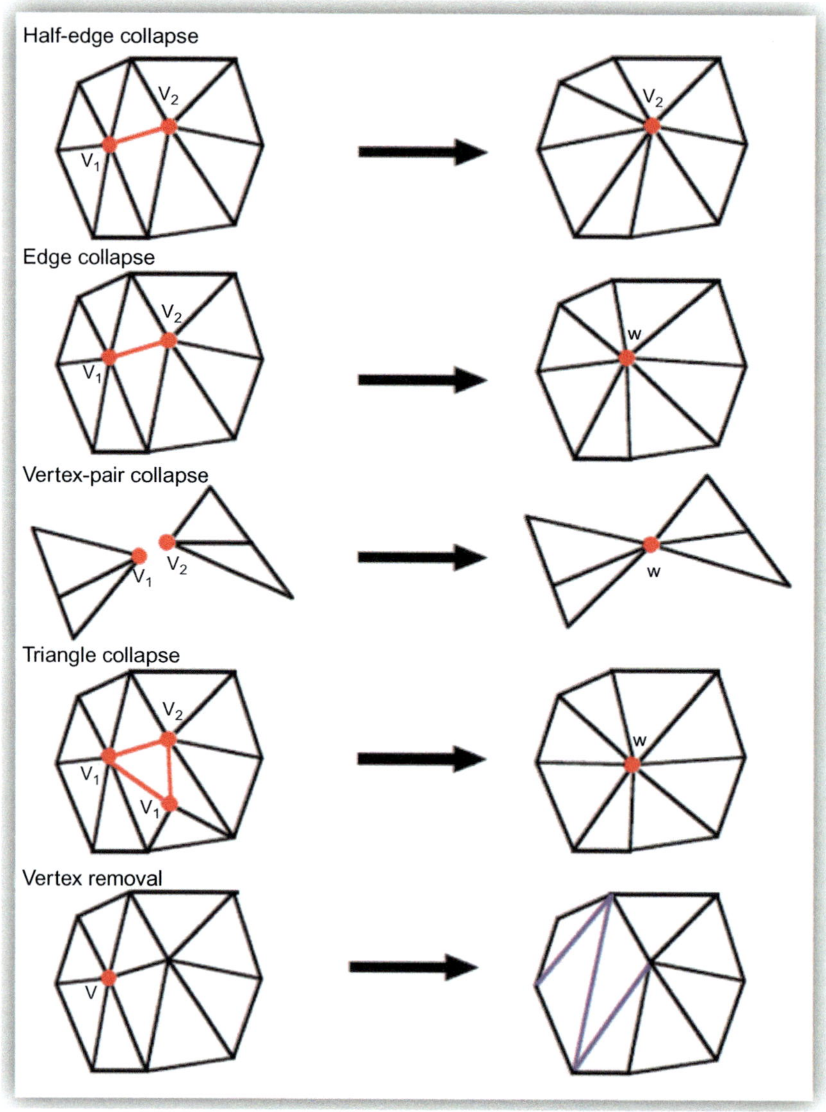

Fig. 9.13 Different mesh simplification algorithms.

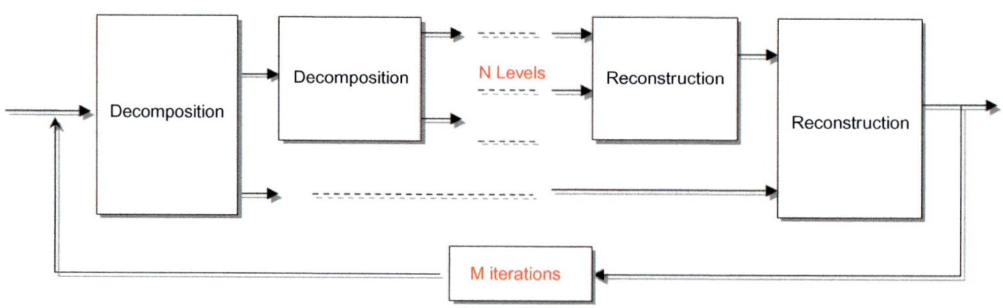

Fig. 9.14 Surface filtration algorithm using the lifting scheme.

decomposed into N-Levels, and the detail coefficients are filtered out and set to zero before the surface is reconstructed again. More decomposition levels mean a smoother output surface. However, the number of decomposition levels is limited, and no further decomposition is possible once we reach the base mesh that could not be simplified anymore. Therefore more filtration could be achieved by reapplying the output-filtered surface as an input to the filtration system and repeat the process as many times as needed, which is represented by the M-iterations in Fig. 9.14.

9.7 Multiscale free-form surface decomposition using a Laplacian mesh relaxation

Filtering a regular profile or areal signal in a time/spatial domain is usually carried out by defining a weighted profile or areal window, usually referred to as the kernel. This weighted kernel is used to calculate a weighted summation that represents a new functional or attribute value at a certain data position, and then, it is translated across the signal to calculate the new values at other positions. The weights inside the kernel are assigned using many different methods depending on the application; for example, the weights could represent a Gaussian filter, Laplacian coefficients, or Sobel coefficients.

In an analogy with regular cases, mesh relaxation could be defined as the extension of the weighted window (kernel) to the irregular cases. In other words, mesh relaxation is the process of calculating a new function value at a particular position of the mesh based on a weighted kernel defined by the mesh relaxation scheme, as shown in Fig. 9.15. In Fig. 9.15A the central vertex of the mesh that has the value of 6 is to be relaxed using the weighted kernel that is represented by smaller circles; applying this kernel will result in new relaxed value of 5.7 that would replace the original value as shown in Fig. 9.15B.

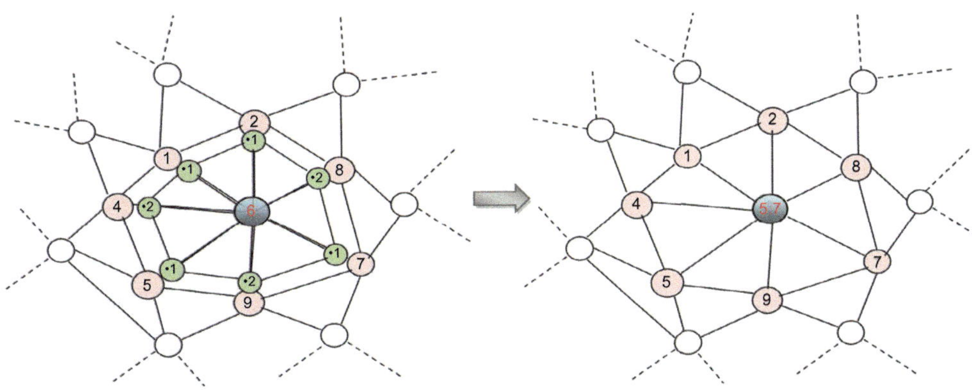

(A) (B)

Fig. 9.15 Demonstration of mesh relaxation using a mesh kernel: (A) the original mesh and the kernel associated with the center vertex and (B) the relaxed value of the center vertex calculated using the kernel.

Unlike the regular cases the kernel in the irregular data sets has to be adaptive in the size (number of kernel nodes) and the weights. These two parameters have to be changeable depending on geometry and/or connectivity between the data points at each vertex of the mesh.

Mesh relaxation plays an important role in building a multiresolution analysis for 3-D triangular meshes. Guskov et al. [45] were the first to propose a multiresolution signal processing tool for 3-D triangular meshes with a mesh relaxation operator being the central ingredient in the algorithm [45]. They proposed a nonuniform relaxation operator that minimizes the second-order differences (SOD) at every edge of the mesh.

Roy et al. [43] proposed another mesh relaxation algorithm to build a multiresolution analysis for irregular 3-D meshes based on a nonuniform relaxation operator that minimizes the local curvature of the mesh [43]. Roy's curvature relaxation was inspired by the work of Meyer et al. [38] on the local and accurate discrete differential geometry operators that were previously applied for surface denoising.

The two relaxation schemes were compared [43], and comparisons showed that the two relaxation operators give similar smoothing results. However, the curvature relaxation is four times faster than the SOD operator [42]. The SOD relaxation uses the two-ring neighborhood to compute a new relaxed value of the mesh; the curvature relaxation requires only the one-ring neighborhood. The one-ring neighborhood of a vertex ν is defined as the set of vertices that are only one edge away from that vertex ν. The two-ring neighborhood, on the other hand, is defined as the set of vertices that are less than two edges away from the vertex ν.

In this chapter the curvature relaxation is adopted to build our multiscale filtering technique as will be explained in the following subsections.

9.7.1 Laplacian mesh relaxation

The Laplacian relaxation operator defines the new function or attribute value of a particular vertex of the mesh based on the discrete Laplace-Beltrami operator that minimizes the local curvature of that mesh at a particular position [Meyer 2002].

The relaxed function value $\mathcal{R}\mathrm{f}(v_i)$ of the vertex v_i using the Laplacian relaxation operator is given by

$$\mathcal{R}\mathrm{f}(v_i) = \sum_{j \in \mathcal{N}_1(v_i)} \mathcal{W}_{i,j}.f(v_j) \tag{9.36}$$

where $f(v_i)$ could be any function or attribute defined over the vertex v_i and in our case, it represents the surface texture; $\mathcal{N}_1(v_i)$ represents the one-ring neighborhood of the vertex v_i; $\mathcal{W}_{i,j}$ represent the weights of the kernel associated with the vertex v_i and are defined by the Laplace-Beltrami operator as shown previously in Eq. (9.31), and it is given by

$$w_{i,j} = \frac{\cot(\alpha_{i,j}) + \cot(\beta_{i,j})}{\sum_{l \in \mathcal{N}_1(v_i)} \{\cot(\alpha_{i,l}) + \cot(\beta_{i,l})\}}.$$

The weights $w_{i,j}$ of the relaxation operator minimizes the curvature energy of an edge $e_{i,j}$ where $\alpha_{i,j}$ and $\beta_{i,l}$ are the angles opposite to the edge $e_{i,j}$ as shown in the Fig. 9.12. Once the weights are calculated, the new relaxed value can be computed as demonstrated in the example shown in the Fig. 9.15.

9.7.2 Multiscale decomposition using a relaxation scheme

The manufacture of a surface usually leaves multiscale features on the surface that is a fingerprint of the manufacturing process. A traditionally manufactured surface will contain three different scales of irregularities that together represent the surface as discussed earlier in this chapter: the short-scale texture (roughness) that includes tool marks and machine vibrations, the middle-scale texture (waviness) that results from the low variation of the machine, and the long-scale (form) that represents the nominal surface form. Surface filtering and decomposing into different scales are very important to characterize and parameterize that surface. Decomposing a free-form surface can be achieved using multiple levels of the mesh curvature relaxation scheme described in the previous section. Each level of decomposition contains a mesh relaxation operator that smooths the mesh and a difference operator that calculates the difference between the smoothed and original mesh as shown in Fig. 9.16. The mesh relaxation operator acts like a low-pass smoothing filter, and the difference operator acts like a high-pass filter that makes this technique similar in concept to the wavelet transform.

Decomposing a free-form surface is carried out using multiple levels of mesh relaxation as shown in Fig. 9.17. The input of the first decomposition level of this decomposition ladder is the original surface, represented by a triangular mesh; and the two outputs

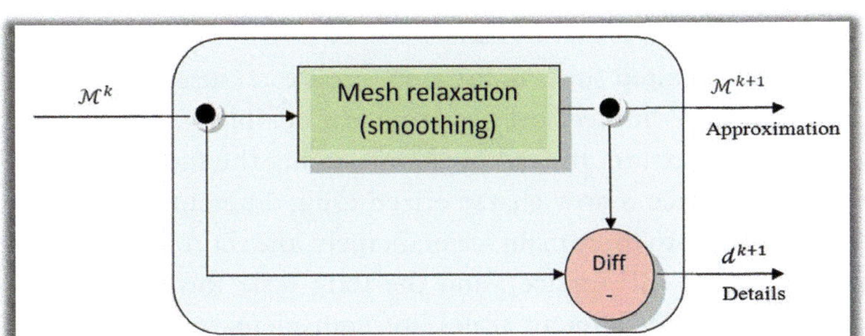

Fig. 9.16 One decomposition level of a mesh relaxation scheme.

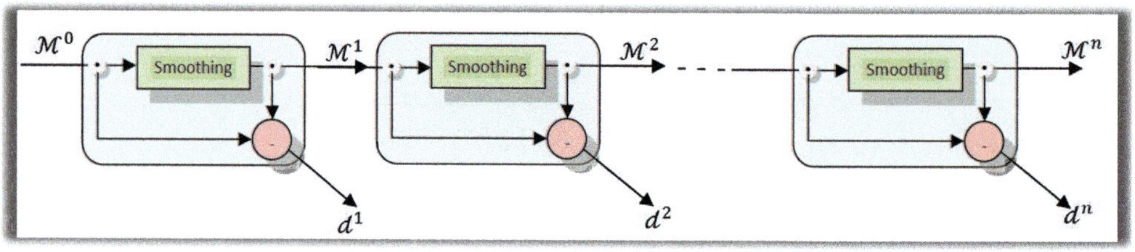

Fig. 9.17 Multiple decomposition levels of a mesh relaxation scheme.

are a smoother version of the input mesh (the approximation) and the detail coefficients that represent the difference between the input and the approximated mesh (the details). The details are necessary to reconstruct the input surface from its approximation. Each subsequent level consists of further splitting of the smooth (approximated) surface into new two components, namely, the approximation that will be even smoother than the input and the details that contain information about the difference between the two levels (the input and output levels) as shown in the Fig. 9.17. The approximation output at each level becomes the input of the next level and so on. Each level of this decomposition ladder represents a different scale of the original input surface.

The original input mesh can be perfectly reconstructed by adding all of the details to the final approximation output of the final decomposition level $\mathcal{M}^n$. The reconstruction can be described mathematically as

$$\mathcal{M}^0 = \mathcal{M}^n + \sum_{i=0}^{n} d^i. \tag{9.37}$$

To obtain different scale components of the surface (e.g., roughness, waviness, and form), all that is necessary is to set the detail coefficients of some levels to zero and then reconstruct the surface using Eq. (9.37). By setting the details of some levels to zero, the reconstructed surface will consist of only the required scales and will filter out all other scales as will be explained in the next subsection.

9.7.3 Scale-limited surfaces using mesh relaxation

The concept of scale-limited surfaces for areal surface texture has recently been introduced [20, 31, 46]. Scale-limited surfaces provide a flexible way to identify various different scales of surface texture. Instead of characterizing the surface using the traditional decomposition, the surface is now characterized using different filters and operators that decompose the surface into three main scales, namely, the short-scale surface (SL surface), the middle-scale surface (SF surface), and the long-scale form surface (F-operator) as shown in Fig. 9.18; these different scales are equivalent to the roughness, waviness, and the form in the traditional definition.

Obtaining different scale-limited surfaces using the mesh relaxation scheme is explained with the help of Fig. 9.19. Fig. 9.19A shows a spectral-like decomposition

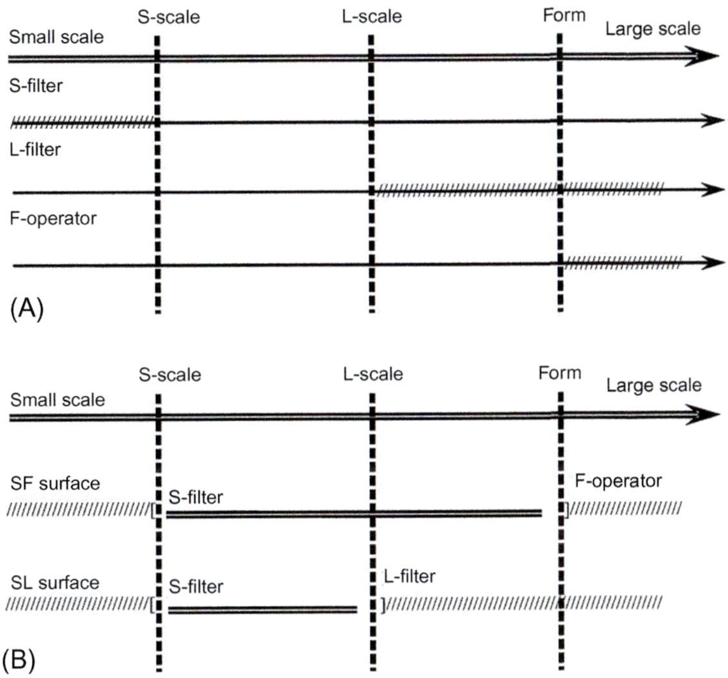

Fig. 9.18 The concept of the scale-limited surfaces: (A) filters and operators used in surface texture and (B) scale-limited surface used in surface texture.

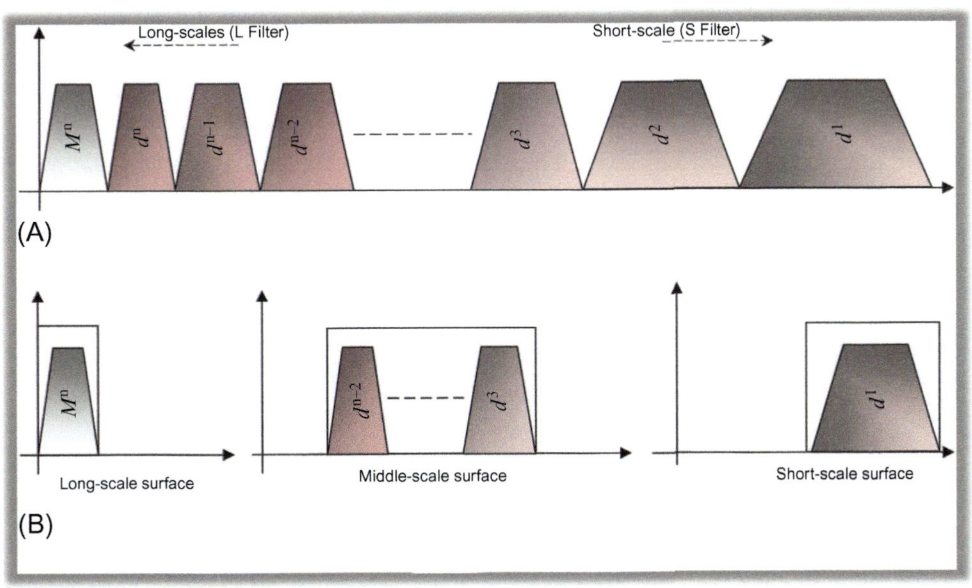

Fig. 9.19 Spectral-like analysis of a free-form surface using mesh relaxation scheme: (A) spectral-like analysis of a free-form surface and (B) the extraction of different scale-limited surfaces: (right) the long-scale surface extracted using low-pass filtering of the spectrum, (middle) the middle-scale surface extracted by band-pass filtering, and (left) the short-scale surface extracted using high-pass filtering.

of a free-form surface using n decompostion levels. As shown in the figure, the surface can be represented using the smoothest approxiamtion $\mathcal{M}^n$ and a series of the detail coefficients from all decomposition levels as explained mathematically in the Eq. (9.37). Lower indices in the figure represent smaller scales, and higher indices represent longer scales.

The long-scale form surface can be extracted by selecting the smoothest approximation $\mathcal{M}^n$ and the details of some higher indexed levels and filtering out all other levels. This operation is equivalent to a low-pass filtering of the given spectrum. The middle-scale surface can be extracted by choosing the details of some of the middle levels that are equivalent to a band-pass filtering of the spectrum. The short-scale surface is extracted by choosing the details of lower indexed levels that are considered as a high-pass filtering of the spectrum. The extraction of these different scale-limited surfaces is shown in Fig. 9.19B.

9.8 Case studies

The aforementioned lifting wavelet and multiscale relaxation algorithms have been implemented and tested to filter and decompose measured and computer-generated surfaces represented by 3-D irregular triangular meshes, and the results are shown, discussed, and compared in this section.

9.8.1 Computer-generated surfaces

To test the performance of the proposed algorithms to decompose different types of free-form surfaces, two computer-generated free-form surfaces have been used. These surfaces are designed to cover a wide range of free-form non-Euclidean surfaces with different topological types. The first surface is a saddle-shaped surface that is considered to be a typical example of non-Euclidean surface with negative curvature. The second surface is a sphere that represents a positive curvature non-Euclidean geometry. These two surfaces are shown in Fig. 9.20. These surfaces are represented by triangular meshes,

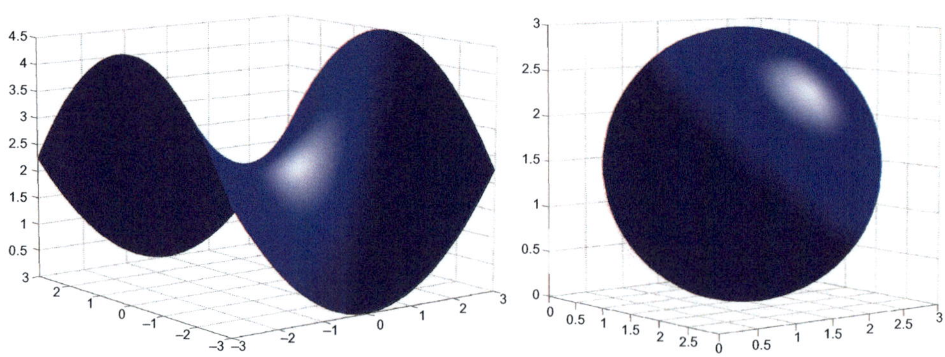

Fig. 9.20 Computer-generated free-form non-Euclidean surfaces with (A) negative curvature surface "saddle-shaped surface" and (B) positive curvature surface "spherical surface."

Table 9.1 Mesh details of the computer-generated surfaces.

Surface	Faces	Vertices
Saddle shaped	28,800	14,641
Sphere	4608	2306

and the number of faces and vertices for each of these surfaces is shown in Table 9.1. An artificial Gaussian noise has been added to these surfaces to represent the texture that is required to be filtered out using the lifting wavelet scheme and the multiscale decomposition algorithms. The results are discussed for each of the algorithms in the following subsections.

9.8.1.1 Lifting wavelets filtration results

The results of applying the lifting wavelet algorithm on the saddle-shaped surface are shown in Fig. 9.21. The original surface is shown in Fig. 9.21A. Fig. 9.21B–F show the results using 1, 4, 8, 16, and 24 decomposition levels, respectively. Fig. 9.21G–I show the results of using 24 decomposition levels and 2, 3, and 4 iterations, respectively.

Surface texture has been encoded into the figure using a color map. Vertices with the highest texture value are represented using the red color; however, vertices with the lowest texture value are represented using the blue color. Other vertices and other texture values are assigned to different colors depending on their texture as shown Fig. 9.21.

The proposed algorithm has also been applied to filter the spherical surface, and the results are shown in Fig. 9.22.

Fig. 9.23 shows the filtering results of the lifting wavelets using the Laplacian prediction operator and the Gaussian prediction operator, as explained earlier in this chapter, of the spherical surface. In this particular example the Gaussian prediction operator produces smoother results at the same decomposition levels.

Figs. 9.24 and 9.25 show the output of the mesh simplification or mesh downsampling operation using the half-edge collapse algorithm as discussed earlier in this chapter. In these figures the coarser meshes (wavelet approximations) after 1, 4, 8, 16, and 24 decomposition levels, for the saddle and sphere surfaces, are shown in figures (A)–(F), respectively.

Fig. 9.26 shows the output of the split operator using the QEM algorithm as explained in the preceding text. The figure demonstrates how the odd and even vertices are distributed, and number of even and odd vertices are shown at the first, second, fifth, eighth, 16th, and 24th levels of the decomposition. The odd vertices are represented by blue dots, while the even vertices are the red dots in the figure. In this algorithm, we do not allow any boundary vertex to be an odd vertex; this ensures that all odd vertices have a complete ring of neighbors and thus will produce more accurate predictions. This

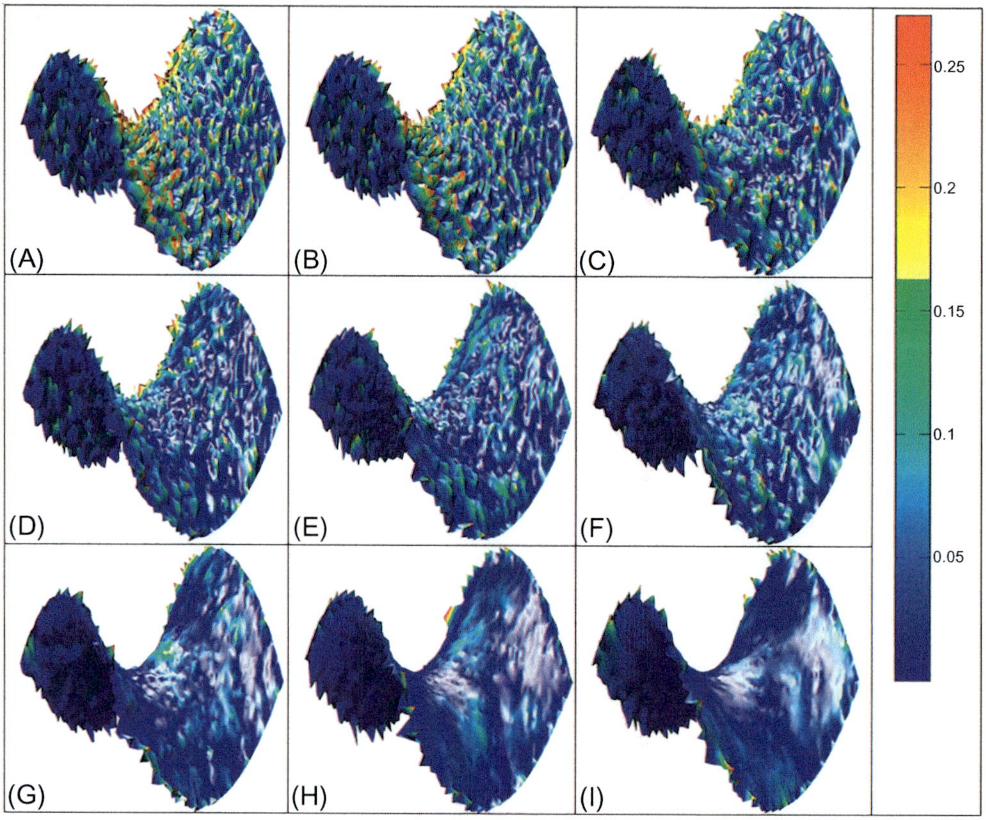

Fig. 9.21 Texture filtration results of the saddle-shaped surface: (A) the original textured surface. (B)–(F) The filtration results using 1, 4, 8, 16, and 24 decomposition levels, respectively. (G)–(I) The filtration results using 24 decomposition levels and 2, 3, and 4 iterations, respectively.

improves the filtration process and also eliminates the boundary problem that will occur when predicting at the boundary.

Fig. 9.27 compares the three different split operators, discussed earlier in the chapter, that is, the random, shortest edge, and QEM algorithms. The figure shows that the random and the shortest edges give higher odd vertices percentage in the first decomposition levels but quickly drop and give very low percentage at higher levels. On the other hand the QEM has the smallest percentage in the beginning, but it has slow decay and therefore covers more decomposition levels. This means that the QEM split operator gives the best mesh approximation output, and the mesh gradually becomes coarser at each decomposition level as could be shown in Figs. 9.24–9.26.

9.8.1.2 Laplacian mesh decomposition results

To demonstrate the ability of the multiscale decomposition algorithm to decompose free-form surfaces into different components, an artificial multiscale texture has been

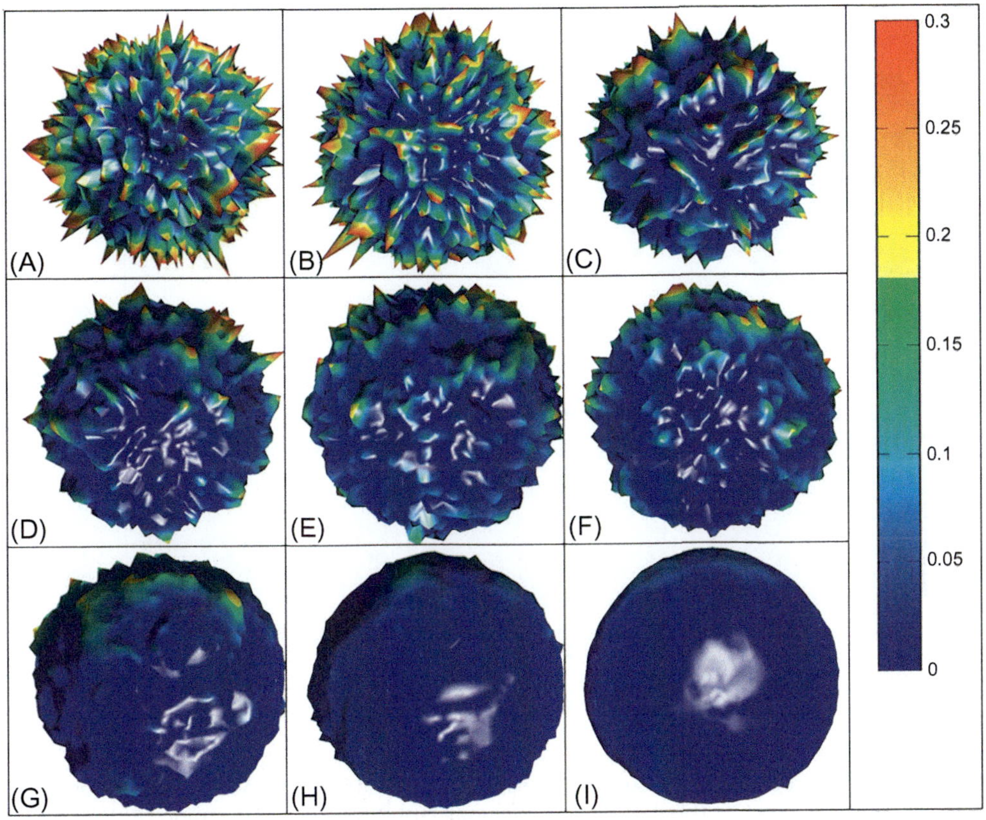

Fig. 9.22 Texture filtration results of the spherical surface: (A) the original textured surface. (B)–(F) The filtration results using 1, 4, 8, 16, and 24 decomposition levels, respectively. (G)–(I) The filtration results using 24 decomposition levels and 2, 3, and 4 iterations, respectively.

imposed onto the computer-generated surfaces. The texture is generated by adding two scale components to the surface's form; the first is a high varying Gaussian noise representing the short-scale texture or surface roughness, and the second is a low varying sinusoidal surface that would represent the middle-scale texture or surface waviness. The results of decomposing these surfaces are shown in Figs. 9.28 and 9.29.

Fig. 9.28 shows the decomposition results of applying the proposed algorithm on the saddle-shaped surface. The original textured surface is shown in Fig. 9.28A. Three different components can be easily distinguished in the textured surface: the nominal surface or form surface, the middle-scale surface or surface waviness, and finally the short-scale surface or the surface roughness. The proposed technique has been applied to decompose this surface into its basic components, and it successfully manages to disintegrate this surface into different components as shown in Fig. 9.28B–D.

In this example, 32 decomposition levels are used to filter the surface into different scales: a short-scale, middle-scale, and long-scale equivalent to roughness, waviness, and

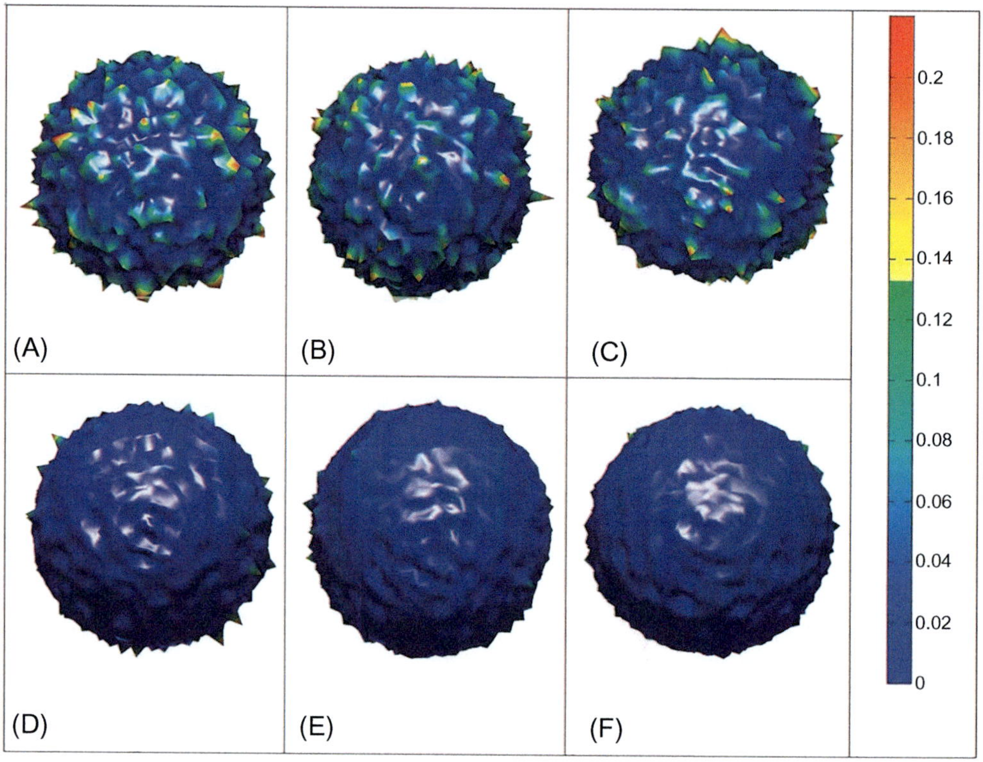

Fig. 9.23 Comparisons between curvature relaxation prediction and Gaussian prediction for the spherical surface. (A)–(C) The results of curvature relaxation filtering using 8, 16, and 24 decomposition levels and (D)–(F) the results of Gaussian prediction using 8, 16, and 24 decomposition levels, respectively.

form, respectively. The short-scale surface or the roughness is obtained using only the detail coefficients of the first two decomposition levels and setting all other detail coefficients of all other levels to zero as discussed earlier in this chapter, whereas the middle-scale surface is obtained using the detail coefficients from the middle decomposition levels, the 3rd to the 30th decomposition levels, and setting all the other detail coefficients in the other decomposition levels to zero. Finally the form error surface is obtained using only the approximation output of the last decomposition levels.

Laplacian mesh decomposition algorithm has also been applied to filter the spherical simulated surface, and the results are shown in Fig. 9.29. The figure shows the results of the decomposition of the spherical surface into short-scale (roughness), middle-scale (waviness), and long-scale (form error) surfaces. As in the previous example, 32 decomposition levels are used; the short-scale surface is resulted from considering only the detail coefficients of the 1st and the 2nd levels, the middle-scale surface is resulted from the

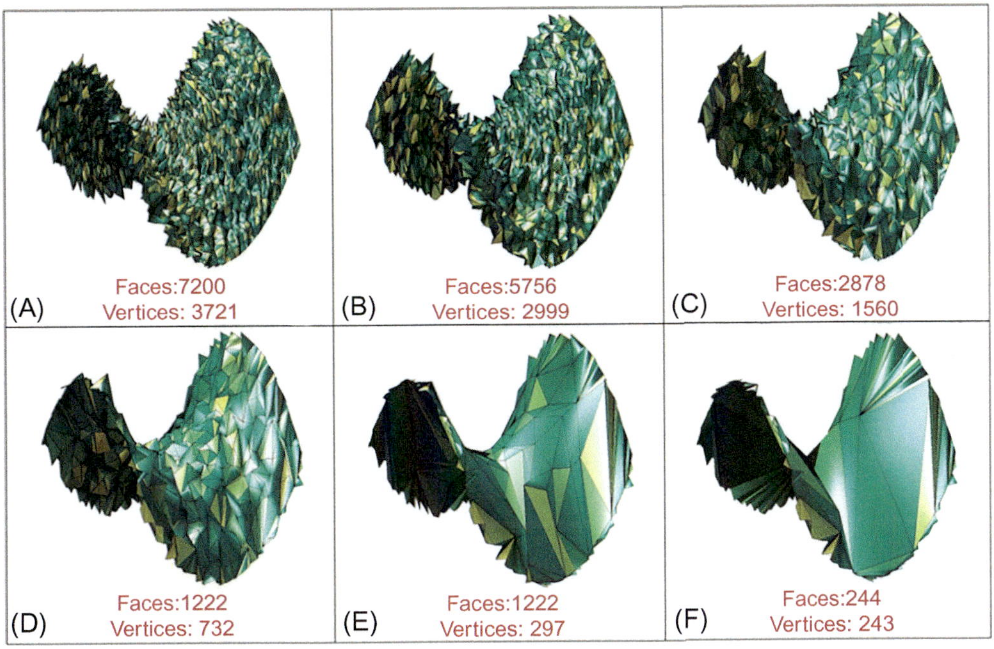

Fig. 9.24 The approximated output meshes (wavelet approximations) for the saddle-shaped surface at different decomposition levels. (A) The original finest input mesh and (B)–(F) the approximated coarse mesh after 1, 4, 8, 16, and 24 decomposition levels, respectively.

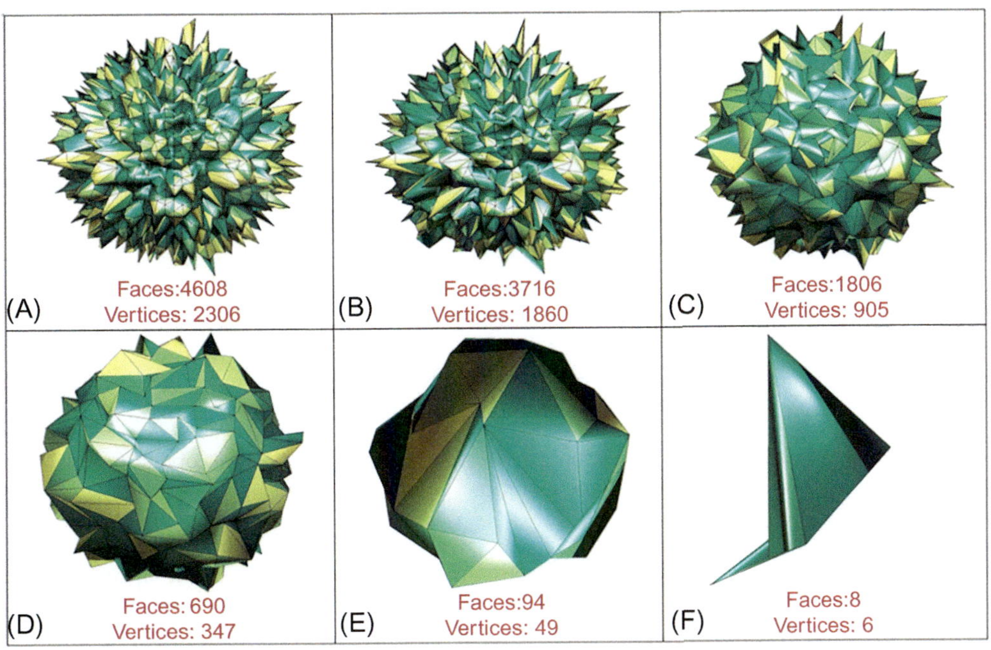

Fig. 9.25 The approximated output meshes (wavelet approximations) for the spherical surface at different decomposition levels. (A) The original finest input mesh and (B)–(F) the approximated coarse mesh after 1, 4, 8, 16, and 24 decomposition levels, respectively.

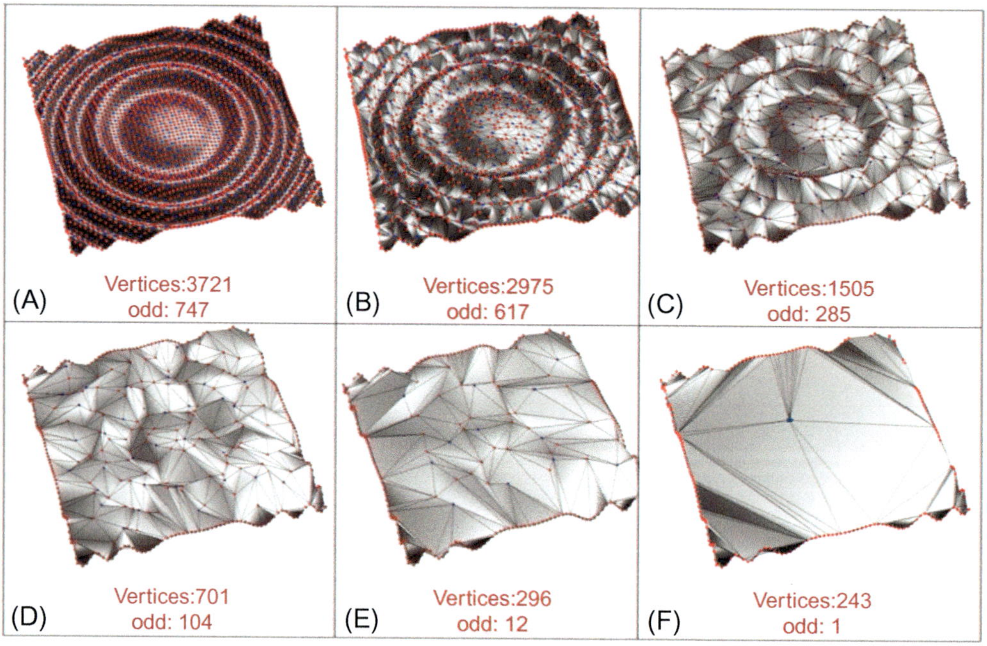

Fig. 9.26 A demonstration of the split operator using the QEM algorithm on computer-generated surface. The distribution of odd vertices at the first, second, fourth, eighth, sixteenth, and twenty-fourth are shown in (A)–(F), respectively. Boundary vertices are not allowed to be odd vertices.

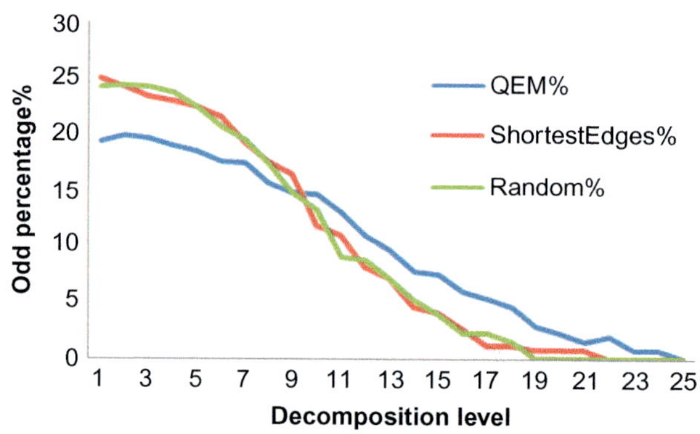

Fig. 9.27 Comparison between the random, shortest edge, and the QEM split operators at different decomposition levels for the saddle-shaped surface.

detail coefficients from the 3rd to the 30th levels, and long-scale is the approximation output of the 32nd level.

The previous simulated examples show that the lifting wavelet and multiscale decomposition algorithms successfully manage to decompose the computer-generated free-form

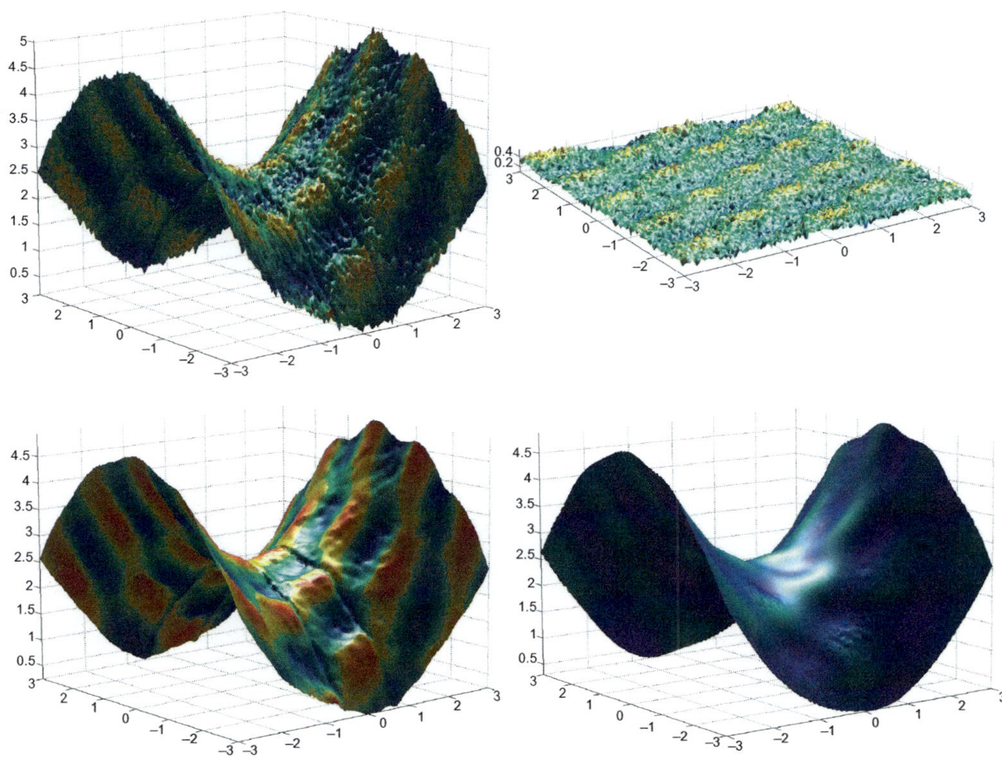

Fig. 9.28 Decomposition results of a saddle-shaped surface using curvature relaxation scheme using 32 decomposition levels: (A) the original textured surface, (B) roughness surface (the short-scale surface), (C) wavy surface (the middle-scale surface), and (D) form surface (the long-scale form surface).

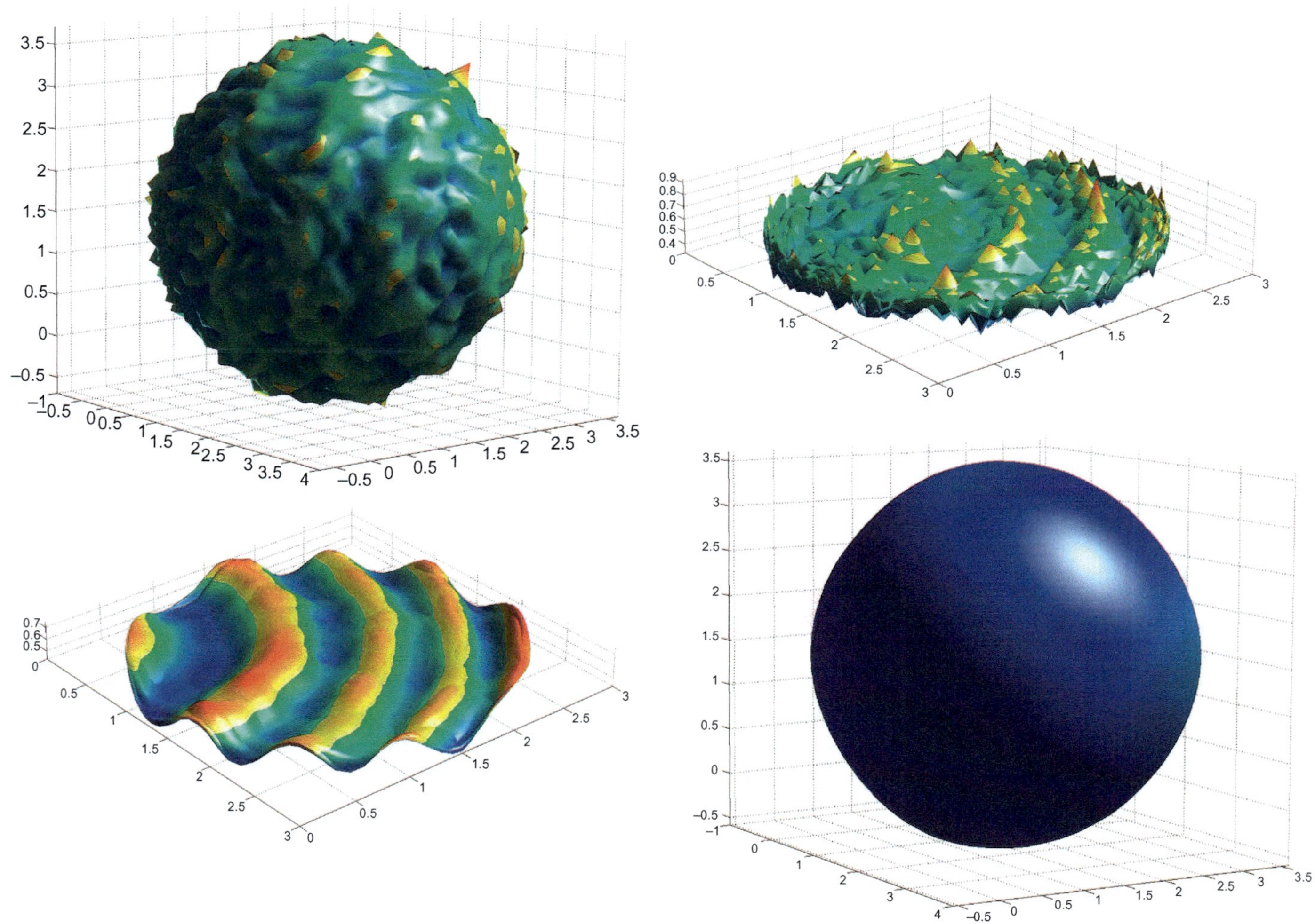

Fig. 9.29 Decomposition results of a sphere surface using curvature relaxation scheme using 32 decomposition levels: (A) the original textured surface, (B) roughness surface (the short-scale surface), (C) wavy surface (the middle-scale surface), and (D) form surface (the long-scale form surface).

surface, represented by 3-D triangular meshes, into different bands and scales. Computer-generated meshes are more likely to be regular or semiregular types of meshes, so it is very important to test the performance of the algorithm on real measured data meshes that are irregular in nature.

9.8.2 Bioengineering surfaces

After the initial application of the aforementioned algorithms to computer-generated free-form surfaces, these algorithms were applied on real surface measurement data. The data were obtained from a coordinate measuring machine (CMM) measurement representing a portion of a hip replacement component. Two surfaces were acquired from the CMM and represented by 3-D triangular meshes; the first surface has 3380 vertices and 6591 faces, and second surface has 7182 vertices and 14,108 faces (triangles) as shown in Table 9.2. These two measured surfaces are shown in Fig. 9.30; we refer to these two surfaces as HipJoint-pt1 and HipJoint-pt2 as shown in Fig. 9.30A and B, respectively. As with the computer-generated surfaces, an extra artificial Gaussian noise is added to these surfaces, and then the surfaces are filtrated using the proposed algorithms.

9.8.2.1 Lifting wavelets filtration results

Figs. 9.31 and 9.32 show the results of applying the lifting technique to filter the surface texture at different decomposition levels and iterations. As with the simulated results, the proposed algorithm successfully smoothed the texture of different scales according to the

Table 9.2 Mesh details of the real measured surfaces.

Surface	Faces	Vertices
HipJoint-Pt1	6591	3380
HipJoint-Pt2	14,108	7182

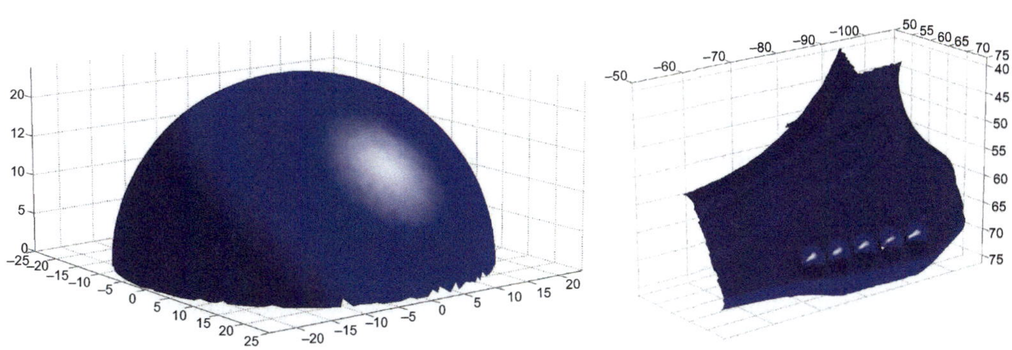

Fig. 9.30 Two bioengineering-measured surfaces obtained with CMM for hip replacement components. We refer to these surfaces as (A) HipJoint-Pt1 and (B) HipJoint-Pt2.

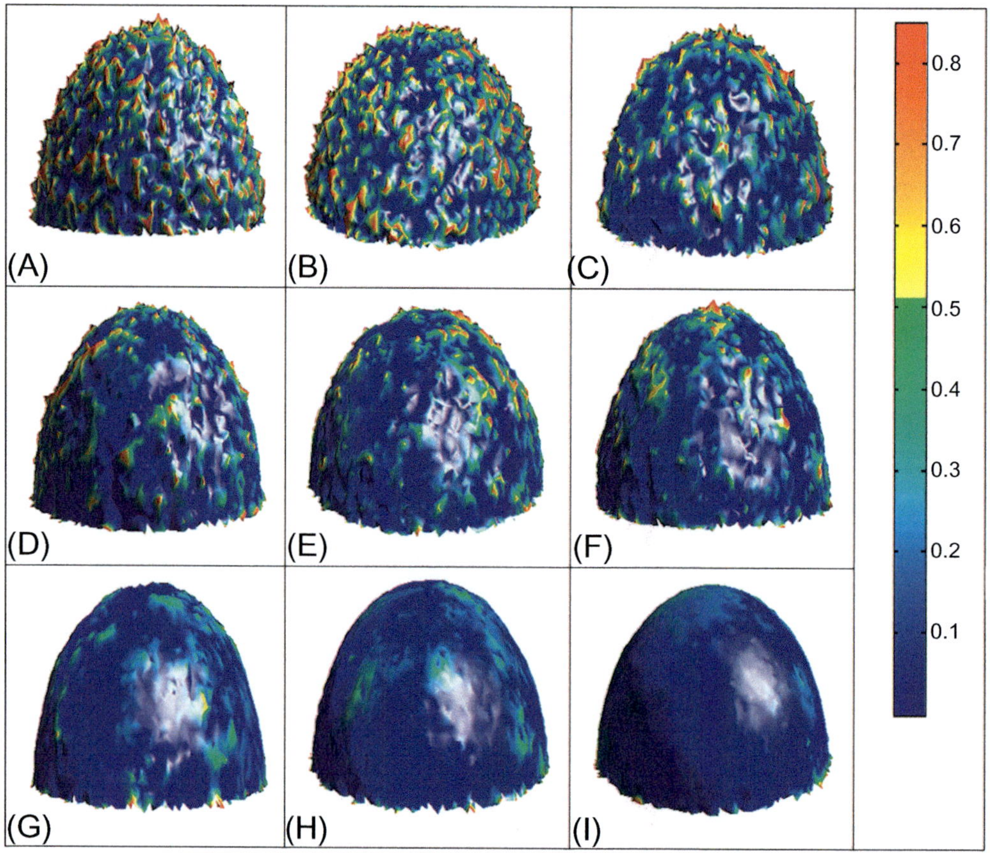

Fig. 9.31 Texture filtration results of the hip-part1 surface: (A) the original textured surface. (B)–(F) The filtration results using 1, 4, 8, 16, and 24 decomposition levels, respectively. (G)–(I) The filtration results using 24 decomposition levels and 2, 3, and 4 iterations, respectively.

decomposition levels and number of iterations. Fig. 9.33 shows the outputs of the mesh simplification process (wavelet approximations) at different decomposition levels for the hip-part1 surface shown in Fig. 9.30A.

9.8.2.2 Laplacian mesh decomposition results

Figs. 9.34 and 9.35 show the results of applying the Laplacian mesh decomposition algorithm to decompose the HipJoint-Pt1 and HipJoint-Pt2 surfaces, respectively, into different scales. The original textured surfaces are shown in Figs. 9.34A and 9.35A. As with the simulated surfaces, 32 decomposition levels are used to analyze the measured surfaces. The short-scale surfaces (roughness) are extracted using the detail coefficients of the first and second levels and filtering out all other levels; the short-scale surfaces for

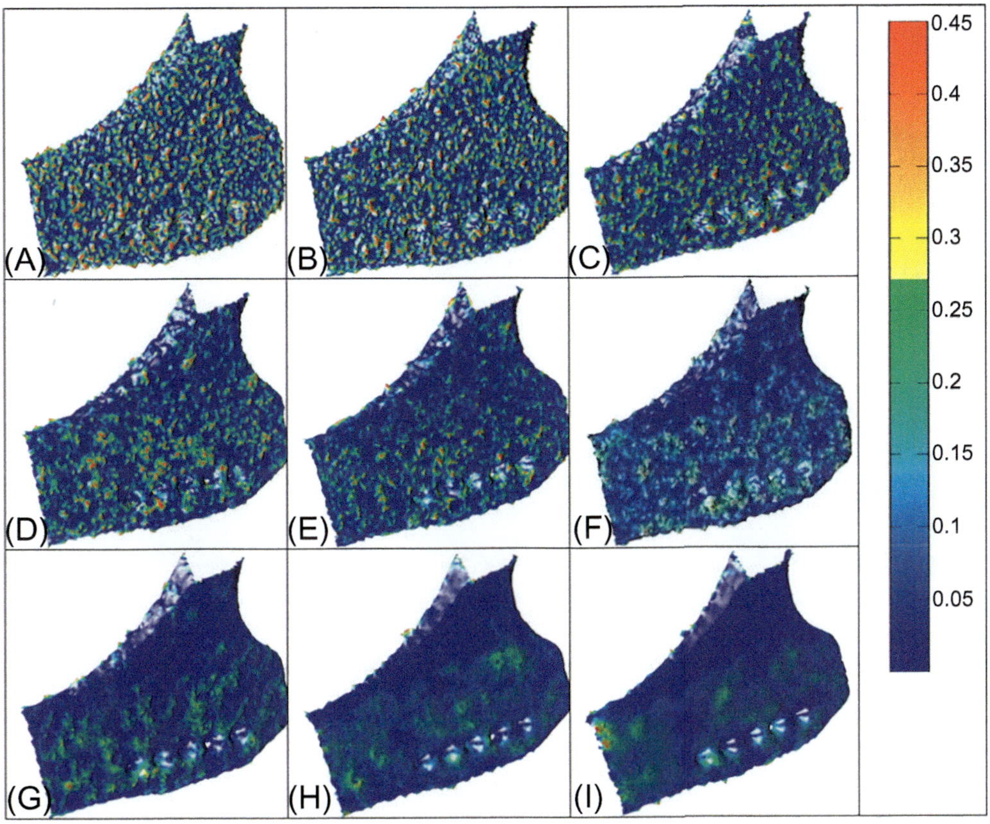

Fig. 9.32 Texture filtration results of the hip-part2 surface: (A) the original textured surface. (B)–(F) The filtration results using 1, 4, 8, 16, and 24 decomposition levels, respectively. (G)–(I) The filtration results using 24 decomposition levels and 2, 3, and 4 iterations, respectively.

HipJoint-Pt1 and HipJoint-Pt2 surfaces are shown in Figs. 9.34B and 9.35B, respectively. The middle-scale surfaces (waviness) that resulted from only considering the detailed coefficients of the middle levels for the two measured surfaces are shown in Figs. 9.34C and 9.35C. The long-scale surfaces (form) are shown in Figs. 9.34D and 9.35D, and as mentioned earlier the form surface is obtained by only considering the output of the final approximation level and filtering out all the detail coefficients of all other levels.

The aforementioned examples of the filtering and decomposing computer-generated and real measured surfaces demonstrate that the lifting wavelets and multiscale mesh decomposition algorithms are capable of filtering the texture of various types of free-form surfaces represented by regular or irregular 3-D triangular meshes.

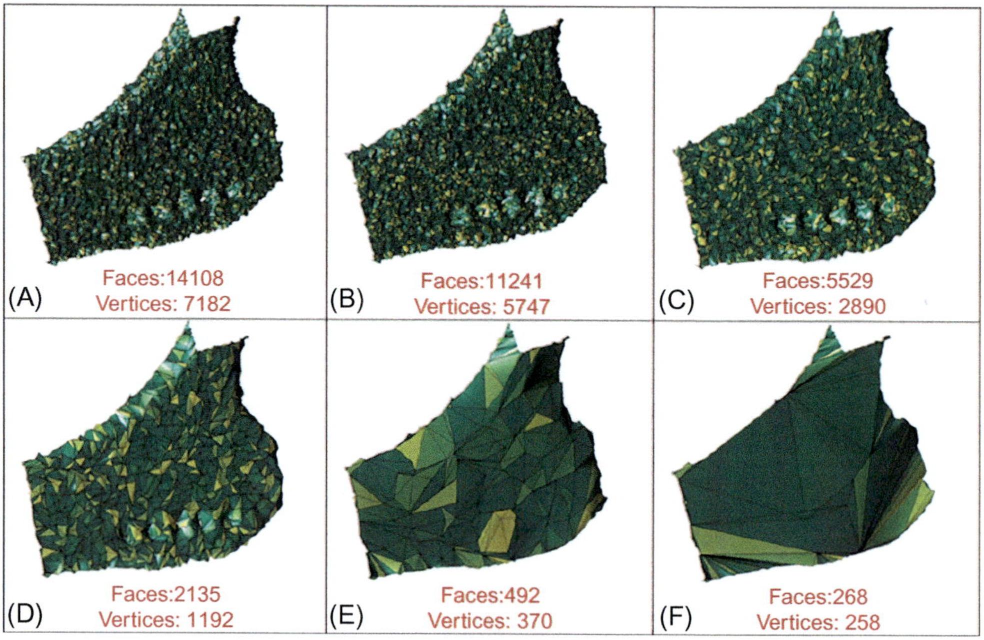

Fig. 9.33 The approximated output meshes (wavelet approximations) for the hip-part2 surface at different decomposition levels. (A) The original finest input mesh and (B)–(F) the approximated coarse mesh after 1, 4, 8, 16, and 24 decomposition levels, respectively.

9.8.3 Comparisons

The lifting wavelet technique is compared with the Laplacian mesh decomposition algorithm and the results are shown in Figs. 9.36–9.38.

Fig. 9.36 compares the filtering results of the computer-generated spherical surface using the lifting-based algorithm with the proposed techniques. The filtered forms, long-scale surfaces, using the lifting algorithm with 4, 16, and 24 decomposition levels are shown in Fig. 9.36A, C, and E respectively. On the other hand the long-scale surfaces resulted from the proposed algorithm using 4, 16, and 24 decomposition levels are shown in Fig 9.36B, D, and F, respectively.

Figs. 9.37 and 9.38 show the comparison results between the two algorithms for the real measured surfaces, namely, HipJoint-Pt1 and HipJoint-Pt2.

As shown in these figures, the relaxation scheme produces smoother form results compared with the lifting-based algorithm. A smoother form (long-scale) result means

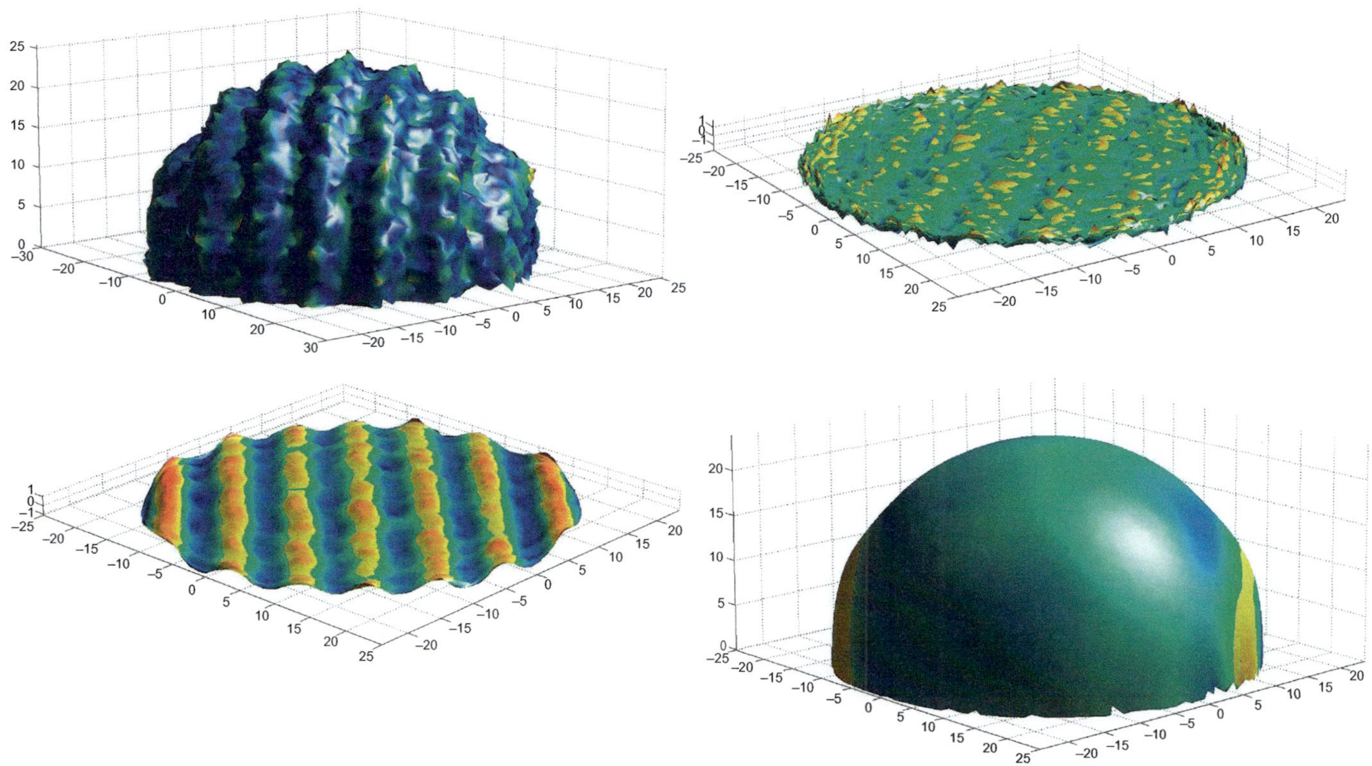

Fig. 9.34 Decomposition results of a HipJoint-Pt1 surface using curvature relaxation scheme using 32 decomposition levels: (A) the original textured surface, (B) roughness surface (the short-scale surface), (C) wavy surface (the middle-scale surface), and (D) form surface (the long-scale form surface).

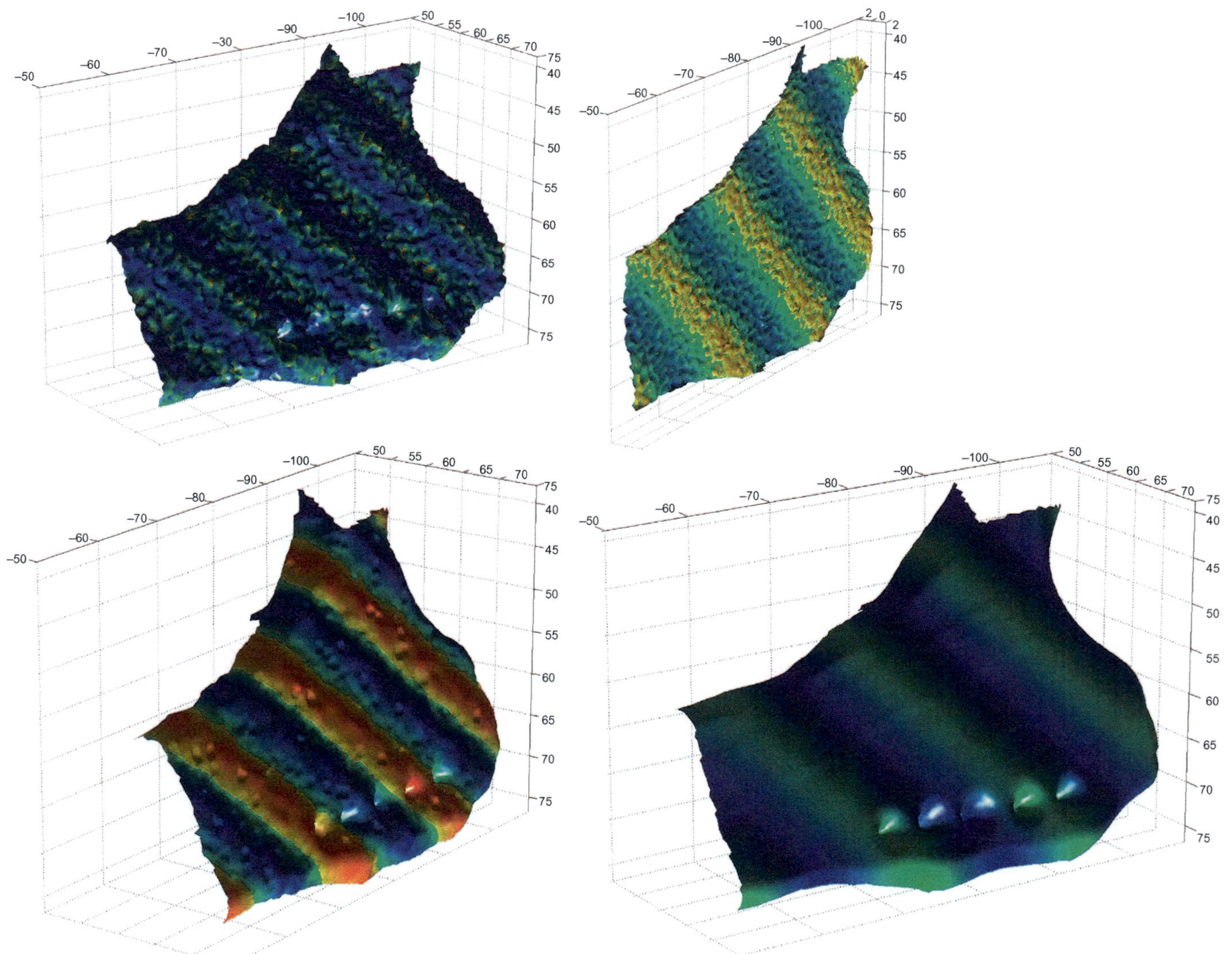

Fig. 9.35 Decomposition results of a HipJoint-Pt2 surface using curvature relaxation scheme using 32 decomposition levels: (A) the original textured surface, (B) roughness surface (the short-scale surface), (C) wavy surface (the middle-scale surface), and (D) form surface (the long-scale form surface).

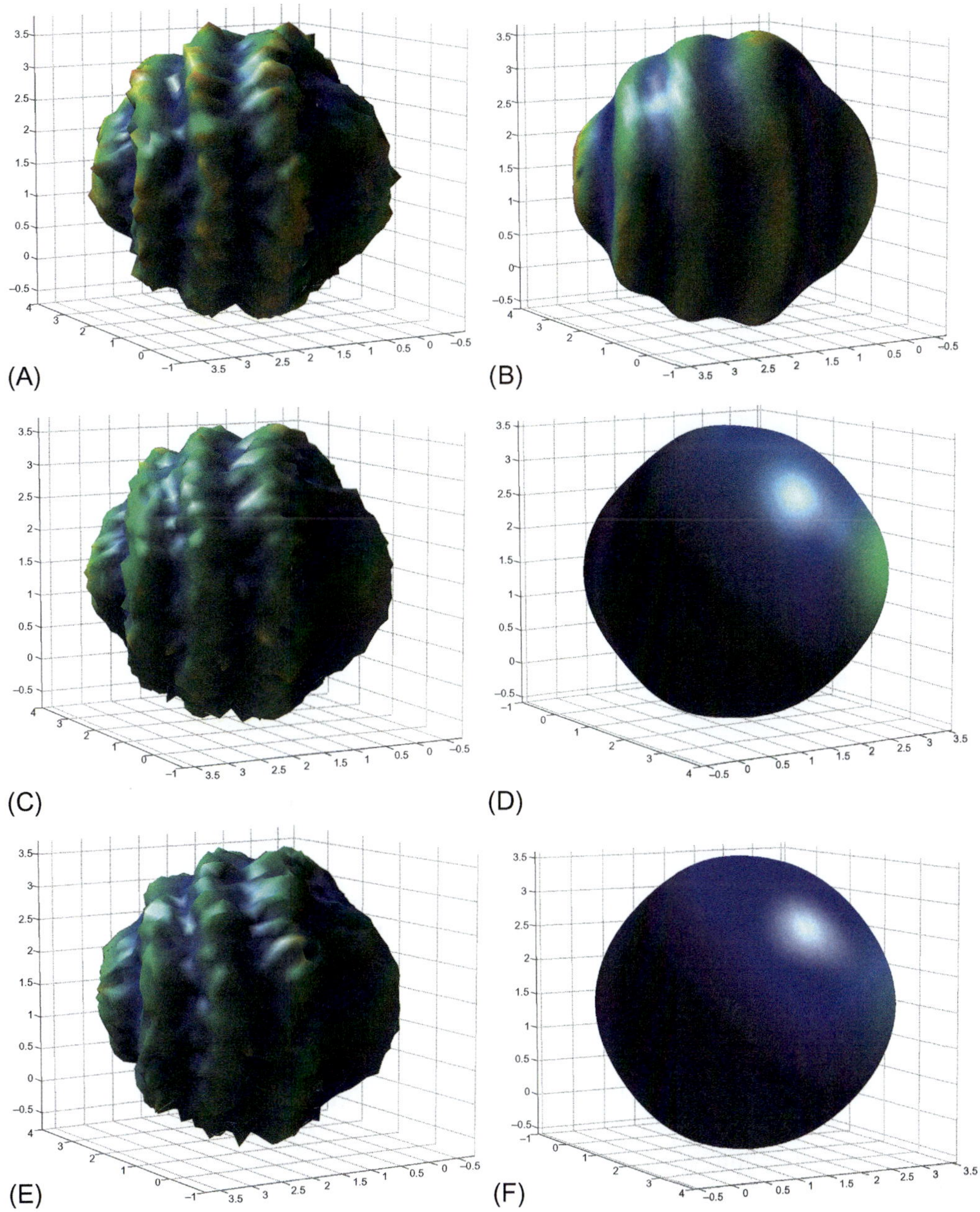

Fig. 9.36 Comparisons between lifting scheme and relaxation scheme filtering methods for the spherical surface: (A), (C), and (E) the results of lifting algorithm using 4, 16, and 24 decomposition levels, respectively; (B), (D), and (F) the results of relaxation algorithm using 4, 16,and 24 decomposition levels, respectively.

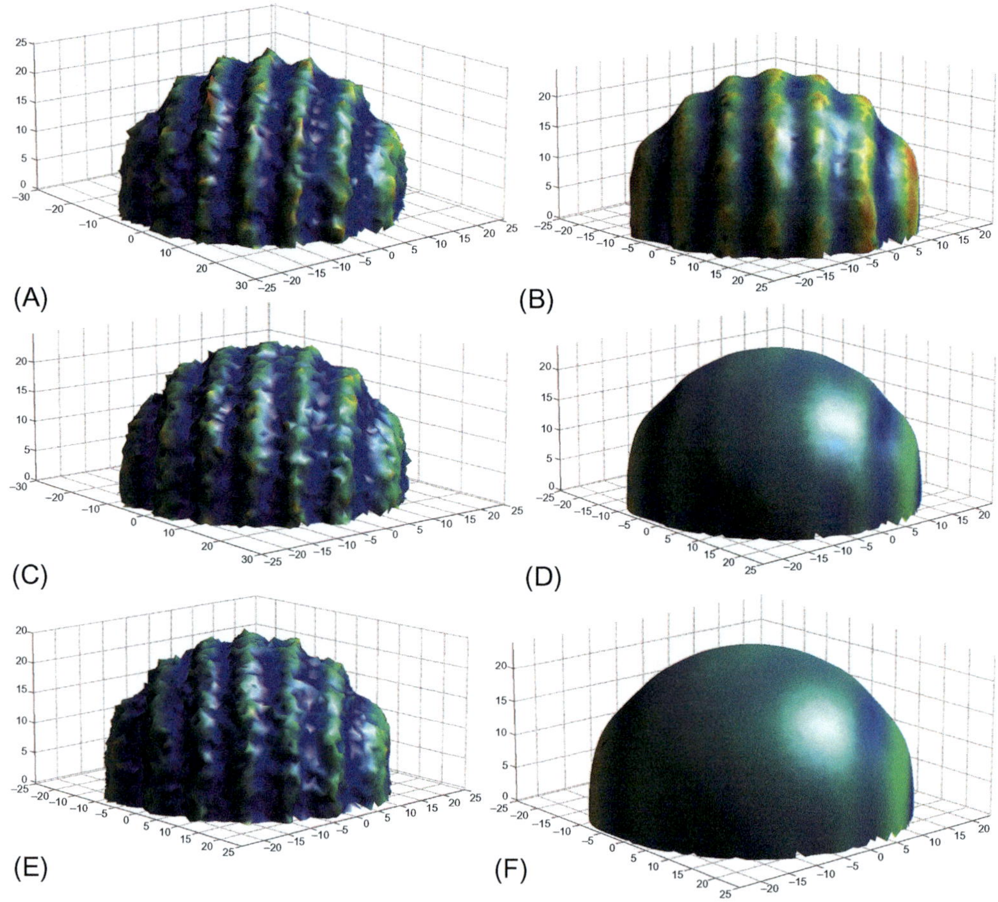

Fig. 9.37 Comparisons between lifting scheme and relaxation scheme filtering methods for the HipJoint-Pt1 surface: (A), (C), and (E) the results of lifting algorithm using 4, 16, and 24 decomposition levels, respectively; (B), (D), and (F) the results of relaxation algorithm using 4, 16, and 24 decomposition levels, respectively.

that the other short-scale and middle-scale surface components (surface textured) are included in the detail coefficients and isolated from the form that makes the decomposition of the surface into different scales easier and more accurate as shown earlier in the chapter.

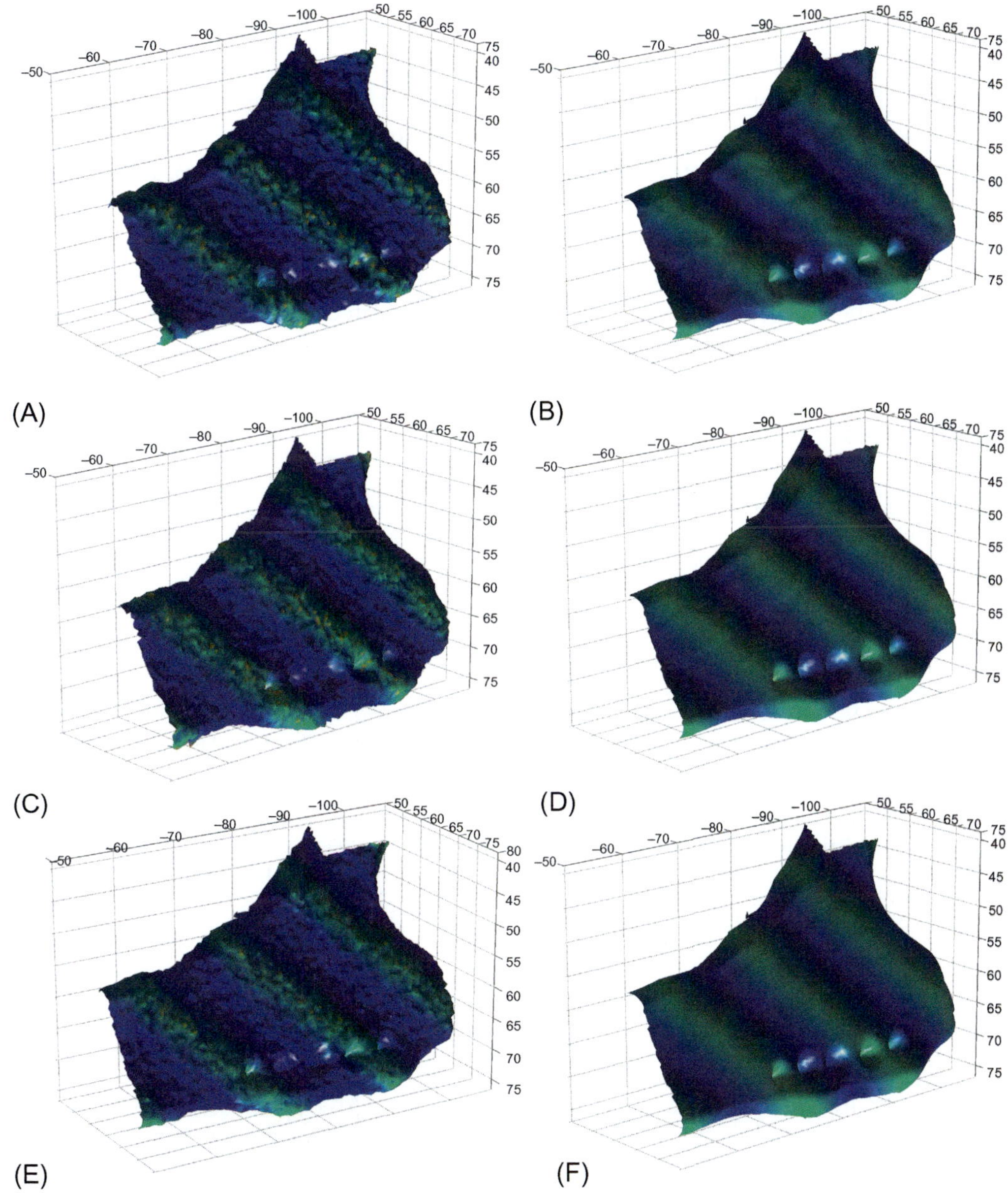

Fig. 9.38 Comparisons between lifting scheme and relaxation scheme filtering methods for the HipJoint-Pt2 surface: (A), (C), and (E) the results of lifting algorithm using 4, 16, and 24 decomposition levels, respectively; (B), (D), and (F) the results of relaxation algorithm using 4, 16, and 24 decomposition levels, respectively.

9.9 Summary

In this chapter, two algorithms for filtering and decomposing engineering surfaces represented by 3-D triangular meshes are presented. The first filtering algorithm is based on the lifting wavelet scheme, where the input surface mesh is filtered using five major operators, namely, the split, the prediction, the update, the mesh simplifications, and the merge operators. The second algorithm decomposes the input surface into three parts: the long-scale, the middle-scale, and the short-scale surface. The decomposing is based on a Laplacian mesh relaxation scheme using the discrete Laplace-Beltrami operator.

The proposed algorithms were applied to filter simulated and bioengineering surfaces, and the results show that both algorithms successfully filter and decompose the surfaces.

References

[1] Daubechies I. Ten lectures on wavelets. SIAM; 1992, ISBN: 0-89871-274-2.
[2] Jansen M, Patrick O. Second generation wavelets and applications. Springer; 2005. ISBN 1-85233-916-0.
[3] Mallat S. A wavelet tour of signal processing. 3rd ed. Academic Press; 2008.
[4] Swelden W. The lifting scheme: A new philosophy in Biorthogonal wavelets reconstructions, In: Wavelet applications in signal and image processing III, pp. 68–79; 1995.
[5] Lou S, Zeng W, Jiang X, Scott PJ. Comparison of Robust filtration techniques in geometrical metrology. In: 18th International Conference on Automation and Computing (ICAC) 2012. Leicestershire, UK: IEEE; 2012.
[6] Swelden W. Wavelets and the lifting scheme: a 5 minute tour. Z Angew Math Mech 1996;76(2):4–7.
[7] Jansen M, Nason GP, Silverman BW. Multiscale methods for data on graphs and irregular multidimensional situations. J Roy Stat Soc 2009;71(1):97–125.
[8] Narang SK, Ortega A. Lifting based wavelet transforms on graphs. In: APSIPA ASC'09; 2009.
[9] Schroder P, Sweldens W. Spherical wavelets: texture processing. In: Hanrahan P, Purgathofer W, editors. Rendering techniques 95. New York: Springer Verlag; 1995.
[10] Blunt L. The development of a basis for 3D surface roughness standards; E.C. contract no SMT4-CT98-2256.
[11] Lou S, Jiang X, Scott PJ. Algorithms for morphological profile filters and their comparison. Precis Eng 2012;36(3):414–23.
[12] Zeng W, Jiang X, Scott P. A generalised linear and nonlinear spline filter. Wear 2011;271(3/4):544–7.
[13] Zeng W, Jiang X, Scott P. Fast algorithm of the Robust Gaussian regression filter for areal surface analysis. Measure Sci Technol 2010;21(5) p. 055108.
[14] Zeng W, Jiang X, Scott P. Metrological characteristics of dual-tree complex wavelet transform for surface analysis. Measure Sci Technol 2005;16(7):1410–7.
[15] Zeng W, Jiang X, Scott P, Blunt L. Metrological characteristics of complex wavelet transform for surface topography analysis. In: 5th euspen International Conference, Montpellier, France; 2005.
[16] ISO 11562: 1996 Geometric product specification (GPS) Surface Texture: profile method metrological characteristics of phase correct filters, 1996.
[17] ISO 16610-21: 2010 Geometrical product specifications (GPS)—Filtration—Part 21: Linear profile filters: Gaussian filters, 2010.
[18] ISO 16610-61: 2010 Geometrical product specifications (GPS)—Filtration—Part 61: Linear areal filters: Gaussian Filters, 2010.
[19] ISO 16610-31: 2010 Geometrical Product Specification (GPS)—Filtration, 2010.
[20] ISO 25178-2: 2012, Geometrical Product Specification (GPS)—Surface Texture: Areal—Part 2: Terms, definitions and surface texture parameters, 2012.

[21] ISO 16610-29, Geometrical product specifications (GPS)—Filteation Part 29: Linear profile filters: Spline wavelets, 2002.
[22] Jiang X, Blunt L. Third generation wavelet for the extraction of morphological features from micro and nano scalar surfaces. Wear 2004;257:1235–40.
[23] Su Y, Xu Z, Jiang X, Pickering J. Discrete wavelet transform on consumer-level graphics processing unit. In: Proceedings of computing and engineering annual Researchers' conference 2008: CEARC'08. Huddersfield: University of Huddersfield; 2008, ISBN 978-1-86218-067-3 pp. 40–47.
[24] Jiang X, Zeng W, Scott P, Ma J. Linear feature extraction based on complex ridgelet transform. Wear 2008;264(5–6):428–33.
[25] Jiang X, Blunt L, Stout KJ. Development of a lifting wavelet representation for characterisation of surface topography. Proc R Soc Lond A 2000;256:1–31.
[26] Jiang X, Blunt L. Morphological assessment of in vivo Wear of orthopaedic implants using multi-scalar wavelet. Wear 2001;250:217–21.
[27] Chen X, Raja J, Simanapalli S. Multi-scale analysis of engineering surfaces. Int J Mach Tool Manuf 1995;35:231–8.
[28] Jiang X, Blunt L, Stout KJ. Recent development in the characterisation technique for bioengineering surfaces. In: Proceedings 8th Inf Conf. Metrology and properties of engineering surfaces, April 2000, Huddersfield, UK; 2000.
[29] Jiang X, Blunt L, Stout KJ. Three-dimensional surface characterization for orthopaedic joint prostheses. J Inst Mech Eng H 1999;213:49–68.
[30] Swelden W. The lifting scheme: a construction of second generation wavelets. SIAM J Math Anal 1998;29(2):511–46.
[31] Jiang X, Scott PJ, Whitehouse DJ, Blunt L. Paradigm shifts in surface metrology. Part II, The current shift. Proc R Soc A 2007;463:2071–99.
[32] Bonneau GP. Multiresolution analysis on irregular surface meshes. IEEE Trans Vis Comput Graph 1998;4(4):365–78.
[33] Daubechies I, Guskov I, Schroder P, Sweldens W. Wavelets on irregular point sets. Phil Trans R Soc Lond A 1999;357:2387–413.
[34] Eck M, DeRose T, Duchamp T, Hoppe H, Lounsbery M, Stuetzle W. Multiresolution analysis of arbitrary meshes. In: SIGGRAPH 95 conference proceedings, ACM SIGGRAPH. Reading, MA: Addison Wesley; 1995. p. 173–82.
[35] Lounsbery M, DeRose T, Warren J. Multi-resolution analysis for surfaces of arbitrary topological type. ACM Trans Graph 1997;16(1):34–73.
[36] Burt P, Adelson E. The Laplacian pyramid as a compact image code. IEEE Trans Commun 1983; 31(4):532–40.
[37] Desbrun M, Meyer M, Schröder P, Barr A. Implicit fairing of irregular meshes using diffusion and curvature flow, In: Proceedings of ACM SIGGRAPH; 1999 pp. 317–324.
[38] Meyer M, Desbrun M, Schröder P, Barr AH. Discrete differential-geometry operators for triangulated 2-manifolds. In: Hege HC, Polthier K, editors. Visualization and mathematics III. Mathematics and Visualization. Berlin, Heidelberg: Springer; 2003.
[39] Valette S, Prost R. Wavelet-based multiresolution analysis of irregular surface meshes. IEEE Trans Vis Comput Graph 2004;10(2):113–22.
[40] Szczesna A. The multiresolution analysis of triangle surface meshes with lifting scheme. In: Proceedings of MIRAGE conference on computer vision/computer graphics collaboration techniques; 2007.
[41] Garland, M. & Heckbert, P., 1997, Surface simplification using quadric error metrics, Proceedings of ACM SIGGRAPH, 1997, pp. 209–216.
[42] Roy M, Foufou S, Koschan A, Truchetet F, Abidi M. Multiresolution analysis for meshes with appearance attributes. In: Proceedings of IEEE on Image Processing ICIP2005. vol. III. 2005. p. 816–9.
[43] Roy M, Foufou S, Koschan A, Truchetet F, Abidi M. Multiresolution analysis for irregular meshes. In: Proceedings of SPIE wavelet applications in industrial processing, Providence, RI. vol. 5266. 2004. p. 249–59.
[44] Hubeli A, Gross M. Multiresolution methods for non-manifold models. IEEE Trans Vis Comput Graph 2001;7(3):207–21.

[45] Guskov I, Sweldens W, Schroder P. Multiresolution signal processing for meshes, In: Proceedings of ACM SIGGRAPH; 1999 pp. 325–334.
[46] Jiang X, Whitehouse DJ. Technological shifts in surface metrology. CIRP Ann—Manuf Technol 2012;61(2):815–36.

Further reading

Bills P, Lou S, Jiang X. Development of morphological filtering method for specification and characterisation of hip replacement taper junctions. In: 12th International conference of the European Society for Precision Engineering and Nanotechnology, 4th–8th June 2012, Stockholm, Sweden; 2012.
Hoppe, H., 1996, Progressive meshes. Proceedings of the 23rd annual conference on Computer graphics and interactive techniques SIGGRAPH 9630, 1996, pp. 99–108.
Jiang X, Zeng W, Scott P. Wavelet analysis for the extraction of morphological features for Orthopaedic bearing surfaces. In: Progress in molecular and environmental bioengineering—from analysis and modeling to technology applications. London: InTech; 2011.
Schroder P, Sweldens W. Building your own wavelets at home. [ACM SIGGRAPH course notes].
Schroder P, Sweldens W. Spherical wavelets: efficiently representing functions on the sphere, In: EGRW 95; 1995 pp. 252–263.
Szczesna A. Designing lifting scheme for second generation wavelet-based multiresolution processing of irregular surface meshes. In: Proceedings of Computer Graphics and Visualization, 2008; 2008.
Schneider R, Kobbelt L. Geometric fairing of irregular meshes for freeform surface design. Comput Aid Geom Des 2001;18(4):359–79.

CHAPTER 10

Characterization of free-form surfaces

10.1 Introduction

Areal parameters were initially developed by Dong et al. [1–4] during the "Birmingham 14 parameters" project and were developed to describe the functionality of a surface [5]. There were some ambiguities in the definition of the parameters regarding feature characterization and the material ratio parameters needed to be more indicative [6]. Further developments and standardizations, such as a detailed analysis of advanced and robust filtration techniques and the subdivision of surface in hills and dales through the watershed decomposition, were proposed in the subsequent SurfStand project and published in Blunt and Jiang [5]. These definitions were later adopted in ISO 25178-2 [7].

Today, advanced manufacturing technologies enable complicated geometries to be designed and manufactured. For example, additive manufacturing (AM) allows internal and external components with free-form geometries to be manufactured. This has led to a series of challenges in the characterization of complex surfaces.

Computed tomography (CT) is considered one of the most promising measurement systems for certain complex surfaces (subject to its material), as extracting surface information from CT measurement is an imperative operation. Characterization of the surface was originally designed for contact stylus and line-of-sight optical sensors; therefore the available software is designed to analyze height map data. The implemented algorithms were designed to optimize the speed if the data are presented as an image. With irregular 3D meshes, such as the extracted surface from CT measurement, it is not possible to directly perform the integrals based on the definitions from ISO 25178-2. Methods to convert a mesh to a height map have been investigated [8, 9], and they are available in commercial software such as MountainsMap [10]. If the measured portion of the free-form surface is high, the computation of the parameters may have a significant bias. It is then necessary to generalize those parameters to be suitable for both regular and irregular meshes.

Generalization of texture parameters was proposed in Abdul-Rahman et al. [11] and Pagani et al. [12]; in both works the authors proposed to use the triangular mesh as the surface representation. Triangular mesh representation was chosen because it can easily represent complex surfaces and computing a linear approximation of scalar function on a mesh is fast and accurate. The usual workflow of the surface characterization is shown in Fig. 10.1. After a surface is measured, to characterize it, the first step is to estimate

* For more examples and practical implementation, check the website: www.digitalsurf.com/freeform.

Advanced Metrology
https://doi.org/10.1016/B978-0-12-821815-0.00010-1

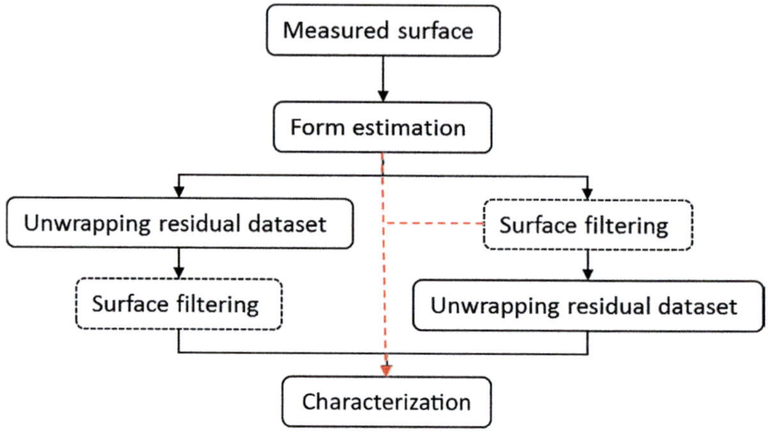

Fig. 10.1 Surface characterization flowchart.

the form. It represents an approximation of the designed surface but is computed using the measured data. After the first decomposition a filter may be applied to compute the so-called scale-limited surface, and after that the surface is unwrapped to convert it to a planar surface, at which point, it can be then finally characterized. A second option is to invert the surface filtering and the unwrapping phase. Using a method that is directly based on the free-form surface allows the elimination of the unwrapping phase, avoiding any distortion of the surface and allowing a better computation of the area and volume values.

Section 10.2 describes the model used to represent the surface; Section 10.3 describes the method used to compute the reference form; Section 10.4 presents the definition and the computation of the texture parameters, and in Section 10.5, test cases are presented.

10.2 Surface representation

To develop the theory of the surface characterization of free-form surface, a manufactured surface is described as a 3D regular parametric surface $\Sigma \in \mathbb{R}^3$ as

$$\boldsymbol{r}(u,v) = \begin{cases} x = x(u,v) \\ y = y(u,v) \\ z = z(u,v) \end{cases}$$

where $\boldsymbol{u} = (u,v)^T \in U$ is the column vector of the parameters and $U \subset \mathbb{R}^2$ is called the parameters' space. This representation is useful both to represent a free-form surface and to define the texture parameters. The computation will be done using a linear approximation, so the parametrization process is not needed during the computation stage. An example of a parametric curve is shown in Fig. 10.2A (blue curve); using this surface representation, it is possible to describe surfaces with reentrant features

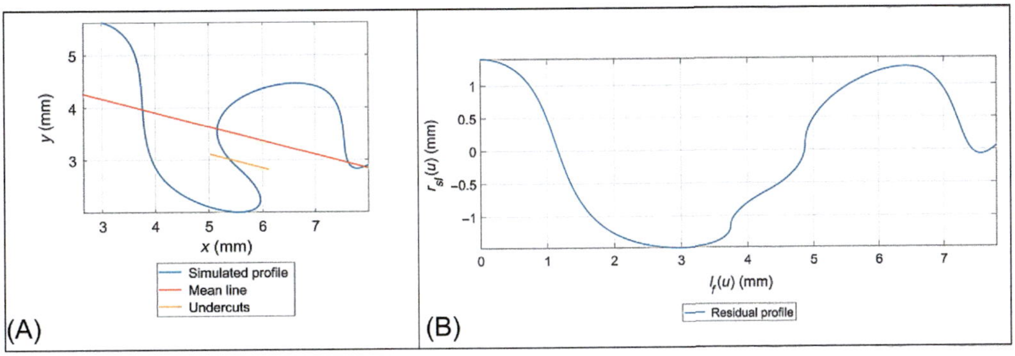

Fig. 10.2 Simulated profile and computed mean line on a portion of the undercuts (A) and residual profile as a function of the mean line (B).

(yellow line in figure). To compute the texture parameters, the surface must be decomposed into two parts:

$$\boldsymbol{r}(u, v) = \boldsymbol{r}_{\text{form}}(u, v) + \boldsymbol{r}_{sl}(u, v)$$

where $\boldsymbol{r}_{\text{form}}(u, v)$ represents the reference form surface and $\boldsymbol{r}_{sl}(u, v)$ the scale-limited surface [7]. All three surfaces share a common parameter space, so there is a unique relation between each coordinate. Since the scale-limited surface may be expressed as an orthogonal signed distance of the form surface, the formula can be rewritten as

$$\boldsymbol{r}(u, v) = \boldsymbol{r}_{\text{form}}(u, v) + r_{sl}(u, v) \cdot \boldsymbol{n}_f(u, v)$$

where $\boldsymbol{n}_f(u, v)$ is the normal of the form surface and $r_{sl}(u, v)$ is a scalar field representing the scale-limited surface. The form surface refers to the large-scale components that may be represented by a primitive, such as a plane and a cylinder, or by a free-form surface. The scale-limited surface refers to the measured surface after removing the reference form and eventually applying a filtering operation to remove the small- and large-scale surface components. In this chapter, no filtration will be applied; the scale-limited surface refers to the measured surface after removing the form surface. A profile example of the decomposition operation is shown in Fig. 10.2: the reference form is represented by a straight line (orange line) in Fig. 10.2A. The scale-limited surface, as a function of the form length, is shown in Fig. 10.2B. It should be noted that, in the parameters' space, the reentrant feature has disappeared and the length of the form is equal to the orange line plus two times the length of the yellow line.

10.3 Reference form computation

The computation of the reference form may be extracted using a filtration method, using a total least squares (TLS) algorithm (also called orthogonal regression), or by using the nominal designed geometry. Some filtration methods were already described in Chapters 6 and 7; however, in this section, the TLS method is presented, and a method to deal with anisotropic meshes is analyzed. In a TLS method the coefficients of the reference form are computed, minimizing the orthogonal distance between the measured points and its orthogonal projection on the reference surface:

$$\hat{\boldsymbol{\beta}} = \arg\min_{\boldsymbol{\beta}} \sum_{i=1}^{n} d^2(\boldsymbol{x}_i, \boldsymbol{f}(\boldsymbol{\beta}, \boldsymbol{x}_i)) = \arg\min_{\boldsymbol{\beta}} \sum_{i=1}^{n} (\boldsymbol{x}_i - \boldsymbol{f}(\boldsymbol{\beta}, \boldsymbol{x}_i))^2 \tag{10.1}$$

where $\boldsymbol{\beta}$ is the vector of parameters that define the reference form, $d(\boldsymbol{a}, \boldsymbol{b})$ is the distance between $\boldsymbol{a}$ and $\boldsymbol{b}$, $\boldsymbol{x}_i$ is a point of the measured surface, and $\boldsymbol{f}(\boldsymbol{\beta}, \boldsymbol{x}_i)$ is the value of the point $\boldsymbol{x}_i$ projected on the function, for example, a plane. The minimization of Eq. (10.1) assigns an equal weight to each of the measured points. A measured surface may be represented by an anisotropic triangular mesh; smaller triangles are used to reconstruct high curvature portions of the surface, while the triangles are bigger in flat portions. If the spacing between each of the measured points is not constant, the minimization of Eq. (10.1) may lead to a biased form estimation; for example, a denser portion of the surface may "pull" the form toward them. To assign the same importance to each portion of the surface, a weighted least squares approach can be used [13, 14]. The vector of the coefficients can be computed as

$$\hat{\boldsymbol{\beta}} = \arg\min_{\boldsymbol{\beta}} \sum_{i=1}^{n} w_i \cdot (\boldsymbol{x}_i - \boldsymbol{f}(\boldsymbol{\beta}, \boldsymbol{x}_i))^2 \tag{10.2}$$

where the weight to assign to each point is computed as

$$w_i = \frac{1}{k} \sum_{j \in N_i} A_j$$

where A_j is the area of the jth triangle, N_i represents the triangles around the vertex i, and k is a normalization constant such that the sum of all the weights adds up to one. Measurement errors may be affected by outliers; these values are anomalies, but they can influence the estimation of the form's parameters. To assign a smaller importance to the outliers, a loss function can be introduced:

$$\hat{\boldsymbol{\beta}} = \arg\min_{\boldsymbol{\beta}} \sum_{i=1}^{n} \rho_i\left(w_i \cdot (\boldsymbol{x}_i - \boldsymbol{f}(\boldsymbol{\beta}, \boldsymbol{x}_i))^2\right) \tag{10.3}$$

where $\rho_i(\cdot)$ is the loss function. Commonly used loss functions are Cauchy, Tukey, Huber, soft L^1, arctan, and tolerant.

If Eqs. (10.1), (10.2) cannot be solved in a closed form or if one wants to solve Eq. (10.3), a numerical method must be used; in this chapter the Ceres solver [15] was used to perform the optimization step.

In the following subsections the equation to minimize to compute the TLS form of plane, cylinder, and sphere will be discussed. To compute the nonweighted version, the weight w_i can be simply omitted.

10.3.1 Plane estimation

The parameters of the plane can be easily estimated in closed form. A plane can be univocally defined by a point lying on it $\boldsymbol{x}_0$ and its normal $\boldsymbol{n}$; the other points are defined by the equation

$$(\boldsymbol{x}-\boldsymbol{x}_0)\cdot\boldsymbol{n}=0$$

A point of the plane can be computed by the weighted average:

$$\boldsymbol{\mu}_w=\sum_{i=1}^{n} w_i\,\boldsymbol{x}_i$$

while the normal vector can be defined after computing the principal vectors of the weighted variance-covariance matrix:

$$\boldsymbol{\Sigma}_w=\sum_{i=1}^{n} w_i\,(\boldsymbol{x}_i-\boldsymbol{\mu}_w)(\boldsymbol{x}_i-\boldsymbol{\mu}_w)^T$$

The normal vector of the plane corresponds with the eigenvector associated to the eigenvalues with smallest value, that is, the normal corresponds to the direction that has the lowest variability. The residual data can then be found on the principal component coordinate system.

10.3.2 Cylinder estimation

A cylinder can be defined by a point on the axis $\boldsymbol{x}_0$, the axis direction $\boldsymbol{a}$, and the radius r, the equation to minimize becomes

$$\min_{r,\boldsymbol{x}_0,\boldsymbol{a}}\sum_{i=1}^{n} w_i\cdot\left[r-\|\boldsymbol{x}_i-\boldsymbol{x}_0+(\boldsymbol{x}_i-\boldsymbol{x}_0)\cdot\boldsymbol{a}\|\right]^2$$

It is not possible to solve it in a closed form, so a numerical method must be used.

The cylinder, not the measured mesh, can then be unwrapped without any stretching; the resulting unwrapped points, $\boldsymbol{x}_i^u$, can be computed as

$$\boldsymbol{x}_i^u=\begin{cases} x_i^u=r\cdot\theta_i=r\cdot\arctan\dfrac{y_i^r}{x_i^r}\\ y_i^u=z_i^r\\ z_i^u=\sqrt{x_i^{r2}+y_i^{r2}}-r\end{cases}$$

where $\boldsymbol{x}_i^r$ represents a measured point translated and rotated such that the axis of the cylinder $\boldsymbol{a}$ coincides with the z-axis versor $(0,0,1)^T$.

10.3.3 Sphere estimation

A sphere can be defined by the center $\boldsymbol{x}_0$ and the radius r; the objective function becomes

$$\min_{r, \boldsymbol{x}_0} \sum_{i=1}^{n} w_i \cdot (r - \|\boldsymbol{x} - \boldsymbol{x}_0\|)^2$$

As in the cylinder scenario, it is not possible to solve it in a closed form, and a numerical method must be used.

The sphere can be unwrapped, but, since the Gaussian curvature is not null, there will be some stretching. The unwrapped point can be computed as

$$\boldsymbol{x}_i^u = \begin{cases} x_i^u = r \cdot \theta_i = r \cdot \arctan \dfrac{y_i^p}{x_i^p} \\ y_i^u = r \cdot \psi_i = r \cdot \arctan z_i^p \\ z_i^u = \|\boldsymbol{x} - \boldsymbol{x}_0\| - r \end{cases}$$

where $\boldsymbol{x}_i^p$ represents a point projected on the sphere with radius 1.

10.4 Areal texture parameters definition

10.4.1 Height parameters

Given the previous assumptions, the areal parameters can be computed as the integral of a scalar field on a surface [12]. The absolute value of the height can be computed as

$$Sa = \frac{1}{A_{\text{form}}} \iint_{\Sigma_{\text{form}}} |r_{sl}(u, v)| \, d\sigma_{\text{form}} \tag{10.4}$$

where

$$d\sigma_{\text{form}} = \|\boldsymbol{r}_{\text{form},u}(u, v) \times \boldsymbol{r}_{\text{form},v}(u, v)\| \, du \, dv$$

and $\boldsymbol{r}_{\text{form},i}(u, v)$ is partial derivative of $\boldsymbol{r}_{\text{form}}(u, v)$ in the i direction, $d\sigma_{\text{form}}$ is the infinitesimal areal element, and $A_{\text{form}} = \iint_{\Sigma_{\text{form}}} d\sigma_{\text{form}}$ is the area of the form surface. With the proposed approach the parameters are computed by weighting the height values with the area element, this allows it to take into account the distortion of the parameters' space. It should be noted that the formula in Eq. (10.4) is similar to the definition in the ISO 25178-2. Since the parameter is a measure of the dispersion of the measured height value around zero (it does not matter if the point is part of a reentrant feature or not), the meaning of the proposed set of parameters coincides with the one described in the standard.

The computation of the form surface when there are reentrant features is shown in Fig. 10.3 using a profile as an example. Fig. 10.3A shows the analyzed profile; the color represents the portion of the profile whose normal forms an angle, to the reference form profile (red dashed line), that is less or equal than $^{\pi}/_{2}$ (blue) or greater (orange).

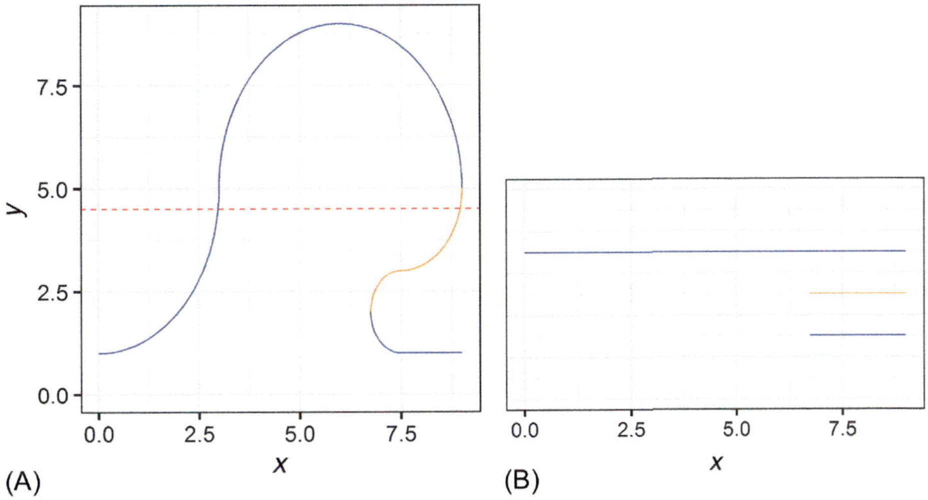

Fig. 10.3 Example of a profile with reentrant features, undercut highlighted in *yellow* (A), and reference profile, measured line projected on the form line (B).

To compute the texture parameters, the measured profile must be projected onto the nominal form, the horizontal line in Fig. 10.3A. The corresponding portions of the profile are shown in Fig. 10.3B. It is possible to observe the real length of the profile if there are reentrant features; it is longer than the corresponding one if the undercuts are removed. It is assumed that the normal of the form surface corresponds to the normal of the nominal surface, not the projected one. That is, in the example, the normal of the points of the form are all pointing upward (or downward).

The remaining height parameters can be computed with similar formulae, that is, the root-mean-square height as

$$Sq = \sqrt{\frac{1}{A_{\text{form}}} \iint_{\Sigma_{\text{form}}} r_{sl}^2(u, v)\, d\sigma_{\text{form}}}$$

the skewness of the heights as

$$Ssk = \frac{1}{A_{\text{form}} \cdot Sq^3} \iint_{\Sigma_{\text{form}}} r_{sl}^3(u, v)\, d\sigma_{\text{form}}$$

and the kurtosis as

$$Sku = \frac{1}{A_{\text{form}} \cdot Sq^4} \iint_{\Sigma_{\text{form}}} r_{sl}^4(u, v)\, d\sigma_{\text{form}}$$

The last class of height parameters that refers to the extreme values, the highest valley, the highest peak, and the range is, respectively,

$$Sv = \left| \min_{u,v} r_{sl}(u,v) \right|, \quad Sp = \left| \max_{u,v} r_{sl}(u,v) \right|, \quad Sz = \left| \min_{u,v} r_{sl}(u,v) \right| + \left| \max_{u,v} r_{sl}(u,v) \right|$$

10.4.2 Hybrid parameters

The root-mean-square gradient can be computed as the weighted mean of the squares of the gradient of a function defined on a surface:

$$Sdq = \sqrt{\frac{1}{A_{\text{form}}} \iint_{\Sigma_{\text{form}}} \left\| \boldsymbol{J}_{\text{form}} \left(\boldsymbol{G}_{\text{form}}^{-1} \right)^{-1} \boldsymbol{\nabla}_U r_{sl}(u,v) \right\| d\sigma_{\text{form}}}$$

where $\boldsymbol{J}_{\text{form}} = \boldsymbol{J}_{\text{form}}(u,v)$ is the Jacobian matrix of the form surface, $\boldsymbol{G}_{\text{form}} = \boldsymbol{G}_{\text{form}}(u,v) = \boldsymbol{J}_{\text{form}}^T \boldsymbol{J}_{\text{form}}$ is the metric tensor, and

$$\boldsymbol{\nabla}_U r_{sl}(u,v) = \begin{pmatrix} r_{sl,u}(u,v) \\ r_{sl,v}(u,v) \end{pmatrix}$$

is the gradient of $r_{sl}(u,v)$ in the parametric space.

A second hybrid parameter measures the ratio between the difference in the surface and the form meshes and the area of the form mesh:

$$Sdr = \frac{A - A_{\text{form}}}{A_{\text{form}}}$$

where $A = \iint_{\Sigma} d\sigma$ is the measured surface area. The *Sdr* parameter directly relates to the area used to compute all the other texture parameters, and it is easy to compute if the form surface is complex.

Another useful parameter may be the percentage of the reentrant features on the form surface. To compute this value the value of the shadow area must be computed (the length of the top line in Fig. 10.3B).

Let θ be the angle between $\boldsymbol{n}(u,v)$ (normal of the measured surface) and $\boldsymbol{n}_{\text{form}}(u,v)$ and

$$k(u,v) = \begin{cases} 1 & \text{if } -\frac{\pi}{2} \leq \theta \leq \frac{\pi}{2} \\ -1 & \text{otherwise} \end{cases}$$

the area of the shadow can be computed as

$$A_{\text{shadow}} = \iint_{\Sigma_{\text{form}}} k(u,v)\, d\sigma_{\text{form}}$$

If there are reentrant features on the border on the surface, the portion of the reentrant features connected to the boundary must be readded to the area of computation.

Using the decomposition of the reference line in Fig. 10.3B, the computation of the length of the shadow of the measured profile coincides with adding the blue lines and subtracting the orange ones.

The percentage of the reentrant features on the form surface can be computed as

$$Srf = \frac{A_{\text{form}} - A_{\text{shadow}}}{2 \cdot A_{\text{shadow}}}$$

It is equal to 0 if the surface has no reentrant features and positive otherwise.

10.4.3 Functional parameters

In this section a method to approximate the density of the Abbott-Firestone (AF), material ratio, and curve is presented. A possible curve, for a surface with reentrant features, is shown in Fig. 10.4; the portion of the volume of the reentrant features is highlighted with a red line. It is possible to observe that the curve does not describe the cumulative of the density of the height value because the sectioned area can be lower for smaller height values if there are reentrant features.

The form surface is first transposed along its normal by a value equal to $r_{\min} = \min_{u,v}\{r_{sl}(u, v)\}$; the volume below the measured surface can therefore be computed as

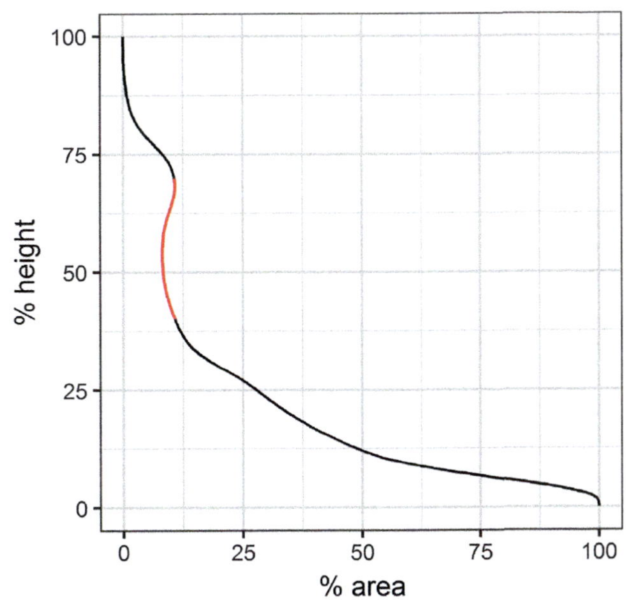

Fig. 10.4 Example of the AF density curve if there are reentrant features.

$$V = \iint_{\Sigma'_{\text{form}}} k(u, v) \cdot h(u, v)\, d\sigma'_{\text{form}}$$

where $h(u, v) = r_{sl}(u, v) - r_{\min}$ and $d\sigma'_{\text{form}}$ are the area element of the form surface translated along the normal $\boldsymbol{r}'_{\text{form}}(u, v) = \boldsymbol{r}_{\text{form}}(u, v) - r_{\min}\ \boldsymbol{n}_{\text{form}}(u, v)$.

The procedure is described in Fig. 10.5; the simulated profile and the reference line, equal to 4.5, are shown in Fig. 10.5A. Assuming that Sv is equal to 4.5, the portion of the area below the profile with $k(u, v)$ equal to 1 is shown in blue in Fig. 10.5B. It is possible to observe that, when there is a reentrant feature, the volume below the lowest curve is

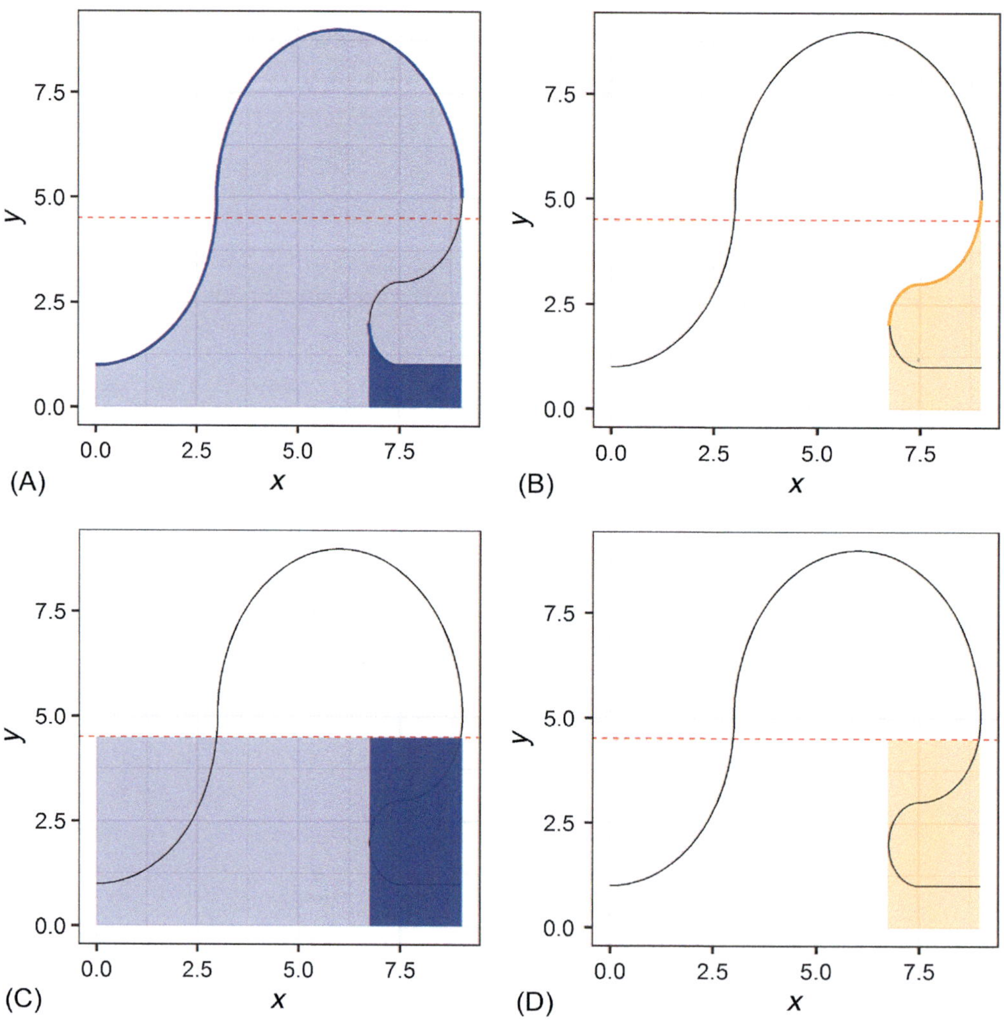

Fig. 10.5 Computation of the volume below a curve.

computed twice (the dark blue area in the figure). To remove the volume of the reentrant feature, the orange area must be removed; this is achieved by setting $k(u, v)$ to -1 (see Fig. 10.5B). The area below the form profile is computed in a similar manner; see Fig. 10.5C and D.

If a numeric quadrature rule is used, it is possible to proportionally split the value of a quadrature point along the height. To compute the density curve, a number of bins have to be arbitrarily set, and the portion of volume for each bin can be computed. The height range was divided into 500 bins. Let n be the total number of bins, $b_i \forall i=1, \ldots, n$ the portion of the volume of each bin, $h(u, v)$ the value of the quadrature point, and $\Delta_b = {}^{Sz}/_{n}$ the range of each bin; the number of bins with equal value is

$$n_b = \left\lfloor \frac{h(u, v)}{\Delta_b} \right\rfloor$$

so the value of the n_{b+1}-th bin is

$$b_{n+1} = \frac{h(u, v) - n_b \cdot \Delta_b}{\Delta_b}$$

while

$$b_i = \frac{h(u, v) - b_{n_b+1}}{n_b} \forall i = 1, \ldots, n_b$$

Summing the contribution of each quadrature point, it is possible to compute the density of the volume as a function of the height (that ranges from 0 to Sz). To compute the AF curve, the volume must be transformed in area, dividing the volume previously computed by Δ_b. It should be noted that if the translation of the form surface is not possible without postprocessing, for example, to eliminate some autointersection, the conversion from volume to surface is not possible with the aforementioned method.

It is not possible to compute the parameters Vxx (functional parameters) according to ISO 25178-2 because the bearing curve is not a bijective function. It was then proposed to compute the parameters directly using the percentage of the height. Let $f_V(h^\star)$ be the density distribution of the volume as a function of the percentage of the height; a possible definition of $Vm(p)$, with $0 \leq p \leq 1$, can be

$$Vm(p) = \frac{h_{\max} - h_{\min}}{A_{\max}} \int_p^1 f_V(h^\star)\, dh^\star$$

while

$$Vv(p) = \frac{h_{\max} - h_{\min}}{A_{\max}} \int_0^p f_{V,\max} - f_V(h^\star)\, dh^\star$$

where $A_{\max}$ is the maximum section area and

$$h^\star = h^\star(u,v) = \frac{h(u,v) - h_{\min}}{h_{\max} - h_{\min}}, \quad h_{\max} = \max_{u,v} h(u,v),$$
$$h_{\min} = \min_{u,v} h(u,v), \quad f_{V,\max} = \max_{h^\star} f_V(h^\star)$$

The volume-related parameters can be computed as

$$Vmp = Vm(p) - Vm(1), \quad Vmc = Vm(q) - Vm(p)$$
$$Vvc = Vv(q) - Vv(p), \quad Vvv = Vv(p) - Vv(0)$$

where p and q, according to the ISO 25178-3 [16], are set to 0.1 and 0.8, respectively.

10.4.4 Comparison with ISO 25178-2

Assume that the form operator was applied on a surface and the scale-limited surface computed. According to the ISO standard, to compute the parameters, the form surface becomes a plane with constant height, $\boldsymbol{r}_{\text{form}}(x,y) = (x,y,0)^T$, while the scale-limited surface is a function of the two parameters, $r_{sl}(x,y) = (x,y,z(x,y))^T$. One should note that the nominal form surface is not taken into account during the parameter's computation, for example, the stretching due to the projection of the points on a sphere is not considered. If the form surface is isometric to a plane, that is, the Gaussian curvature is null everywhere, it is possible to compute the transformation

$$\|\boldsymbol{r}_{\text{form},u}(u,v) \times \boldsymbol{r}_{\text{form},v}(u,v)\|\, du\, dv = du^\star\, dv^\star$$

without stretching [17]. If it is possible to analytically compute this change of variable, the height parameters of the proposed method and the ISO 25178-2 coincide. For example, for a cylinder,

$$\boldsymbol{r}_{\text{form}}(\rho,\theta) = \begin{cases} x = \rho\cos\theta \\ y = \rho\sin\theta \\ z = z \end{cases}$$

a transformation equal to $u^\star = \rho\,\theta$ (the arc length) and $v^\star = z$ (the height) allow an unwrapping of the surface without stretching.

Using the previous assumptions, it is possible to compute the following quantities:

$$\boldsymbol{n}_{\text{form}}(u^\star, y^\star) = (0,0,1)^T$$
$$r_{sl}(u^\star, y^\star) = z(x,y)$$
$$d\sigma_{\text{form}} = dx\, dy$$
$$\boldsymbol{J}_{\text{form}} = \begin{bmatrix} 1 & 0 \\ 0 & 1 \\ 0 & 0 \end{bmatrix}$$
$$\boldsymbol{G}_{\text{form}} = \begin{bmatrix} 1 & 0 \\ 0 & 1 \end{bmatrix}$$
$$\boldsymbol{\nabla}_U r_{sl}(u,v) = \left(z_x, z_y\right)^T$$

the areal parameters of ISO 24178-2 become a special case of the proposed general case, for example,

$$A_{\text{form}} = \iint_{\Sigma_{\text{form}}} d\sigma_{\text{form}} \Rightarrow \iint_{\Sigma_{\text{form}}} dx\, dy$$

$$Sa = \frac{1}{A_{\text{form}}} \iint_{\Sigma_{\text{form}}} |r_{sl}(u, v)|\, d\sigma_{\text{form}} \Rightarrow \frac{1}{A_{\text{form}}} \iint_{\Sigma_{\text{form}}} |z(x, y)|\, dx\, dy$$

$$Sdq = \sqrt{\frac{1}{A_{\text{form}}} \iint_{\Sigma_{\text{form}}} \left\| \boldsymbol{J}_{\text{form}} \left(\boldsymbol{G}_{\text{form}}^{-1}\right)^T \nabla_U r_{sl}(u, v) \right\| d\sigma_{\text{form}}} \Rightarrow \sqrt{\frac{1}{A_{\text{form}}} \iint_{\Sigma_{\text{form}}} z_x^2 + z_y^2\, dx\, dy}$$

The computation of the volume of the surface coincides with the ISO definition if the form surface is a plane.

If the form surface is not a plane, with the method described in the norm, it is not possible to compute the exact volume because the unwrapping of the surface causes a distortion of the volume. For example, the volume of a cylinder of radius r and height h is equal to $\pi\, r^2\, h$, but if the cylinder is unwrapped, the volume becomes double.

10.4.5 Triangular mesh approximation

A possible approximation of the surface is the triangular mesh. To compute the texture parameters, a height value has to be associated for each vertex of the mesh; then the integration of a scalar function on a mesh can be performed. The surface integral is usually computed using the barycentric coordinate system [18]. The integral of a function $f(b_1, b_2, b_3)$ can be computed as

$$\sum_{i=1}^{n_{\text{tri}}} A_i^{\text{tri, form}} \sum_{j=1}^{n_q} f_{ij}\, w_j$$

where b_1, b_2, and b_3 are the barycentric coordinates of a generic triangles; n_{tri} is the number of the triangles in the mesh; $A_i^{\text{tri, form}}$ is the area of the ith triangle of the form surface; n_q is the number of the quadrature points; f_{ij} is the value of the function in the ith triangle at the jth quadrature point; and w_j is the quadrature weight.

An example of a liner function over a mesh is shown in Fig. 10.6. The value of the function is equal to f at the vertex, and it linearly decays up to zero at the opposite edges.

The barycentric coordinates (b_1, b_2, b_3) correspond to the normalized masses (the sum is equal to 1) placed at each vertex of the triangle; a point on the triangle is determined as the geometric centroid [18]. Each point of a linear function defined on a triangle can be expressed as

$$f(u, v) = f_1\, u + f_2\, v + f_3\, (1 - u - v)$$

where $b_1 = u$, $b_2 = v$, and $b_3 = 1 - u - v$, the value of the triangle, given a barycentric coordinate, can be computed as a linear combination of the vertices:

$$\boldsymbol{p} = \boldsymbol{v}_1\, b_1 + \boldsymbol{v}_2\, b_2 + \boldsymbol{v}_3\, b_3.$$

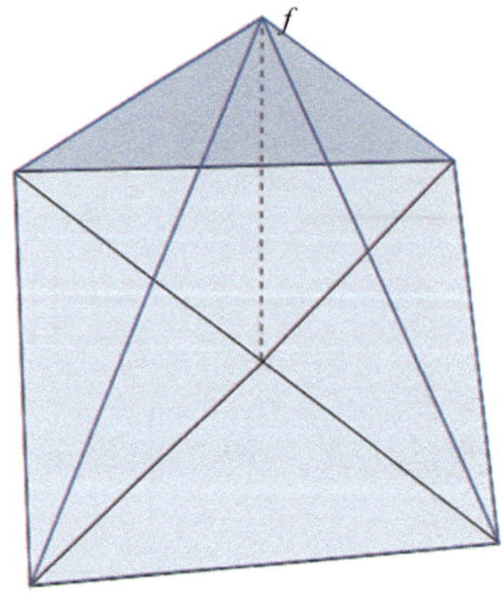

Fig. 10.6 Example of a linear implementation of a function defined on a triangular mesh.

The number of quadrature points depends on the degree of the function that has to be integrated. The numerical quadrature described in Xiao and Gimbutas [19] was used, and they allow the computation of the minimum number of quadrature points to evaluate the integral on a triangular domain.

The area of each triangle can be computed using the definition of the cross product:

$$A_i^{\text{tri, form}} = \frac{(\boldsymbol{v}_2 - \boldsymbol{v}_1) \times (\boldsymbol{v}_3 - \boldsymbol{v}_1)}{2}$$

where $\boldsymbol{v}_i$ is the ith vertex of the triangle, while surface area is the sum of the triangle's areas:

$$A_{\text{form}} = \sum_{i=1}^{n_{\text{tri}}} A_i^{\text{tri, form}}$$

Since a linear approximation was used, the value of the gradient is constant on each element (triangle). For each triangle, it can be computed starting from the definition as

$$\boldsymbol{\nabla}_i^{\text{tri, form}} = \boldsymbol{J}_i^{\text{tri, form}} \left(\mathbf{G}_i^{\text{tri, form}}\right)^{-T} \boldsymbol{\nabla}_U^i r_{sl}(u, v)$$

where

$$\boldsymbol{J}_i^{\text{tri, form}} = [\boldsymbol{v}_1 - \boldsymbol{v}_3, \boldsymbol{v}_2 - \boldsymbol{v}_3]$$

$$\boldsymbol{\nabla}_U^i r_{sl}(u, v) = (f_1 - f_3, f_2 - f_3)^T$$

The mean square gradient can therefore be computed as

$$Sdq = \sqrt{\frac{1}{A_{\text{form}}} \sum_{i=1}^{n_{\text{tri}}} A_i \left\| \boldsymbol{\nabla}_i^{\text{tri, form}} \right\|^2}$$

There is only one quadrature point because the gradient is a constant function on each element. The value of the parameter *Sdr* can be approximated by evaluating both the area of the measured or scale-limited mesh and the area of the form surface.

10.5 Test cases

In this section, four free-form surfaces are reconstructed, and the areal texture parameters are computed. The analyzed meshes are a portion of a lattice structure, a ball bearing, a heavy-duty stick-on Velcro, and a polymer impeller blade.

The portion of the lattice structure has both reentrant features and a complex form, due to the spatter particles on the top of the surface. It was additively manufactured from Ti6Al4V ELI using an electron beam melting (EBM) system. Lattice structures are used in the mechanical field for weight reduction [20] and in the medical field for bioattachment and making implants more lightweight [21].

The second surface is a plain ball bearing that presents some scratches due to the wear. It is shown that it is possible to analyze the whole sphere and the differences between the method presented in Section 10.4 and the one described in ISO 25178-2.

The third mesh represents a structured mesh designed with reentrant features with the goal of maximizing the locking between the two parts of the product. A structured surface is analyzed to show an application of the proposed parameters.

The last test case represents a blade of an impeller manufactured from a polymer using a selective laser sintering (SLS) system. It is shown how it is possible to characterize generic free-form surfaces.

Two surface approximation methods are analyzed:

- Projection method (grid): the measured mesh is projected on a grid to simulate the acquisition with an optical device; it is a so-called 2.5-D surface, and it is possible to compute the areal parameters per ISO 25178-2.
- Triangular mesh (mesh): the measured surface is reconstructed using linear triangular elements.

All the samples were scanned on a Nikon XT H 225 CT; the reconstruction was performed using Nikon CT Pro 3D [22].

10.5.1 Portion of lattice structure

In this section a portion of a lattice structure, manually segmented, is analyzed. The reconstructed surfaces are shown in Fig. 10.10. The lattice structure was manufactured from Ti6Al4V ELI using an Arcam Q10 electron beam melting (EBM) system. The nominal powder particle size was 45–100 μm. The scanning parameters are shown in Table 10.1. The surface determination was performed using the geodesic active contour method implemented in ITK [23]; two volume-to-surface methods were used: the marching cube (MC) method [24] and the adaptive Delaunay isosurface reconstruction

Table 10.1 XT H 225 settings for the lattice structure scan.

Parameter	Value	Parameter	Value
Filter material	None	Voxel size	3.6 μm
Acceleration voltage	60 kV	Magnification	56.1
Filament current	100 μA	Detector size (pixels)	1008 × 1008
Exposure time	1000 ms	Number of projections	1583

(AR) implemented in CGAL [25, 26]. The AR algorithm allows the reconstruction of an implicit surface with a desired approximation, and it is possible to control the manifoldness of a mesh. The maximum distance between the implicit function and the reconstructed mesh was set to 1 μm. If the mesh is reconstructed with the marching cube method, it must be cleaned; that is, nonmanifold edges and faces must be removed. To perform this operation, RameshCleaner was used [27]. The number of reconstructed faces with the MC method was 654039 and 140971 with the AR algorithm. The form was estimated with a TLS fitting using the Ceres solver [15] to minimize the objective function. The radii of the estimated cylinders were 209.35 and 205.08 μm for the MC and the AR method, respectively.

In Section 10.5.1.1 the mesh reconstructed using the AR method is analyzed to show the differences between the grid and the mesh method, while in Section 10.5.1.2, both the reconstructed meshes are used to show the differences due to the surface extraction algorithm and the form estimation method used.

10.5.1.1 Comparison between the grid and mesh methods

In this section the areal parameters are compared with the ones described in ISO 25718-2. To remove any possible source of discrepancy due to the boundaries of the mesh, the first step consists of filling the holes of the mesh. The CGAL algorithm for closing the holes with fairing [28] was used; to obtain a better reconstruction, the mesh was set to be a Delaunay triangulation. After that the form must be removed so that the residuals lay on a planar domain. The residual must then be cropped to create a rectangular border (see Fig. 10.7A). The mesh must be finally converted into a height map; the distance between each point in the x–y plane was chosen to be equal to half of the average edge length of the mesh (2 μm). Clipping and ray-casting algorithms implemented in VTK [29] were used for these purposes. A magnification showing the features lost due to the conversion process is shown in Fig. 10.7B. The red mesh represents the height map, while the transparent mesh is the measured mesh; it is possible to observe that almost the whole feature under the surface cannot be analyzed. The triangles along the unwrapping direction, if they are not on a border, must be finally removed.

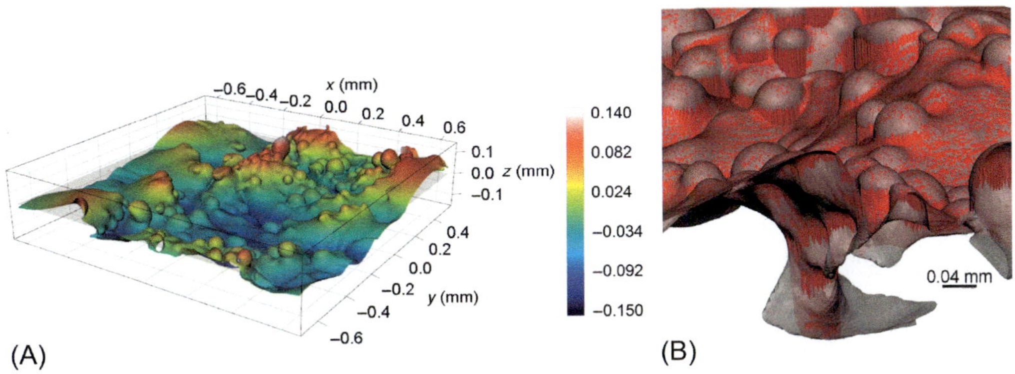

Fig. 10.7 Residual dataset (A), magnification of a reentrant feature *(gray mesh)*, and (B) mesh projected on a grid *(red mesh)*.

To estimate the effect of the stretching due to the unwrapping phase, two methods are annotated for the mesh model:

- Unwrap: the form is used to estimate the parameters corresponding to a plane.
- Wrap: the estimated form surface is used to compute the parameters.

To evaluate the difference between the unwrapped mesh and the measured one, the mesh was rewrapped using the coefficients of the estimated form surface. The mesh, along with the form surface, after the rewrapping operation, is shown in Fig. 10.8. The effect of cropping the mesh can be observed on the missing triangles along the tangential direction.

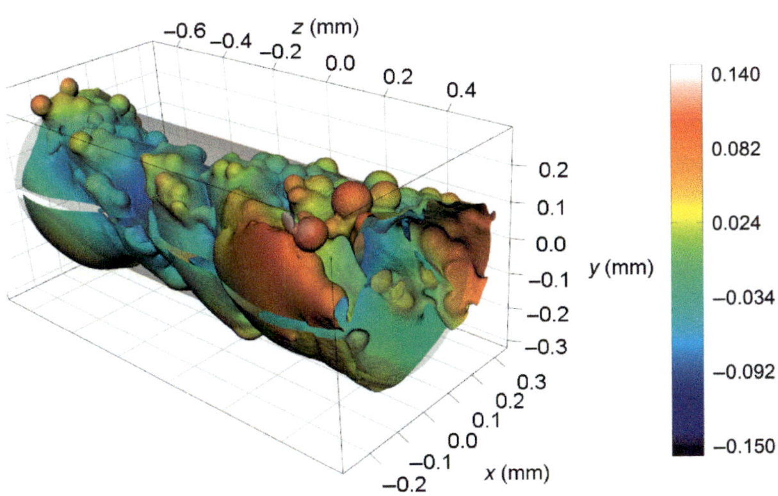

Fig. 10.8 Mesh rewrapped using the estimated cylindrical form.

Table 10.2 shows the estimated texture parameters; the differences are computed taking the grid method as reference. The height parameters (*Sa*, *Sq*, *Ssk*, and *Sku*) show a good agreement between all the analyzed methods. The biggest discrepancies can be observed on the estimation of *Sv* (≈11%) due to the missing reentrant features after the projection into a grid. Another difference can be observed in the estimation of the *Sdr* parameter. It corresponds to −8% and −17% for the unwrapped and rewrapped mesh, respectively. The percentage of the form surface with reentrant features corresponds to 5%, and it can be computed only with the mesh-based method. The bias on the computation of the volume-related parameters of the projection method corresponds to 15% and 1% for the *Vmc* and *Vvc*, respectively. It should be noted that the area of the form surface (A_{form}) of the unwrapped surfaces (mesh and height map) differs due to the projection of the triangles of the reentrant features. The differences between the unwrapped and the rewrapped mesh can be explained by the computation of the area of the scale-limited surface; the curvature of the measured surface is not null everywhere, so the Gauss Egregium theorem does not hold. Due to the unwrapping process, the difference is higher if the compared values of the volume are below the scale-limited and the form surfaces; that is, if the object of the analysis is the volume, the real form surface must be taken into account.

The distribution of the volume below the surface and the Abbott-Firestone curve are shown in Fig. 10.9. It is possible to observe both the effect of the unwrapping and the undercut removal in Fig. 10.9A:

- The unwrapping generates wrong estimation of the volume computation, so the two curves have different behavior (see the cyan curve compared with the others that are overlapping).
- Removing the undercuts does not allow to correctly compute *Sv*; the yellow curve ends before the black one. So the limits used to compute the functional parameters differ.

10.5.1.2 Comparison of different reconstruction and form estimation methods

In this section, both the mesh reconstructed with the MC and the AR methods are analyzed. The effect of the form estimation applying the LS and the WLS methods is discussed too. The two reconstructed meshes are shown in Fig. 10.10 though it is not possible to visually see any big differences between the two reconstruction methods.

Though there is a high difference in the number of generated faces (see Fig. 10.10C and D), the distances between the two meshes are negligible. The AR reconstruction adds more triangles where there is a high value of the curvature, while the triangles get bigger in flat portions of the surface. The form was estimated using the LS method, and it is possible to observe that the range of the signed distance between the form and the mesh (color scale) does not match; this mean that the coefficients of the form differ. As it is possible to observe in Fig. 10.10E and F and in Table 10.3, if the LS method is used,

Table 10.2 Estimated areal texture parameters of the portion of lattice surface.

Method	*Sa* (μm)	Δ*Sa* (%)	*Sq* (μm)	Δ*Sq* (%)	*Ssk*	Δ*Ssk*	*Sku*	Δ*Sku*
Grid	38.44	–	46.79	–	−0.49	–	2.39	–
Mesh	38.97	1.38	47.29	1.09	−0.57	−0.08	2.38	−0.01
Mesh wrap	38.98	1.40	47.30	1.11	−0.57	−0.08	2.38	−0.01
Method	***Sp* (μm)**	**Δ*Sp* (%)**	***Sv* (μm)**	**Δ*Sv* (%)**	***Sz* (μm)**	**Δ*Sz* (%)**	***Sdq***	**Δ*Sdq***
Grid	138.00	–	132.89	–	270.89	–	2.07	–
Mesh	139.91	1.39	148.59	11.81	288.50	6.50	2.50	0.43
Mesh wrap	139.91	1.39	148.59	11.81	288.50	6.50	2.14	0.06
Method	***Vmp* (μm)**	**Δ*Vmp* (%)**	***Vmc* (μm)**	**Δ*Vmc* (%)**	***Vvv* (μm)**	**Δ*Vvv* (%)**	***Vvc* (μm)**	**Δ*Vvc* (%)**
Grid	0.05	–	68.11	–	0.66	–	121.64	–
Mesh	0.04	−4.15	78.86	15.78	0.20	−69.40	123.25	1.32
Mesh wrap	0.04	−4.15	78.86	15.78	0.20	−69.40	123.25	1.32
Method	***Sdr* (%)**	**Δ*Sdr* (%)**	***Srf* (%)**	**Δ*Srf* (%)**	**A_{sl} (mm^2)**	**ΔA_{sl} (%)**	**A_{form} (mm^2)**	**ΔA_{form} (%)**
Grid	52.21	–	0.00	–	2.50	–	1.64	–
Mesh	43.51	−8.70	4.90	−4.90	2.60	4.02	1.81	10.08
Mesh wrap	35.15	−17.06	4.90	−4.90	2.44	−2.05	1.81	10.08

Method	A_{shadow} (mm^2)	ΔA_{shadow} (%)	V_{sl} (mm^3)	ΔV_{sl} (%)	V_{form} (mm^3)	ΔV_{form} (%)
Grid	1.64	–	0.20	–	0.22	–
Mesh	1.65	0.26	0.22	12.73	0.24	12.10
Mesh wrap	1.65	0.26	0.07	−66.39	0.07	−66.57

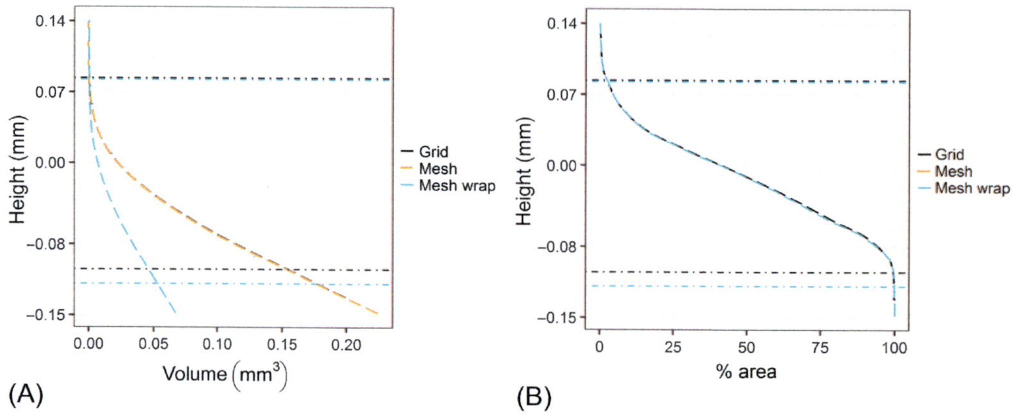

Fig. 10.9 Volume below the scale-limited surface as function of the height (A) and Abbott-Firestone curve (B).

there is a discrepancy in both the radius and the axis direction of the cylinder; while applying a WLS method, the two cylinder almost overlaps.

According to the values of the texture parameters reported in Table 10.3, the major difference due to the estimation using the LS method can be observed in the computation of the height parameters that involve the extreme values (*Sp*, *Sv*, and *Sz*), the parameters related to the area of the form surface, and the functional parameters.

10.5.2 Ball bearing surface

In this section a plain ball bearing is measured using the CT device, and the scanning parameters are reported in Table 10.4. A curvature-based anisotropic diffusion filter implemented in ITK [23] was performed to enhance the edge, then the local Chan-Vese algorithm was applied to reconstruct the surface [30]. The geodesic active contour was not used because it is more sensitive to the initial conditions. The reconstructed mesh, using the marching cube method, was then remeshed using the anisotropic remeshing algorithm proposed in Botsch and Kobbelt [31] implemented in OpenFlipper [32]. A different approach was used compared with the previous section to show that the optimal reconstruction process of a CT dataset is not unique. There are different alternatives in the preprocessing (filters to remove the noise), surface determination, surface reconstruction, and postprocessing (decimation or remeshing) steps.

A magnification of the mesh after the MC cube algorithm (2786113 points and 5572222 triangles) is shown in Fig. 10.11D; the surface presents some step effect due to the voxels of the volume. The remeshing was performed, applying 10 iterations of the algorithm setting a minimum edge length of 0.007 μm, a maximum edge length of 0.020 μm, and a maximum error of 0.001 μm. The total number of points was 562125 with 1124246 triangles. The remeshing algorithm is used both to remove some

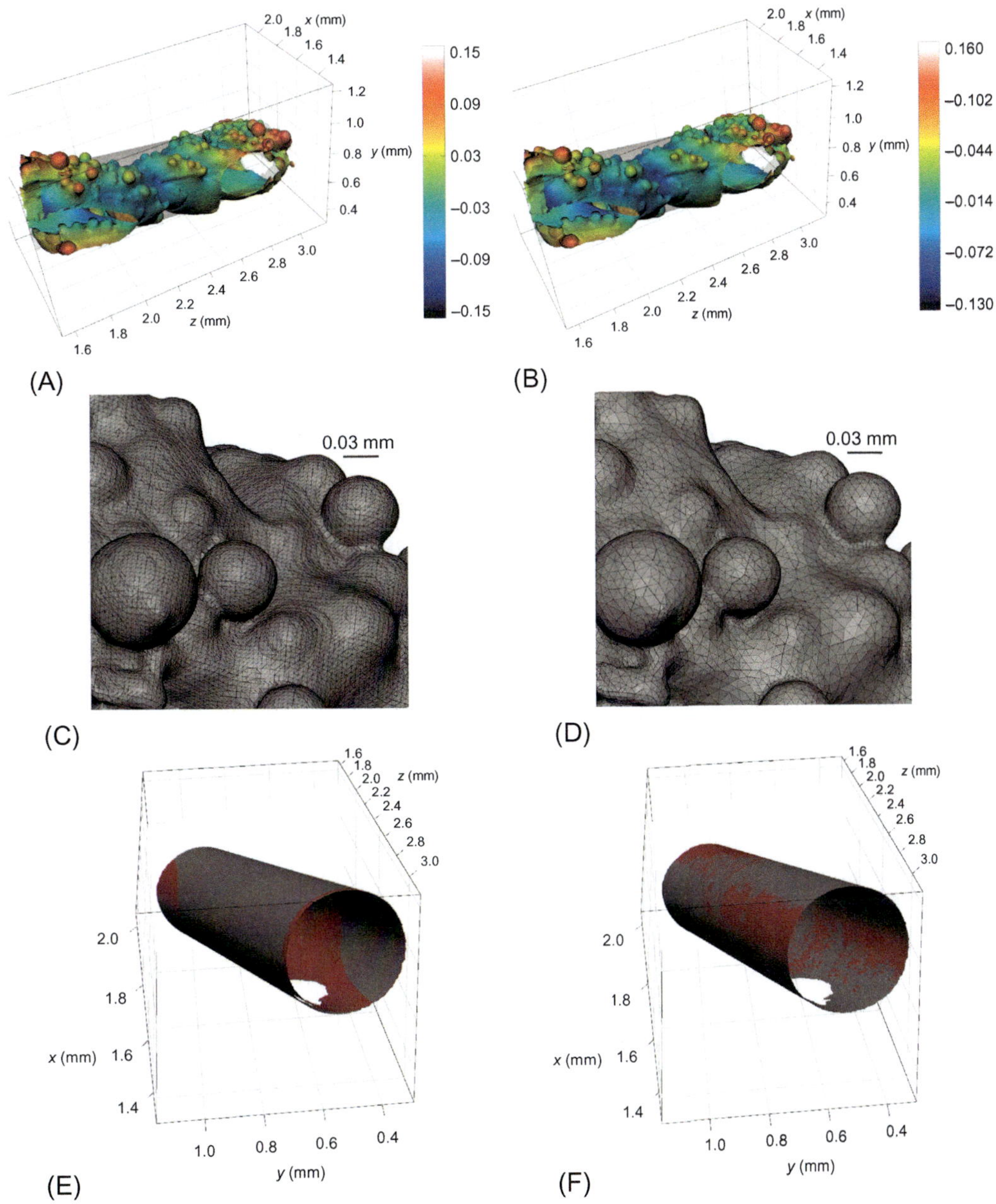

Fig. 10.10 Analyzed portion of lattice structure: reconstruction using the marching cube method (A) and adaptive Delaunay reconstruction (B), magnification of the meshes reconstructed using the MC (C) and AR (D) methods, and reference forms estimated using the LS (E) and WLS (F) algorithms (*gray* form MC and *red* form AR generated meshes).

Table 10.3 Estimated radii and texture parameters.

Method	Radius (μm)	*Sa* (μm)	*Sq* (μm)	*Ssk*	*Sku*	*Sp* (μm)	*Sv* (μm)	*Sz* (μm)
Mesh MC LS	209.35	38.85	47.31	−0.43	2.47	148.35	145.06	293.40
Mesh AR LS	205.08	38.45	46.75	−0.17	2.52	157.09	134.76	291.85
Mesh MC WLS	209.23	38.85	47.31	−0.43	2.47	148.64	145.01	293.66
Mesh AR WLS	209.28	38.82	47.28	−0.43	2.47	147.78	144.30	292.09

Method	*Sdq*	*Sdr* (%)	*Srf* (%)	A_{sl} (mm^2)	A_{form} (mm^2)	A_{shadow} (mm^2)
Mesh MC LS	2.37	40.63	5.43	2.82	2.00	1.81
Mesh AR LS	2.26	42.97	5.22	2.81	1.96	1.78
Mesh MC WLS	2.28	40.69	5.43	2.82	2.00	1.81
Mesh AR WLS	2.31	40.24	5.28	2.81	2.00	1.81

Method	*Vmp* (μm)	*Vmc* (μm)	*Vvv* (μm)	*Vvc* (μm)	V_{sl} (mm^3)	V_{form} (mm^3)
Mesh MC LS	0.03	77.22	0.38	128.31	0.075	0.081
Mesh AR LS	0.02	71.33	0.65	133.11	0.079	0.082
Mesh MC WLS	0.02	77.22	0.38	128.49	0.075	0.080
Mesh AR WLS	0.02	76.82	0.39	127.80	0.076	0.081

Table 10.4 XT H 225 settings for ball bearing scan.

Parameter	Value	Parameter	Value
Filter material	Copper (0.5 mm)	Voxel size	9.33 μm
Acceleration voltage	200 kV	Magnification	21.43
Filament current	48 μA	Detector size (pixels)	1008 × 1008
Exposure time	2000 ms	Number of projections	1583

effects during the reconstruction process and to reduce the number of triangles to allow fastest mesh operations, such as the form estimation and the parameters' computation.

The remeshed surface with the computed form surface, colored with the residual dataset, and a magnification of a groove on the ball bearing are shown in Fig. 10.11A and B. The form was computed using the LS algorithm, since the anisotropy of the surface was not high; similar coefficients of the reference sphere were found. The scale-limited surface after the unwrapping and the rasterization is shown in Fig. 10.12A. It shows a big groove on the surface and two walls at the border of the surface. These walls are the effect of the cone beam error due to the CT measurement (see the top of Fig. 10.12C). As it is possible to observe on the magnification of the mesh in Fig. 10.12B, the error covers only a small portion of the surface. In the grid method, it appears bigger in the unwrapping process; the two errors due to the cone beam effect are located at the top and the bottom of the ball bearing.

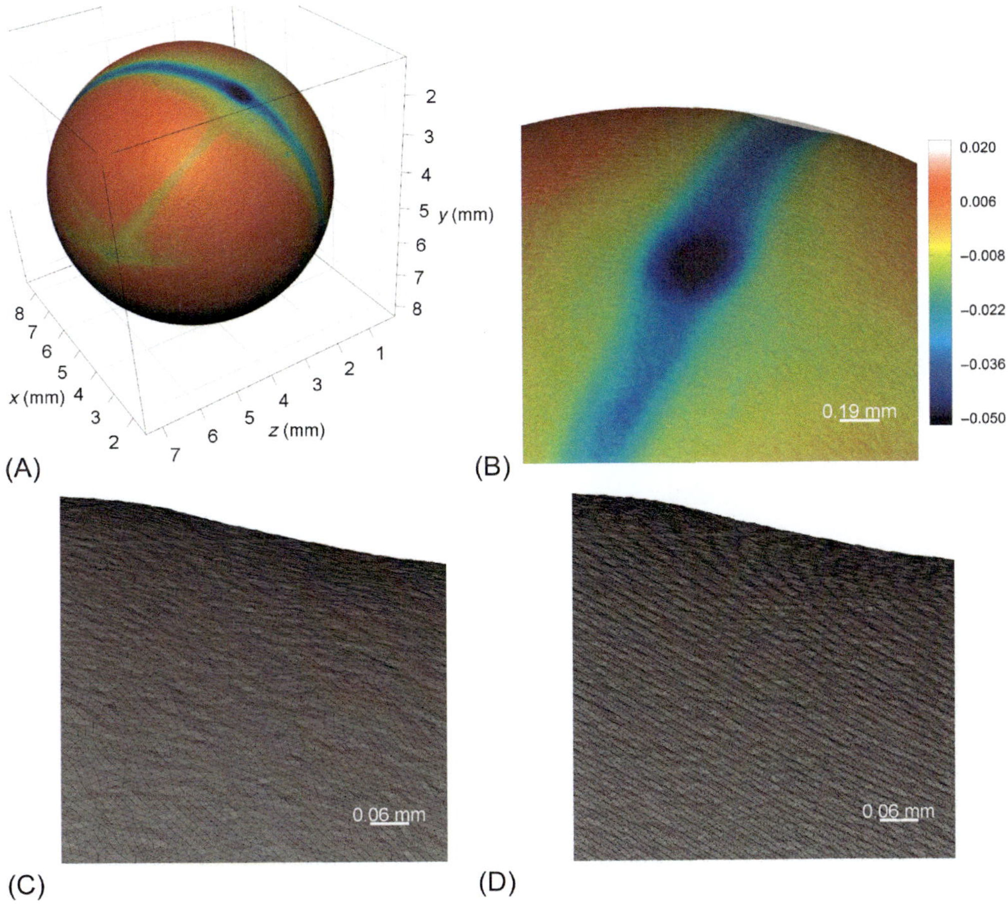

Fig. 10.11 Analyzed mesh and estimated form (A), magnification for the mesh (B), anisotropic remeshed mesh (C), and mesh after the surface reconstruction (D).

Table 10.5 shows the estimated parameters; the word rem refers to the remeshed mesh. The differences were computed taking different references:

- Mesh: the differences were computed using the grid method as reference to show the effect of the unwrapping and projection operations.
- Grid rem: the differences were computed using the grid method as reference to show the effect of the remeshing process.
- Mesh rem: the differences were computed using the mesh method as reference to show the effect of the remeshing process.

The unwrapping phase influences all the values related to both the areas and the volumes regarding both the scale-limited and the form meshes. The area of the form mesh varies because the Gaussian curvature of the sphere is not null, so the unwrapping process generated some stretching. The remeshing process has a small effect on the parameters

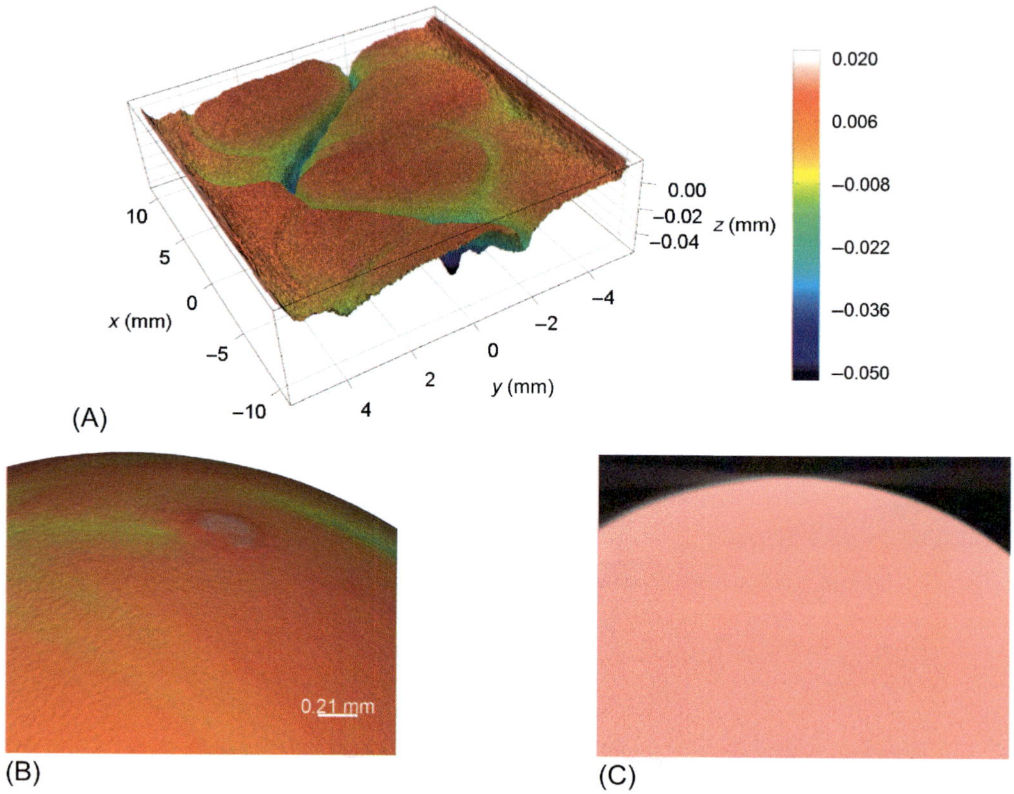

Fig. 10.12 Mesh after the projection operation (A) and cone beam error of the CT measurement: mesh (B) and *gray* level image (C).

relating to the extreme values (*Sp*, *Sv*, *Sz*, and *Vmp*), *Sdq* and *Sdr*, due to the smoothing effect. It should be noted that the area of the form surface of the mesh and mesh rem method are equal; it means that the coefficients of the two reference forms coincide.

The distribution of the volume as a function of the height and the Abbott-Firestone curve is shown in Fig. 10.13. The effect of the remeshing was small; there is a difference due the different estimated *Sv*. The Abbott-Firestone curves are reported as a function of the area, not the percentage to show the effect of the unwrapping process. It should be noted, at the top of the curve, that the cone beam error has a high effect on the unwrapped surface, while it has no effect of the two mesh surfaces.

10.5.3 Heavy-duty stick-on Velcro surface

In this section a portion of a heavy-duty stick-on Velcro surface shown in Fig. 10.14 is analyzed, and the scanning parameters are shown in Table 10.6. The surface was reconstructed using the AR algorithm setting the maximum distance parameter to 10 μm. This

Table 10.5 Estimated texture parameters of the ball bearing surface.

Method	*Sa* (μm)	Δ*Sa* (%)	*Sq* (μm)	Δ*Sq* (%)	*Ssk*	Δ*Ssk*	*Sku*	Δ*Sku*
Grid	5.13	–	7.35	–	−1.63	–	8.88	–
Mesh	5.34	4.05	7.84	6.64	−2.21	−0.58	9.45	0.57
Grid rem	5.07	−1.11	7.29	−0.77	−1.69	−0.06	9.07	0.19
Mesh rem	5.32	−0.27	7.82	−0.23	−2.21	−0.01	9.47	0.02

Method	*Sp* (μm)	Δ*Sp* (%)	*Sv* (μm)	Δ*Sv* (%)	*Sz* (μm)	Δ*Sz* (%)	*Sdq*	Δ*Sdq*
Grid	16.19	–	54.41	–	70.59	–	0.04	–
Mesh	16.21	0.15	55.63	2.24	71.84	1.76	0.13	0.08
Grid rem	15.78	−2.57	54.13	−0.51	69.91	−0.97	0.03	−0.01
Mesh rem	16.02	−1.19	54.63	−1.65	70.65	−1.65	0.08	−0.05

Method	*Vmp* (μm)	Δ*Vmp* (%)	*Vmc* (μm)	Δ*Vmc* (%)	*Vvv* (μm)	Δ*Vvv* (%)	*Vvc* (μm)	Δ*Vvc* (%)
Grid	0.12	–	41.11	–	<0.01	–	8.34	–
Mesh	0.01	−88.07	41.30	0.46	<0.01	–	9.02	8.23
Grid rem	0.12	−2.20	40.94	−0.43	<0.01	–	8.04	−3.75
Mesh rem	0.01	−10.59	40.55	−1.82	<0.01	–	8.94	−0.90

Method	*Sdr* (%)	Δ*Sdr* (%)	*Srf* (%)	Δ*Srf* (%)	A_{sl} (mm²)	ΔA_{sl} (%)	A_{form} (mm²)	ΔA_{form} (%)
Grid	2.22	–	0.00	–	243.60	–	243.07	–
Mesh	8.11	0.59	0.00	–	157.05	−35.53	155.79	35.91
Grid rem	1.24	−0.10	0.00	–	242.10	−0.62	241.80	−0.52
Mesh rem	3.37	−0.47	0.00	–	156.31	−0.47	155.79	0.00

Method	A_{shadow} (mm²)	ΔA_{shadow} (%)	V_{sl} (mm³)	ΔV_{sl} (%)	V_{form} (mm³)	ΔV_{form} (%)
Grid	243.07	–	13.436	–	13.225	–
Mesh	155.79	−35.91	8.398	−37.50	8.394	−36.52
Grid rem	241.80	−0.52	13.306	−0.98	13.089	−1.03
Mesh rem	155.79	0.00	8.253	−1.72	8.249	−1.73

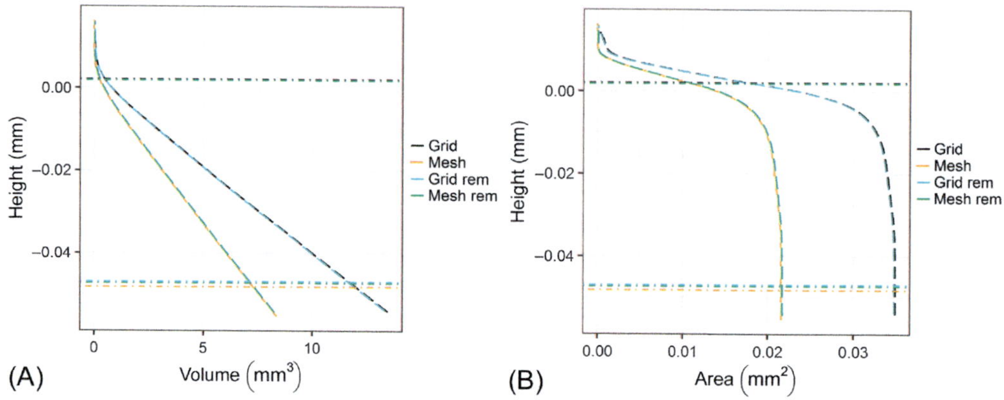

Fig. 10.13 Volume below the scale-limited surface as function of the height (A) and Abbott-Firestone curve (B).

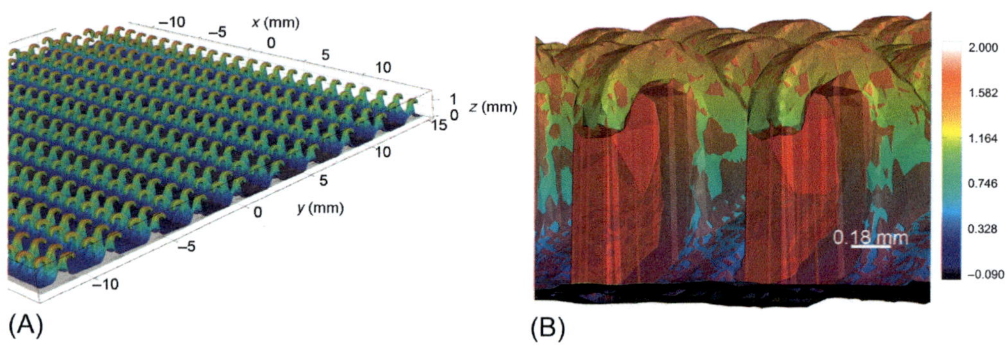

Fig. 10.14 Heavy-duty stick-on Velcro surface: reconstructed surface and estimated form surface in *gray* (A) and reconstructed mesh and projected mesh in *red* (B).

Table 10.6 XT H 225 settings for Velcro surface.

Parameter	Value	Parameter	Value
Filter material	None	Voxel size	32.30 μm
Acceleration voltage	140 kV	Magnification	6.19
Filament current	71 μA	Detector size (pixels)	1008 × 1008
Exposure time	708 ms	Number of projections	1583

product presents a lot of hooks, whose aim is to maximize the adhesion between two surfaces.

The form surface was estimated using the plane random sample consensus algorithm implemented in the PCL library [33] with the intention of not including the hooks in the plane estimation. Texture parameter estimation is reported in Table 10.7. It is possible to

Table 10.7 Estimated texture parameters of the Velcro surface.

Method	*Sa* (μm)	Δ*Sa* (%)	*Sq* (μm)	Δ*Sq* (%)	*Ssk*	Δ*Ssk*	*Sku*	Δ*Sku*
Grid	343.88	–	530.58	–	1.95	–	4.16	–
Mesh	381.33	10.89	563.80	6.26	1.80	−0.15	3.48	−0.67
Method	***Sp* (mm)**	**Δ*Sp* (%)**	***Sv* (mm)**	**Δ*Sv* (%)**	***Sz* (mm)**	**Δ*Sz* (%)**	***Sdq***	**Δ*Sdq***
Grid	1.57	–	0.09	–	1.66	–	5.41	–
Mesh	1.57	0.07	0.09	0.11	1.66	0.07	7.59	2.18
Method	***Vmp* (μm)**	**Δ*Vmp* (%)**	***Vmc* (μm)**	**Δ*Vmc* (%)**	***Vvv* (μm)**	**Δ*Vvv* (%)**	***Vvc* (μm)**	**Δ*Vvc* (%)**
Grid	0.18	–	201.05	–	106.94	–	962.61	–
Mesh	0.18	−2.72	131.53	−34.58	121.40	13.53	1032.68	7.28
Method	***Sdr* (%)**	**Δ*Sdr* (%)**	***Srf* (%)**	**Δ*Srf* (%)**	**A_{sl} (mm^2)**	**ΔA_{sl} (%)**	**A_{form} (mm^2)**	**ΔA_{form} (%)**
Grid	145.47	–	0.00	–	2095.26	–	853.57	–
Mesh	73.12	−72.35	10.76	10.76	1796.75	−14.25	1037.85	−21.59

Method	A_{shadow} (mm^2)	ΔA_{shadow} (%)	V_{sl} (mm^3)	ΔV_{sl} (%)	V_{form} (mm^3)	ΔV_{form} (%)
Grid	853.57	–	365.06	–	77.44	–
Mesh	854.05	0.06	294.47	−19.34	77.57	0.17

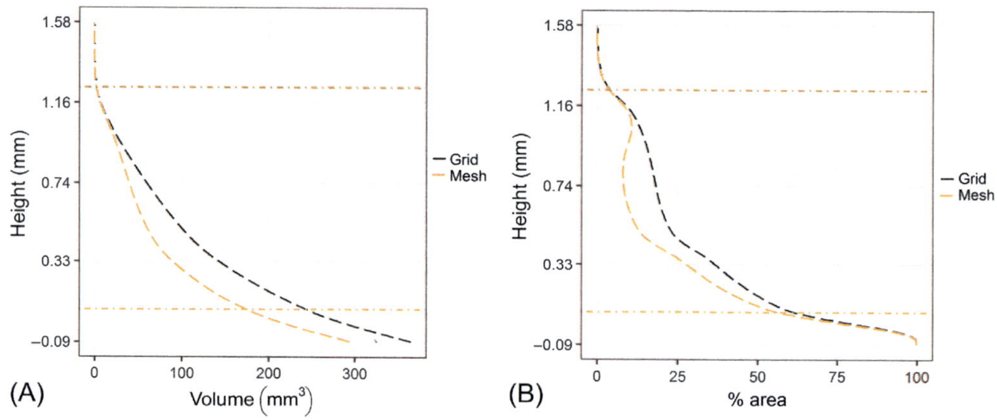

Fig. 10.15 Volume below the scale-limited surface as a function of the height (A) and Abbott-Firestone curve (B).

observe that the difference between the mesh-based methods and the height map is higher compared with the previous case. The most relevant high differences can be found on the estimation of *Sdq* (≈2) and *Sdr* (≈−70%) and the volume-related parameters *Vmc* (≈−35%) and *Vvc* (≈7%). The projection method leads to high bias in structured surface with reentrant features in the estimation of the density of the volume and the Abbott-Firestone curve (see Fig. 10.15).

10.5.4 Portion of turbine blade

In this section a blade of a turbocharger impeller is analyzed. The CT-scanning parameters are reported in Table 10.8. The analyzed blade, reconstructed with the AR algorithm setting the maximum distance of 10 μm, is shown in Fig. 10.16A. S magnification showing the estimated form surface is shown in Fig. 10.16B. The reference form is computed using the fast and effective feature-preserving mesh denoising [34] implemented in guided denoising [35]. This method was used to remove the low-frequency component of the measured mesh to estimate the roughness data set. The roughness data set was computed with the signed distance using the pseudonormal method implemented in LibIGL [36]. To allow the computation of the parameters per ISO 25178-2, the

Table 10.8 XT H 225 settings for turbine blade surface.

Parameter	Value	Parameter	Value
Filter material	Copper (0.25 mm)	Voxel size	93.69 μm
Acceleration voltage	120 kV	Magnification	5.85
Filament current	150 μA	Detector size (pixels)	1008 × 1008
Exposure time	1000 ms	Number of projections	1583

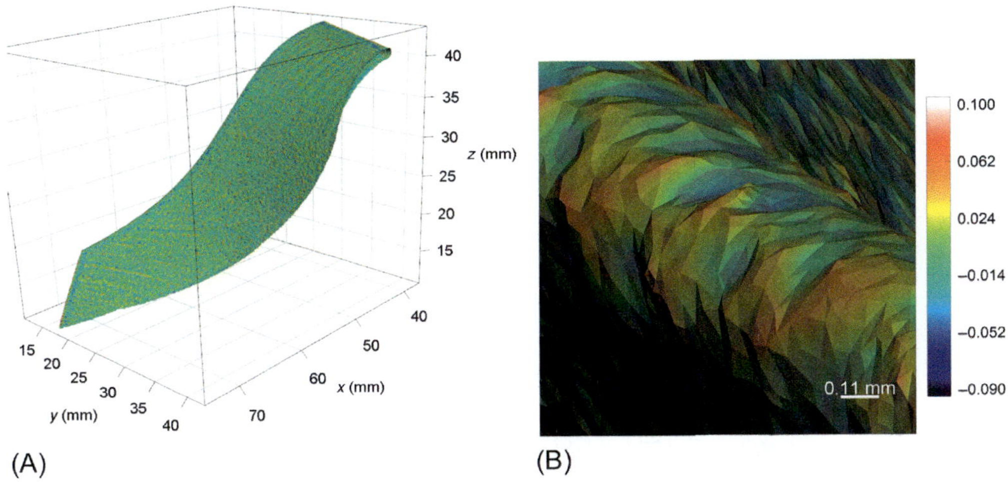

Fig. 10.16 Measured portion of turbine blade and estimated form surface (A) and magnification of the mesh (B).

measured mesh was parametrized using the minimum area distortion algorithm implemented in OpenGI [37].

The parameters per ISO 25718-2 are computed projecting onto a grid of 500 × 500 points; the residual data set is then expressed as a function of the parameters' space. Since it is an abstract space, the values of the parameters regarding the area cannot be properly estimated. The estimated parameters are reported in Table 10.9. The differences between the estimation of the height parameters are negligible; except *Sv* and *Sz* the difference is the effect of the sampling size used in projecting the data set onto a grid. Due to the parametrization, it is not possible to compute the parameters that depend on the metric of form surface, *Sdq*, *Sdr* and *Srf* and the functional parameters, with the projection method. It is possible to observe that the percentage of the reentrant features is almost null. Fig. 10.17 shows volume's distribution, and the Abbott-Firestone curve and the grid-based curve present a shifting due to the difference in the estimation of *Sv*.

10.6 Feature parameters

The last type of parameters analyzed is the so-called feature parameters. The idea is to first subdivide the measured surface into different regions, each one characterized by a feature, then the parameters can be computed. The first step of a feature-based characterization is called segmentation; the surface is subdivided into nonintersecting regions using the watershed segmentation [38]. The watershed algorithm simulates the rain; that is, it assumes that the water is falling on a (nonflat) surface. Due to gravity the water flow will be along the steepest descendent slope, until it reaches a local minimum. The subset of all

Table 10.9 Estimated texture parameters of the turbine blade surface.

Method	*Sa* (μm)	Δ*Sa* (%)	*Sq* (μm)	Δ*Sq* (%)	*Ssk*	Δ*Ssk*	*Sku*	Δ*Sku*
Grid	10.66	–	13.77	–	−0.08	–	3.70	–
Mesh	10.70	0.41	13.78	0.05	−0.12	−0.04	3.65	−0.05

Method	*Sp* (μm)	Δ*Sp* (%)	*Sv* (μm)	Δ*Sv* (%)	*Sz* (μm)	Δ*Sz* (%)	*Sdq*	Δ*Sdq*
Grid	96.74	–	72.67	–	169.41	–	5.21	–
Mesh	97.68	0.97	89.24	22.81	186.92	10.34	0.20	−5.01

Method	*Vmp* (μm)	Δ*Vmp* (%)	*Vmc* (μm)	Δ*Vmc* (%)	*Vvv* (μm)	Δ*Vvv* (%)	*Vvc* (μm)	Δ*Vvc* (%)
Grid	< 0.01	–	38.58	–	0.028	–	80.10	–
Mesh	< 0.01	–	51.27	32.88	0.003	−90.63	79.69	−0.52

Method	*Sdr* (%)	Δ*Sdr* (%)	*Srf* (%)	A_{sl} (mm^2)	A_{form} (mm^2)	A_{shadow} (mm^2)	V_{sl} (mm^3)	V_{form} (mm^3)
Grid	469.12	–	0.00	5.69	1.00	1.00	0.07	0.07
Mesh	21.15	−447.97	0.00	1438.16	1408.37	1408.37	123.28	124.21

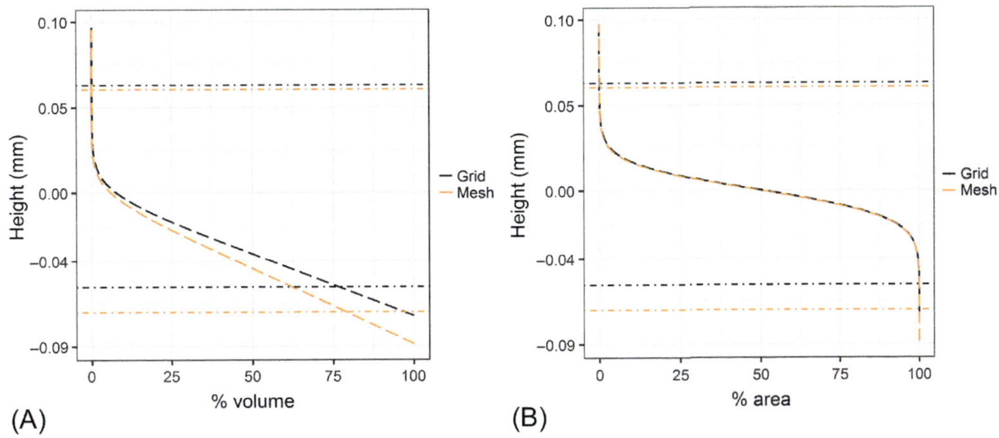

Fig. 10.17 Volume below the scale-limited surface as function of the height (A) and Abbott-Firestone curve (B).

the points whose steepest slope reaches the minimum is called a catchment basin. The lines dividing each region are called watershed lines. The watershed algorithm can be easily generalized to a triangular mesh [39] associating a scalar value to each vertex value. The scalar value in the texture characterization is the signed distance function between the measured point and its projection on the form mesh. According to this definition the minima of the scalar field are the pits, so it is called pit-based watershed. If the value of the scalar field is multiplied by −1, another segmentation can be computed. The minima become the peaks, so the watershed is called peak-based watershed. Fig. 10.18 shows both the pit-based (A) and peak-based (C) watersheds. A common drawback of the watershed algorithm, due to the noise, is the over segmentation; that is, if the number of regions is high, the features of interest cannot be identified. The number of segmented regions is reduced by an operation called Wolf pruning [40]; the pits (peaks) with the smallest difference are merged. The operation is then repeated until the smallest difference is higher than a given threshold. The threshold is usually defined as a percentage of *Sz* value. The segmentation after a 5% Wolf pruning of the pit- and peak-based watersheds are shown in Fig. 10.18B and D, respectively.

As described in Scott [40] and in ISO 25178-2 [7], the feature-related parameters can be divided into spacing/hybrid, peak, and material. The hybrid parameters are divided into the following:

- density of the summits (*Sds*): computed as the number of peaks per unit area, and
- arithmetic mean of the peak curvature (*Spc*).

The peak parameters are related to the feature heights:

- five points peak height (*S5p*): computed as the average value of the five peaks with the largest global peak heights,

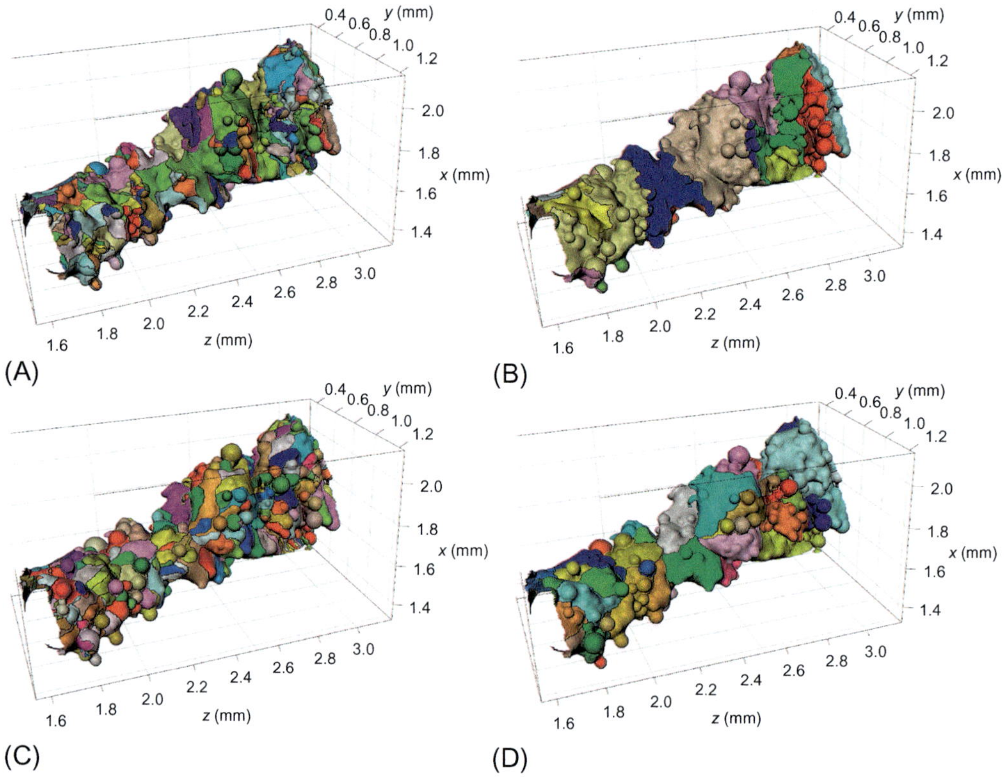

Fig. 10.18 Watershed segmentation: pit-based watershed without (A) and with 10% Wolf pruning (B) and peak-based watershed without (C) and with 10% Wolf pruning (D).

- five points pit height (*S5v*): computed as the average value of the five pits with the largest global pit heights, and
- ten points height of the surface (*S10z*): computed as the sum of *S5p* and *S5v*.

The material parameters describe the area/volume properties of the segmented features:

- mean dale area (*Sda*)
- mean hill area (*Sha*)
- mean dale volume (*Sdv*)
- mean hill volume (*Shv*)

The values of the feature parameters are reported in Table 10.10. The values of the hybrid and material parameters with and without pruning should be noted.

Table 10.10 Feature parameters.

Method	*Spd*	*Spc*	*S5p* (μm)	*S5v* (μm)	*S10z* (μm)	*Sda* (mm^2)	*Sha* (mm^2)	*Sdv* (mm^3)	*Shv* (mm^3)
No pruning	203.05	0.38	143.34	133.48	276.82	0.0058	0.0049	0.0001	0.0001
5% pruning	16.32	6.10	131.92	127.66	259.57	0.1043	0.0612	0.0028	0.0016

10.7 Summary

In this chapter a generalization of the texture parameters to free-form surfaces, with the triangular mesh representation, was discussed. New manufacturing processes and measuring systems pose new challenges in the surface characterization area. The set of parameters allow the characterization of surfaces, which would not be possible using the standard height map representation.

References

[1] Dong WP, Sullivan PJ, Stout KJ. Comprehensive study of parameters for characterizing three-dimensional surface topography I: some inherent properties of parameter variation. Wear 1992; 159:161–71.
[2] Dong WP, Sullivan PJ, Stout KJ. Comprehensive study of parameters for characterizing three-dimensional surface topography II: statistical properties of parameter variation. Wear 1993;167: 9–21.
[3] Dong WP, Sullivan PJ, Stout KJ. Comprehensive study of parameters for characterizing three-dimensional surface topography III: parameters for characterising amplitude and some functional properties. Wear 1994;178:29–43.
[4] Dong WP, Sullivan PJ, Stout KJ. Comprehensive study of parameters for characterizing three-dimensional surface topography IV: parameters for characterising spatial and hybrid properties. Wear 1994;178:45–60.
[5] Blunt L, Jiang X. Advanced techniques for assessment surface topography: development of a basis for 3D surface texture standards "surfstand". Elsevier; 2003.
[6] Jiang X, Scott PJ, Whitehouse DJ, Blunt L, editors. Paradigm shifts in surface metrology. Part II. The current shift. Proceedings of the Royal Society of London A: Mathematical, Physical and Engineering Sciences; 2007.
[7] ISO. ISO 25178-2, Geometrical product specification (GPS)—surface texture: areal—part 2: terms, definitions and surface texture parameters; 2012.
[8] Townsend A, Pagani L, Scott P, Blunt L. Areal surface texture data extraction from X-ray computed tomography reconstructions of metal additively manufactured parts. Precis Eng 2017;48(Supplement C):254–64.
[9] Thompson A, Senin N, Giusca C, RJCA L. Topography of selectively laser melted surfaces: a comparison of different measurement methods. CIRP Ann Manf Technol 2017;66(1):543–6.
[10] DigitalSurf. MountainsMap; 2017.
[11] Abdul-Rahman HS, Zeng W, Lou S, Jiang X, Scott PJ. Freeform texture representation and characterisation based on triangular mesh projection techniques. Measurement 2016;92:172–92.
[12] Pagani L, Qi Q, Jiang X, Scott PJ. Towards a new definition of areal surface texture parameters on freeform surface. Measurement 2017;109:281–91.
[13] Pagani L, Jiang X, Scott PJ. Investigation on the effect of sampling on areal texture parameters. Measurement 2018;128:306–13.
[14] Pagani L, Lou S, Zeng W, Jiang X, Scott PJ, editors. Effect of the form estimation on the areal texture parameters for X-ray computed tomography measurement. 8th Conference on Industrial Computed Tomography (iCT); Wels, Austria; 2018.
[15] Agarwal S, Mierle K. Others. Ceres Solver, http://ceres-solver.org.
[16] ISO. ISO 25178-3, Geometrical product specifications (GPS)—surface texture: areal—part 3: specification operators; 2012.
[17] do Carmo MP. Differential geometry of curves and surfaces. Prentice-Hall; 1976.
[18] Weisstein EW. Barycentric coordinates: MathWorld-A Wolfram web resource. Available from: http://mathworld.wolfram.com/BarycentricCoordinates.html.
[19] Xiao H, Gimbutas Z. A numerical algorithm for the construction of efficient quadrature rules in two and higher dimensions. Comput Math Appl 2010;59(2):663–76.

[20] Tang Y, Kurtz A, Zhao YF. Bidirectional evolutionary structural optimization (BESO) based design method for lattice structure to be fabricated by additive manufacturing. Comput Aided Des 2015;69:91–101.
[21] Emmelmann C, Scheinemann P, Munsch M, Seyda V. Laser additive manufacturing of modified implant surfaces with osseointegrative characteristics. Phys Procedia 2011;12(Part A):375–84.
[22] NV NM. Nikon CT-Pro; 2016.
[23] Yoo TS, Ackerman MJ, Lorensen WE, Schroeder W, Chalana V, Aylward S, et al. Engineering and algorithm design for an image processing Api: a technical report on ITK—the Insight Toolkit. Stud Health Technol Inform 2002;85:586–92.
[24] Lorensen WE, Cline HE. Marching cubes: a high resolution 3D surface construction algorithm. ACM Comput Graph 1987;21(4):163–9.
[25] Boissonnat J-D, Oudot S. Provably good sampling and meshing of surfaces. Graph Model 2005;67:405–51.
[26] Rineau L, Yvinec M. A generic software design for Delaunay refinement meshing. Comput Geom Theory Appl 2007;38:100–10.
[27] Centin M, Signoroni A, editors. RameshCleaner: conservative fixing of triangular meshes. Smart tools and apps for graphics—Eurographics Italian chapter conference. The Eurographics Association; 2015.
[28] Botsch M, Kobbelt L, Pauly M, Alliez P, Lévy B. Polygon mesh processing. A K Peters; 2010.
[29] Schroeder W, Martin K, Lorensen B. The visualization toolkit; 2006.
[30] Wang X-F, Huang D-S, Xu H. An efficient local Chan–Vese model for image segmentation. Pattern Recogn 2010;43(3):603–18.
[31] Botsch M, Kobbelt L. A remeshing approach to multiresolution modeling. In: Proceedings of the 2004 Eurographics/ACM SIGGRAPH symposium on geometry processing; Nice, France. 1057457: ACM; 2004. p. 185–92.
[32] Mobius J, Kobbelt L. OpenFlipper: an open source geometry processing and rendering framework. In: Proceedings of the 7th international conference on curves and surfaces; Avignon, France. 2187629. Springer-Verlag; 2012. p. 488–500.
[33] Rusu RB, Cousins S, editors. 3D is here: point cloud library (PCL). IEEE international conference on robotics and automation (ICRA); May 9–13; Shanghai, China; 2011.
[34] Sun X, Rosin PL, Martin RR, Langbein FC. Fast and effective feature-preserving mesh denoising. IEEE Trans Vis Comput Graph 2007;13(5):925–38.
[35] Deng B. Guided denoising; 2015.
[36] Jacobson A, Panozzo D. Others. Libigl: A simple C++ geometry processing library; 2016.
[37] Rau C. OpenGI; 2011.
[38] Soille P. Morphological image analysis: principles and applications. Springer Science & Business Media; 2013.
[39] Čomić L, De Floriani L, Iuricich F, Magillo P. Computing a discrete Morse gradient from a watershed decomposition. Comput Graph 2016;58:43–52.
[40] Scott PJ. Feature parameters. Wear 2009;266(5):548–51.

CHAPTER 11

Characterization of free-form structured surfaces

11.1 Introduction

Surface metrology as a discipline is currently undergoing a significant paradigm shift, from stochastic to structured surfaces [1]. Structured surfaces can be defined as "surfaces with deterministic patterns of usually high aspect ratio geometric features designed to give a specific function." All structured surfaces are specifically designed to meet a highly defined functional requirement such as the specific optical response of a Fresnel lens. A key common feature of structured surfaces is that they are normally high-aspect ratio surfaces where the surface geometry is the critical determination of the component function. The scale of structured surfaces ranges from the macroscale down to the nanoscale. Specific examples include MEMS/NEMS devices, micromolding, microfluidic systems (lab on a chip), structured coatings, microlenses and optic elements, biointegration coatings, self-cleaning coatings, and sheet metal products.

The ability to adequately characterize the geometry of these free-form structured surface is crucial in the optimization and control of such functional surfaces. However, despite the rapid development of structured surface-related products, the suitable metrology theory and practical methodologies are still far behind the requirements, because traditional surface texture parameters, such as roughness, waviness, and statistical descriptions, are not appropriate for the description of deterministic geometries and add little information.

MEMS surfaces are one type of typical structured surface. From the functional point of view, the accuracy of the geometrical feature (line width, step height, etc.) of this type of surfaces is generally more important than the roughness. Removing noise from the measured surface data is often the first step for the accurate separation of these geometrical features. It's therefore the basic requirement of the denoising techniques that they should not only reduce the noise but also do so without blurring or changing the location of the features. However, traditional filtration techniques lack the ability to preserve the feature boundary with high accuracy. For example, when using the Gaussian filter for the MEMS surface, the boundary of the lines and steps are smoothed. Therefore it is very difficult to exactly evaluate the width of lines and heights of steps. In recent years, several techniques for the analysis of structured surfaces have emerged, including robust filters [2, 3], wavelet

* For more examples and practical implementation, check the website: www.digitalsurf.com/freeform.

Advanced Metrology
https://doi.org/10.1016/B978-0-12-821815-0.00011-3

filters, morphological filters, and pattern analysis. Among them, based on Scott's fundamental work [4], new data segmentation methods and feature parameters have been successfully used in the characterization of the structured surface [5, 6]. Our recent work has included approaches based on nonlinear diffusion. These techniques, based on the use of partial differential equations, have been extensively studied since the early work of Perona and Malik in 1987 [7].

Tessellation surfaces are another typical structured surface; these are surfaces that have a repeated structure comprising of a collection of tiles that fill the plane with no overlaps and no gaps. The following terms refer to the same concept: tessellated structured surface/tessellated surface/tessellations. The most widely used tessellated engineering surfaces are optical lenses, such as the retroreflection used on road surfaces, road signs and safe clothing, and microlens array for back light optics. Another widely used tessellation surface in engineering is the electron beam textured steel sheet surface, which has a repeated pattern on its surface, and also abrasive surfaces (typical example is the 3M TRI-ZAC abrasive paper surface), which consists of an array of triangular-based micron-sized pyramids. Structured surfaces are made up of discrete objects (primitive cells). Hence the definitions and characterization of these nanoscale structured surfaces are at the interface between the discrete and continuous worlds, which are very difficult to evaluate using traditional stochastic models. Traditional surface texture parameters, such as roughness and waviness parameters, cannot be used for discrete geometries and cannot be related to the function for structured surfaces, and statistical descriptions add little information. For example, diffraction grating behavior is basically determined by the ratio of groove spacing to optical wavelength and by the groove geometry. Such surface finish parameters as both profile parameters (Rt, Rq, etc.) and areal parameters (Sa, Sq, etc.), by contrast, provide little information in this case. For tessellation surfaces, it is very often the shape and geometry information that is more important than surface texture. For example, when evaluating the functional quality of a microlens array (MLA), the geometry of the individual features (individual lens); their maximum, minimum, and mean values; and the distribution of the features (spacing of the lens) are more important than the surface roughness. Conventional linear filtration technologies based on the wavelength/frequency extract the useful spatial information nonefficiently.

11.2 Characterization framework of free-form structured surface

Manufactured parts with complex free-form structured surfaces have also been widely used recently in automobile, bioengineering, medical, consumer electronics, etc. The scale of free-form structured surfaces ranges from the macroscale down to the nanoscale. Specific examples include flexible/wearable electronics, MEMS/NEMS devices, micromolding, microfluidic systems, defined geometry abrasives, and biointegration coatings. Compared with traditional "stochastic surface," "free-form surface," and "structured surface," the free-form structured surfaces have two significant characteristics: (1) they have a complex base surface (reference surface/mean surface), which has complex shape/free-form shape, and (2) they have deterministic features with high aspect ratios on the base surface. The ability to adequately characterize these complex structured surface geometry features is crucial in the optimization and control of such functional devices/components. When characterizing complex structured surfaces, three fundamental issues need to be considered:

(1) How to separate the free-form reference surface from the structured features without any distortion both on the free-form/reference surfaces and on the structured features?

(2) How to separate the structured features from high-frequency noise/surface texture?

(3) How to evaluate the structured features?

In this chapter a general framework for characterization of complex free-form structured surfaces is proposed (Fig. 11.1), which includes four steps:

(1) form/form error separation to find a reference surface and the form-removed structured feature,

(2) denoising/filtration to reduce the effect of instrument noise or separate the features from surface texture,

(3) feature extraction to separate feature regions following by identification of metrological characteristics of surface features and relationships, and

(4) feature characterization to quantitative assessment of feature attributes.

For issue (1), two methods can be used. The first is the robust form filtration including higher-order robust Gaussian regression filter and robust spline filter, which have proven to be very effective and efficient for the form separation when the cutoff has been deliberately chosen. This method can only be used for situations where there is no function for the form. The second method is robust fitting/association. This method can be applied when there is a function that can be used to describe the form.

For issue (2) a partial differential equation (PDE)-based adaptive nonlinear diffusion filter is proposed. The proposed diffusion filter is based on the PDE method and can be seen as a nonlinear heat equation, which describes the distribution of heat (or variation in

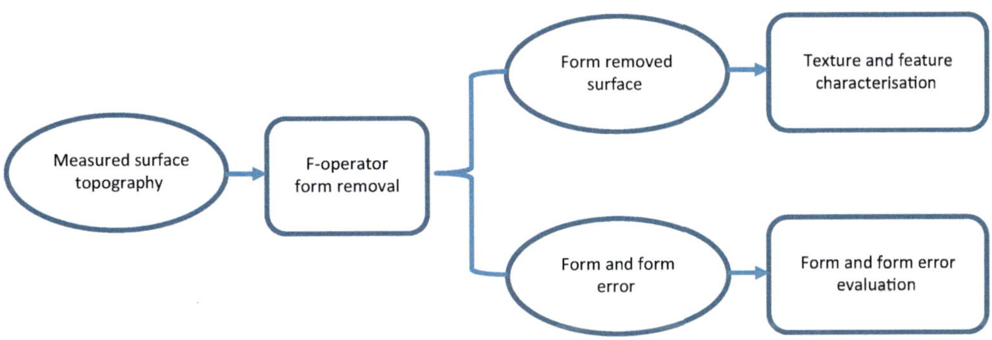

Fig. 11.1 Scheme framework for characterization of complex structured surfaces.

temperature) in a given region overtime. The diffusivity function based on the gradient of the surface can help to separate the internal region area of the boundary area of the measured feature. In this way the diffusion process will take place mainly in the interior regions (line, step, etc.) of the surface, and it will not affect the region boundaries where the magnitude of gradient is large.

For issue (3) a pattern analysis can be used, and a detailed description can be found in the following section.

In this chapter the following section gives an introduction to the use of robust filtration, including robust spline filter and robust Gaussian regression filters, to solve issue (1). In Section 11.4 a newly proposed nonlinear PDE diffusion filter is used to denoise/ remove short-wavelength component to solve issue (2). Then, in Section 11.5, pattern analysis techniques are applied for the characterization of the feature. Section 11.6 gives some case studies on how to apply the proposed framework to evaluate MEMS-type surface. Then in Section 11.7, there is an introduction to the characterization of tessellated features. Finally, in Section 11.8, the summary and conclusions are given.

11.3 F-operator for the form removal of the free-form structured surface

Measured surfaces always contain some irrelevant information that is not of interest for a particular application. Hence, there is need to filter out that irrelevant information, for example, form and form error that are not part of the free-form structured surfaces.

Most widely used F-operators for the separation of form/form error from the surface texture/feature are surface association, or form fitting. For accurate separation, normally, the nominal form of the measured surface should be known, for example, a plane, a cylinder, and sphere. The equation that describes the form needs to be known, then the association/fitting techniques can be used. However, in most cases, the nominal surface form is not known. One of the most widely used methods, when the nominal form is not known, is the assumption that the form can be described by a polynomial function of a fixed order. This is acceptable when the form is simple; however, free-form structured surfaces have, by definition, forms that mean polynomial fitting cannot be used. In this chapter, to separate the form and the surface texture/feature, it is proposed to use filtration.

Conventional linear filters such as the linear Gaussian filter and the linear spline filter are not suitable for the free-form structure surface analysis. This is because free-form structured surfaces always have high-aspect ratio surfaces that are not suitable for linear filtration. When linear filters are used for free-form structured surfaces, not only there is a distortion at the boundary and the departure of the reference surface from the measured surface, but also features (primitive structures) themselves will be distorted.

Consequently, in this chapter, the authors use robust filters to preprocess the free-form structured surfaces.

11.3.1 Robust spline filter

11.3.1.1 Generalized robust spline filter

The ordinary spline filter $s(x_k)$ can be described as follows: let the data points (x_i, z_i), $(i=0\cdots n-1)$ be given, and assume they are sorted in a strictly ascending order in the coordinates x_i, which are equidistant, and of width Δx, that is, $x_0 < x_1 < \ldots x_{n-2} < x_{n-1}$ is valid. To find $s(x_k)$, the residual errors are minimized using the L2-norm together with the condition of minimizing the bending energy, which is proportional to $\int\left(\frac{d^2s(x)}{dx^2}\right)^2 dx$, and the condition of natural boundary condition. According to the well-known method of the variation calculus, the spline filter $s(x_k)$ can be described as the following minimization problem [8–10]:

$$\sum_{k=0}^{n-1}(z_k - s(x_k))^2 + \lambda\int_{x_0}^{x_{n-1}}\left(\frac{d^2s(x)}{dx^2}\right)^2 dx \to \min_{s(x)} \tag{11.1}$$

with λ being the Lagrange parameter. The above minimization problem can be solved by solving the following linear equations:

$$\left[\mathbf{I} + \alpha^4\mathbf{Q}\right]\mathbf{s} = \mathbf{z} \tag{11.2}$$

where $\mathbf{I}$ is the identity matrix, $\alpha = 1/\left(2\sin\frac{\pi\Delta x}{\lambda_c}\right)$, λ_c is the cutoff wavelength, and $\mathbf{Q}$ is the coefficient matrix depending on the boundary conditions.

The robust spline filter is introduced by minimizing the L1-norm of the residual errors under the condition of minimizing the bending energy and the condition of natural boundary condition. It is defined as

$$\sum_{k=0}^{n-1}|z_k - s(x_k)| + \lambda\int_{x_0}^{x_{n-1}}\left(\frac{d^2s(x)}{dx^2}\right)^2 dx \to \min_{s(x)} \tag{11.3}$$

The previous equation can only be solved iteratively using algorithms such as split Bregmann iteration. For full details on this algorithm, see [3].

To unify the theory and algorithms of linear and nonlinear spline filters and to optimize the nonlinear filter by using a better estimator, a generalized spline filter is proposed. It can be defined as the following minimization procedure [11]:

$$\sum\nolimits_{k=0}^{n-1}\rho(z_k - s(x_k)) + \lambda\int_{x_0}^{x_{n-1}}\left(\frac{d^2s(x)}{dx^2}\right)^2 dx \to \min_{s(x_k)} \tag{11.4}$$

Comparing this expression with Expressions (11.1), (11.3), it can be found that the only difference between them is that a more general error metric function $\rho(\cdot)$, which is a symmetric, positive-definite function of the residual error with a unique minimum at zero, is used in the new model.

11.3.1.2 M-estimation

Let r_k be the difference between the kth measurement data z_k and its fitted value s_k. The standard least squares method tries to minimize $\sum_k r_k^2$, which is not robust if there are outliers present in the data because the objective functions increase indefinitely. The M-estimators try to reduce the effect of outliers by replacing the squared residuals r_i^2 by another function of the residuals, yielding

$$\min \sum_k \rho(r_k) \tag{11.5}$$

where ρ is a symmetric, positive-definite function with a unique minimum at zero and is chosen, so its rate of increase is less than square when the residual increases. Instead of solving this problem directly, one can implement it as an iterative reweighted least squares problem. Let $\mathbf{s} = [s_0, \ldots, s_{n-1}]^T$ be the vector to be estimated. The M-estimator of $\mathbf{s}$ based on the function $\rho(r_k)$ is the vector $\mathbf{s}$, which is the solution of the following n equations:

$$\sum_k \psi(r_k) \frac{\partial r_k}{\partial s_j} = 0, \quad \text{for } j = 0, \ldots, n-1 \tag{11.6}$$

where the derivative $\psi(x) = d\rho(x)/dx$ is called the influence function. If we now define a weight function

$$\delta(r) = \psi(r)/r \tag{11.7}$$

Eq. (11.7) becomes

$$\sum_k \delta(r_k) r_k \frac{\partial r_k}{\partial s_j} = 0, \quad \text{for } j = 0, \ldots, n-1 \tag{11.8}$$

This is exactly the system of equations that we obtain if we solve the following iterative reweighted least squares (IRLS) problem:

$$\min \sum_k \delta\left(r_k^{(m-1)}\right) r_k^2 \tag{11.9}$$

where the superscript m indicates the iteration number. The weight $\delta(r_k^{(m-1)})$ should be recomputed after each iteration to be used in the next iteration.

11.3.1.3 Iteratively reweighted least squares solution for generalized spline filter

According to the previous discussion of M-estimation theory, the minimization problem defined in Eq. (11.4) can be generally solved by using the IRLS method as follows [11]:

$$\sum_{k=0}^{n-1} \delta_k (z_k - s(x_k))^2 + \lambda \int_{x_0}^{x_{n-1}} \left(\frac{d^2 s(x)}{dx^2} \right)^2 dx \to \min_{s(x)} \tag{11.10}$$

by using Euler-Lagrange calculus

$$\left[\boldsymbol{\delta}^{(m-1)} + \alpha^4 \mathbf{Q} \right] \mathbf{s}^{(m)} = \boldsymbol{\delta}^{(m-1)} \mathbf{z} \tag{11.11}$$

where

$$\boldsymbol{\delta}^{(m-1)} = \begin{bmatrix} \delta_0^{(m-1)} & & & & 0 \\ & \delta_1^{(m-1)} & & & \\ & & \ddots & & \\ & & & \delta_{n-2}^{(m-1)} & \\ 0 & & & & \delta_{n-1}^{(m-1)} \end{bmatrix} \quad \mathbf{Q} = \begin{bmatrix} 1 & -2 & 1 & & & & 0 \\ -2 & 5 & -4 & 1 & & & \\ 1 & -4 & 6 & -4 & 1 & & \\ & & \ddots & \ddots & \ddots & & \\ & & 1 & -4 & 6 & -4 & 1 \\ & & & 1 & -4 & 5 & -2 \\ 0 & & & & 1 & -2 & 1 \end{bmatrix}$$

Eq (11.12) can be written more concisely and conveniently as [3]

$$\left[\mathbf{I} + \alpha^4 \mathbf{Q} \right] \mathbf{s}^{(m)} = \mathbf{s}^{(m-1)} + \boldsymbol{\delta}^{(m-1)} \left(\mathbf{z} - \mathbf{s}^{(m-1)} \right) \tag{11.12}$$

The other advantage of the Eq. (11.12) is that, at every iteration, the matrix $[\mathbf{I} + \alpha^4 \mathbf{Q}]$ is constant and only needs to be calculated once.

L2 norm—Linear spline filter

Let $\rho(r) = \frac{1}{2} r^2$, then according to Eqs. (11.6), (11.7), $\delta(r) = 1$, and Eq. (11.11), can be simplified as [11]

$$\left[\mathbf{I} + \alpha^4 \mathbf{Q} \right] \mathbf{s}^{(m)} = \mathbf{z} \tag{11.13}$$

thus the superscript m can be removed. Solving Eq. (11.13), only one step is needed; it is equivalent to the linear spline filter as defined in the ISO standard.

Proposed nonlinear spline filter

Other widely used M-estimators including Huber, Cauchy, and Tukey functions. Among them, the Tukey function has been verified to be highly robust against outliers and converges very well [5]:

$$\rho(r)=\begin{cases}\frac{c^2}{6}\left(1-\left[1-(r/c)^2\right]^3\right), & if|r|\le c\\ c^2/6, & if|r|>c\end{cases}\qquad \psi(r)=\begin{cases}r\left[1-\left(\frac{r}{c}\right)^2\right]^2, & if|r|\le c\\ 0, & if|r|>c\end{cases}$$

$$\delta(r)=\begin{cases}\left[1-\left(\frac{r}{c}\right)^2\right]^2, & if|r|\le c\\ 0, & if|r|>c\end{cases}\tag{11.14}$$

In this chapter, Tukey function has been chosen as the standard error metric.

11.3.1.4 Experimental results

The linear and nonlinear robust spline filters are applied to some measured data. The results are shown in Figs. 11.2 and 11.3. Fig. 11.2 shows that the linear spline filter can process data with significant form components very well and with almost no boundary effects. Fig. 11.3 is used to compare the linear and nonlinear spline filters. The left and the right figures are the results from a measured honing surface profile and a measured worn milled surface profile, respectively. The upper figures show the original profile and the mean profile by using linear and nonlinear spline filters, respectively, and the lower figures show the residual profiles. From these figures, one can clearly see that both the linear and nonlinear spline filters can follow large forms well with almost no boundary

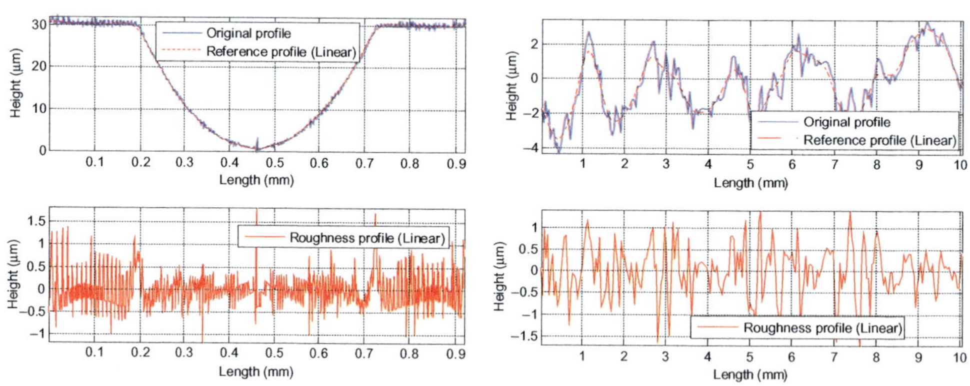

Fig. 11.2 Linear spline filter (left, microlens mold, cutoff = 0.08 mm, and right, chattered milled profile, cutoff = 0.8 mm).

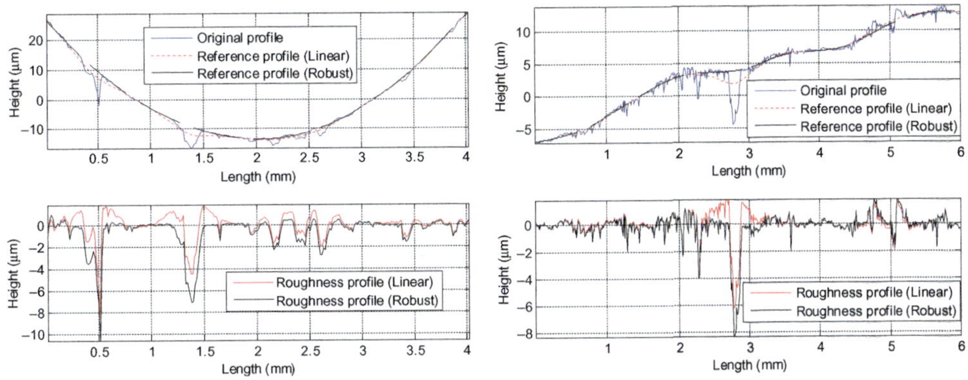

Fig. 11.3 Linear and robust spline filter (left, honing surface, cutoff = 0.8 mm, and right, worn milled surface, cutoff = 1 mm).

effects, while the nonlinear spline filters are robust against data outliers (extremely high peak or deep valley). Also the computation speed is very fast. For a typical 64129 pts data set, the linear spline filter only takes 16 ms, while the nonlinear spline needs only two iterations to get the converge result (32 ms).

11.3.2 Robust Gaussian regression filter

The robust regression filter was originally introduced in [12]. The authors of this paper introduced an M-estimation-based general model and the fast algorithm of the robust regression filter both for profile and areal surface analysis [2]. The general model of the robust regression filters up to second order is as follows:

Let z_{ij}, w_{ij} be the measured surface data and resulting mean surface data, respectively; i, j be the index of the surface coordinate in the x and y directions, respectively; and $i=0, \ldots, m-1$; $j=0, \ldots, n-1$, m, n be the number of points in x and y direction, respectively; and the lateral sample spacing are Δx, Δy. The discrete model of the second-order areal Gaussian regression filter can be described by the following minimization problem:

$$\arg\min_{w_{i,j}} \left\{ \sum_{l=0}^{n-1} \sum_{k=0}^{m-1} \rho\left(z_{k,l} - w_{i,j} - \sum\nolimits_{h=0}^{2} \beta_{h,2-h}(i,j)((k-i)\Delta x)^{h}((l-j)\Delta y)^{2-h} \right) \cdot s_{i,j,k,l} \right\} \tag{11.15}$$

where $s_{i,j,k,l} = (1/\alpha\lambda_{cx})\exp(-\pi((k-i)\Delta x/\alpha\lambda_{cx})^2)\cdot(1/\alpha\lambda_{cy})\exp(-\pi((l-j)\Delta x/\alpha\lambda_{cx})^2)$ and λ_{cx}, λ_{cy} are the selected cutoff wavelength in x and y direction, respectively. The reweighting function $\rho(\cdot)$ is a symmetric, positive-definite function with a unique minimum at zero. The resulting reference surface can be obtained by solving the following linear equations iteratively [2]:

$$w^{(t)}(i,j) = [1\ 0\ 0\ 0\ 0\ 0] \begin{bmatrix} G_{00}^{(t-1)} & G_{10}^{(t-1)} & G_{01}^{(t-1)} & G_{20}^{(t-1)} & G_{11}^{(t-1)} & G_{02}^{(t-1)} \\ G_{10}^{(t-1)} & G_{20}^{(t-1)} & G_{11}^{(t-1)} & G_{30}^{(t-1)} & G_{21}^{(t-1)} & G_{12}^{(t-1)} \\ G_{01}^{(t-1)} & G_{11}^{(t-1)} & G_{02}^{(t-1)} & G_{21}^{(t-1)} & G_{12}^{(t-1)} & G_{03}^{(t-1)} \\ G_{20}^{(t-1)} & G_{30}^{(t-1)} & G_{21}^{(t-1)} & G_{40}^{(t-1)} & G_{31}^{(t-1)} & G_{22}^{(t-1)} \\ G_{11}^{(t-1)} & G_{21}^{(t-1)} & G_{12}^{(t-1)} & G_{31}^{(t-1)} & G_{22}^{(t-1)} & G_{13}^{(t-1)} \\ G_{02}^{(t-1)} & G_{12}^{(t-1)} & G_{03}^{(t-1)} & G_{22}^{(t-1)} & G_{13}^{(t-1)} & G_{04}^{(t-1)} \end{bmatrix}^{-1} \begin{bmatrix} D_{00}^{(t-1)} \\ D_{10}^{(t-1)} \\ D_{01}^{(t-1)} \\ D_{20}^{(t-1)} \\ D_{11}^{(t-1)} \\ D_{02}^{(t-1)} \end{bmatrix} \tag{11.16}$$

where [2]

$$G_{u,v}^{(t-1)}(i,j) = \sum_{l=0}^{n-1} \sum_{k=0}^{m-1} \delta_{k,l}^{(t-1)} \cdot ((k-i)\Delta x)^u \cdot ((l-j)\Delta y)^v \cdot s_{i,j,k,l} \tag{11.17}$$

$$D_{u,v}^{(t-1)}(i,j) = \sum_{l=0}^{n-1} \sum_{k=0}^{m-1} \delta_{k,l}^{(t-1)} z_{k,l} \cdot ((k-i)\Delta x)^u \cdot ((l-j)\Delta y)^v \cdot s_{i,j,k,l} \tag{11.18}$$

and t is the iteration step and $\delta^{(t)}(r) = \rho'(r)/r$ is the reweighting function with r being the residual from last iteration $r_{i,j} = z_{i,j} - w_{i,j}^{(t-1)}$.

Figs. 11.4 and 11.5 show two typical results by using the robust regression filter to process the tessellation surfaces. Compared with the linear filter, the robust regression

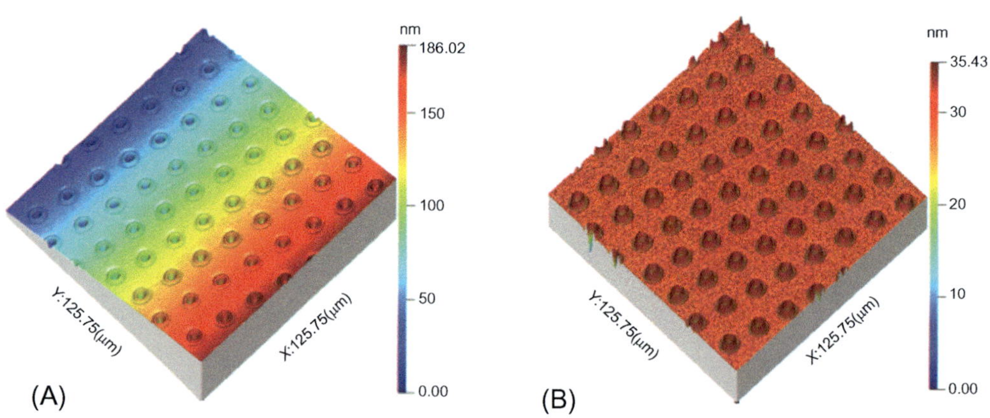

Fig. 11.4 (A) Measured LZT surface and (B) tessellation surface by using the proposed regression filter.

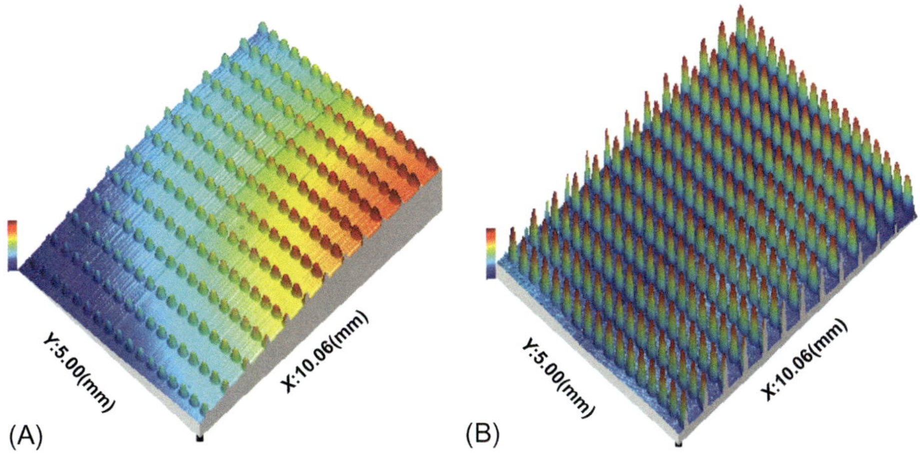

Fig. 11.5 (A) Measured microlens array surface and (B) tessellation surface by using proposed regression filter.

filter has the following advantages: (1) it can process surfaces with significant form and can also act as a form filter; (2) all boundary data can be processed and used with no distortion; and (3) the shape of the feature are well preserved, for example, the shapes and geometries of the individual laser zone spots in the LZT surface in Fig. 11.4 and the individual microlens in the Fig. 11.5.

Fig. 11.6A shows a typical example for free-form structured surface, which is a microlens array with free-form base surface. Fig. 11.6B and C shows the form filtered by using the linear Gaussian filter and the second-order Gaussian regression filter, respectively. One can clearly see that by using the second-order regression filter the reference free-form surface can be separated without distortion, while the linear filter has distortion.

11.4 PDE-based surface characterization

PDE-based linear/nonlinear diffusion methods have proved to be useful in many fields ranging from image processing to computer-aided quality control and postprocessing of noisy data. One typical diffusion method is the model of heat diffusion. The heat equation is an important partial differential equation that describes the distribution of heat (or variation in temperature) in a given region overtime. A more detailed introduction about the heat diffusion can be seen in the previous chapter.

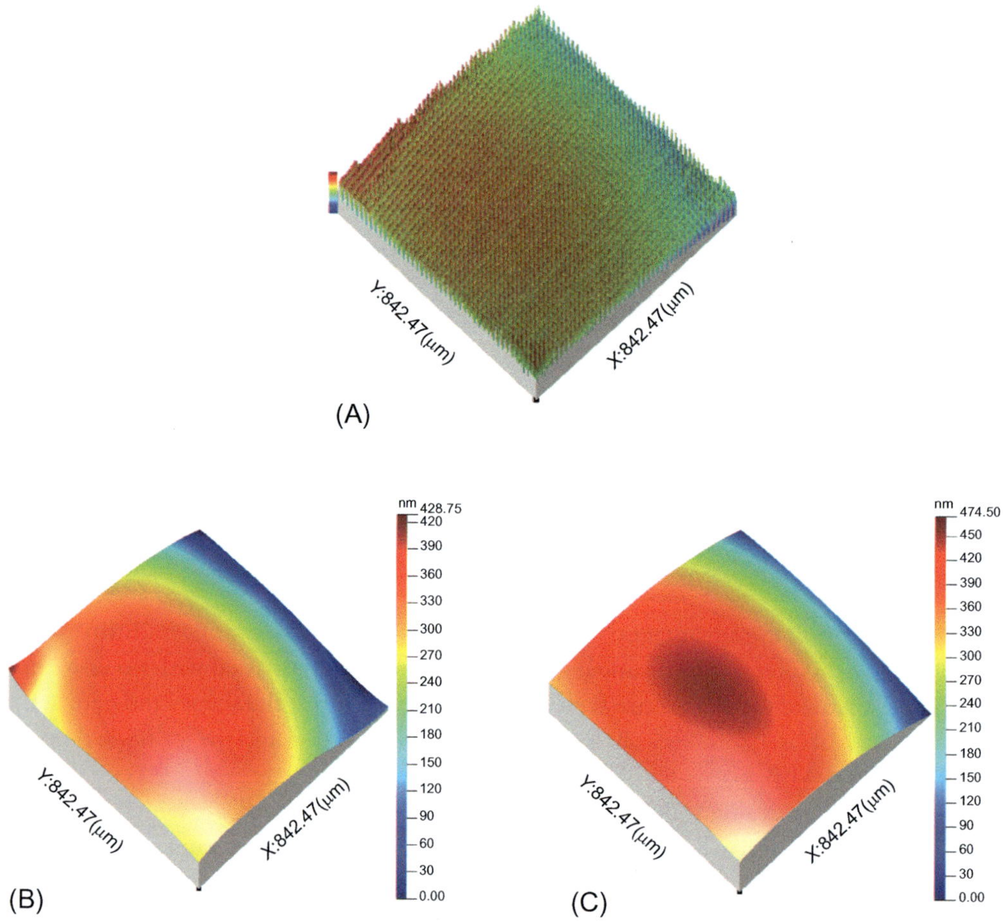

Fig. 11.6 (A) Measured microlens array (B) form by linear Gaussian filter and (C) form by second-order Gaussian regression filter.

11.4.1 Linear diffusion for regular lattice grid surface analysis

For surface characterization, let $f(x)$ being the measured/original profile, with x being the spatial coordinate defined on the domain $(0, lx)$ and lx is the measuring length. The linear diffusion filter for profile surface analysis can be defined as

$$\begin{aligned} &\frac{\partial u}{\partial t} = \operatorname{div}(c\nabla u) = c\Delta u = c \cdot u_{xx} \\ &u(x, 0) = f(x) \end{aligned} \tag{11.19}$$

where $u(x, t)$ is the smoothed/filtered profile from the original profile $f(x)$, with time scale t, which associates with the standard deviation σ and can be related with the cutoff wavelength that is used in general filtration procedure. For linear diffusion filter, c is a constant.

For areal surface analysis, let the surface domain be an open rectangle $\Omega := (0, lx) \times (0, ly)$; the areal surface is represented by a bounded function $f: \Omega \rightarrow \mathbb{R}$. Then a filtered areal surface $u(x, y, t)$ with a scale parameter $t > 0$ may be obtained as the solution of a diffusion equation with the measured surface $f(x, y)$ as initial condition:

$$\begin{aligned} \frac{\partial u(x, y, t)}{\partial t} &= \mathrm{div}\,(c \cdot \nabla u) = c \cdot \Delta u = c \cdot \left(u_{xx} + u_{yy}\right) \\ u(x, y, 0) &= f(x, y) \end{aligned} \tag{11.20}$$

where u_{xx}, u_{yy} represents the second derivative at point (x, y) in two orthogonal measurement direction, respectively.

Linear diffusion filter can smooth/denoise profile/areal surfaces very well. However, two disadvantages have limited its use as a surface filter: (1) the location of the edge of region on the surface will be dislocated with the time scale increase, so the filter results do not give the right position of the surface feature and (2) due to the fact that the linear diffusion is a Gaussian smoothing procedure, it does not only reduce noise but also smooth important features such as edges.

11.4.2 Adaptive diffusion filter for surface analysis

By introducing an adaptive conduction coefficient c, the previous problem can be partly solved. The c can be designed as a function that has small value at the edge while it has a large value within the interior region of the feature on the measured surface. An improved linear diffusion filter described in Eq. (11.21) uses a function of the gradient of the original measured surface f as the conduction coefficient/diffusivity function [7, 13].

$$\frac{\partial u}{\partial t} = \mathrm{div}\left(g\left(\|\nabla f\|^2\right) \cdot \nabla u\right) \tag{11.21}$$

Compared with standard linear diffusion filters, the conduction coefficient in this improved model is substituted with a function, $c = g(\|\nabla f\|^2)$, where $g(\cdot)$ is a nonnegative monotonically decreasing function in respect to $\|\nabla f\|^2$, with $g(0) = 1$. Here, $\|\nabla f\|^2$ is chosen to locate the edge of the feature. In this way the diffusion process (smoothing process) will take place mainly in the interior regions (line, step, etc.) of the surface feature, and it will not affect the region boundaries where the magnitude of $\|\nabla f\|$ is large. According to [14], the optimized $g(\cdot)$ can be chosen as

$$g(x) = e^{\left(-(x/K)^2\right)} \tag{11.22}$$

$$g(x) = \frac{1}{1 + (x/K)^2} \tag{11.23}$$

Here, K is a preset constant that is dependent on the distribution of the gradient of the surface. It can also be set according to the value of integral of histogram of the absolute values of the gradient throughout the surface [7, 13]. For simplicity the K can be preset as the mean absolute value of the gradient of the original surface:

$$K = \text{mean}(\|\nabla f\|) \tag{11.24}$$

Fig. 11.7 gives two types of widely used diffusivity function.

Fig. 11.8 compares the filtering results of a practical measured profile surface by using linear diffusion and a linear adaptive diffusion filter. From these results, one clearly sees that the filtering result of the linear diffusion filter is equivalent to a general Gaussian filter. The profile has been smoothed; however, the edge of the feature has also been blurred. By using the adaptive diffusion filter, the profile has been smoothed very well, while the edges have been retained.

Eq. (11.25) is still a linear procedure, as $c = g(\|\nabla f\|^2)$ is a function of position and does not change with increasing time. A more complex model was proposed by Perona and Malik for image processing and called the nonlinear anisotropic diffusion filter [14]:

$$\frac{\partial u}{\partial t} = \text{div}\left(g\left(\|\nabla u\|^2\right) \cdot \nabla u\right) \tag{11.25}$$

This is different from the linear case in that using the gradient of the original surface as the independent variable, the conduction coefficient/diffusivity function of the nonlinear mode, and using the gradient of the smoothed surface as the independent variable mean that the conduction coefficient varies not only with different location but also with increasing time:

$$c = g\left(\|\nabla u\|^2\right) \tag{11.26}$$

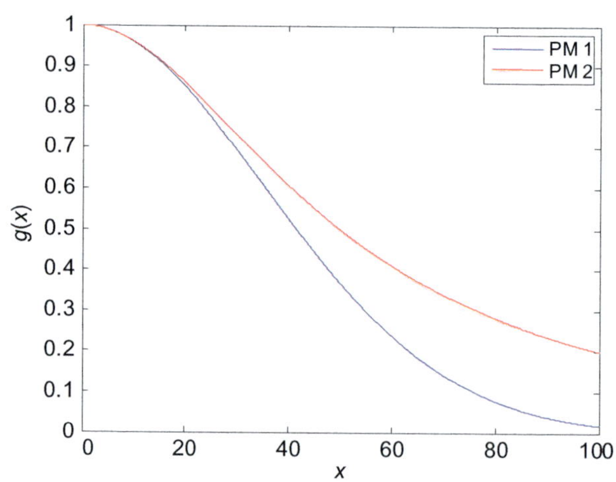

Fig. 11.7 Diffusivity function.

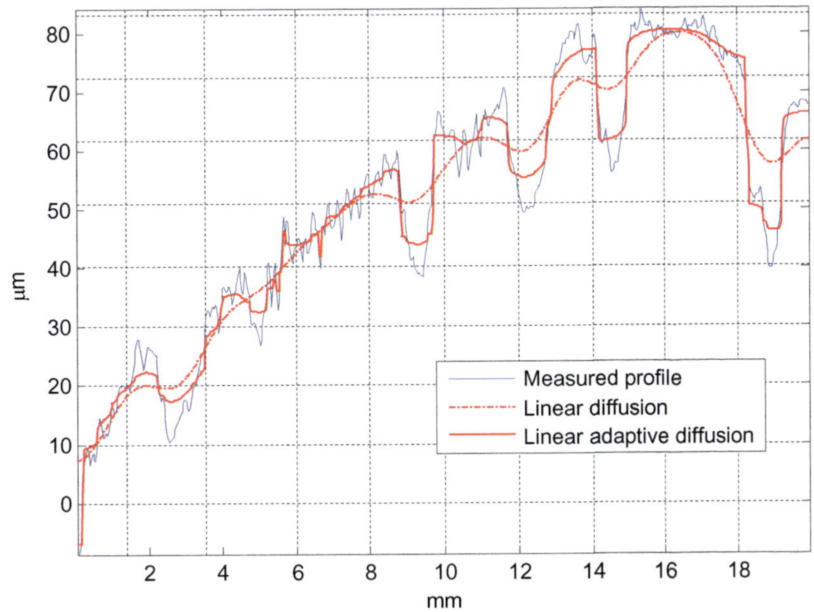

Fig. 11.8 Comparison of the linear diffusion and the linear adaptive diffusion.

And also the K is set as the mean absolute value of the gradient of the updated surface value:

$$K = \text{mean}(\|\nabla u\|) \tag{11.27}$$

11.4.3 Wavelet regularization PDE-based surface characterization

The accuracy of the previous models is highly dependent on the gradient calculation of the surface. One serious problem is that outliers in the measured surface could introduce very large oscillations of the gradient ∇u. Therefore the gradient-based model can possibly miscalculate the true edges and outliers, which lead to undesirable diffusion in regions where there is no true edge. A possible way to improve this is to regularize the model.

Similar to the method of designing a diffusivity function, there are two basic methods to using regularization: one is applying the regularization on the original surface; the other is applying the regularization on the diffused surface at each step. The second one has been widely used for nonlinear diffusion.

A typical improved model is based on Gaussian regularization:

$$\frac{\partial u}{\partial t} = \text{div}\left(g(\|\nabla(u^* G_\sigma)\|)\nabla u\right) \tag{11.28}$$

where G_σ is a Gaussian kernel with variance σ and * denotes the convolution operator. It means that in each iteration there are three steps to calculate the updated diffusivity function: (1) using Gaussian filter to regularize the diffused surface, (2) calculate the gradient using the regularized surface, and (3) calculate the diffusion function.

A wavelet shrinkage method is introduced as the regularization filter in the following session.

The *k*th measured signal value can be written as

$$u = s + n \tag{11.29}$$

where s represents the true signal and n the noise. The aim of denoising is to estimate s as accurately as possible from u in the wavelet domain, at the ith decomposition level:

$$w_{u,i} = w_{s,i} + w_{n,i} \tag{11.30}$$

where $w_{u,i}$ is the noisy wavelet coefficient, $w_{s,i}$ is the true coefficient, and $w_{n,i}$ is the noise wavelet coefficient. The denoising problem is then to estimate $w_{s,i}$ as accurately as possible from $w_{u,i}$, and then reconstruct the signal from the denoised wavelet coefficients $w_{s,i}$. Simple denoising algorithms that use the wavelet transform consist of three steps: (1) calculate the wavelet transform of the noisy signal, (2) modify the noisy wavelet coefficients according to some rules, and (3) compute the inverse wavelet transform using the modified coefficients to reconstruct the denoised signal. One of the well-known rules for the second step is thresholding analysis, which includes soft and hard thresholding. The basic procedure of the improved diffusion filter is described in formulae (11.31), (11.32):

$$\frac{\partial u}{\partial t} = \operatorname{div}\left(g(\|\nabla(u_T)\|)\nabla u\right) \tag{11.31}$$

$$u_T = \mathrm{IWT}(\mathrm{ST}(\mathrm{WT}(u))) \tag{11.32}$$

where u_T represent the shrunken version of u; WT and IWT represent the forward and inverse wavelet, respectively; and ST represent the shrinkage operator.

The selection of the threshold value T is the key issue for the denoising problem. The simplest way is to use the variance of the wavelet coefficients on each level as the threshold value. In this paper the widely used maximum a posteriori (MAP) method is chosen to estimate the thresholding value. The MAP estimates the signal in terms of the probability density function (PDF) of the noise and the PDF of the signal coefficient.

11.4.4 Numerical solution and experiments

To make the model practical for the measured surface analysis, the previous model needs to be discretized.

11.4.4.1 Profile analysis

For profile surface, let z_i be the height of the measured profile surface at position ($i\Delta x$); u_i^t and u_i^{t+1} be the filtered profile data at time scale t and $t+1$, respectively; i be the index of the surface coordinate in the x direction; and $i=0, \ldots, M-1$, M be the number of points in x direction; the lateral sample spacing is Δx. The treatment of the diffusion process can be simplified as [7, 13]

$$\begin{aligned} u_i^{t+1} &= u_i^t + \tau\left[g(\|\nabla_L u\|) \cdot \nabla_L u + g(\|\nabla_R u\|) \cdot \nabla_R u\right]_i^t \\ u_i^0 &= z_i \end{aligned} \tag{11.33}$$

where τ is the time spacing and ∇_L and ∇_R represent the forward and backward difference of the neighbor points that are defined as

$$\nabla_L u_{i,j} = u_{i-1} - u_i \quad \nabla_R u_i = u_{i+1} - u_i \tag{11.34}$$

As the steps are typical feature types in MEMS surface, Figs. 11.9 and 11.10 use two simulated steps to evaluate the performance of the introduced diffusion filter. Fig. 11.9 is a single-sided step, and Fig. 11.10 is a double-sided step. Gaussian distributed noise, and also, some heavy noises have been added on the test data. From these examples, it is very clear that the profiles have been smoothed very well, and also the most important feature, the edge of the steps, is well preserved and not distorted.

Figs. 11.11 and 11.12 use two practical measured profile surfaces to demonstrate the ability of the diffusion filter for processing surfaces with multiple feature regions. Fig. 11.11 is a standard calibration gauge with multisteps, and Fig. 11.12 is an aspheric diffractive lens profile.

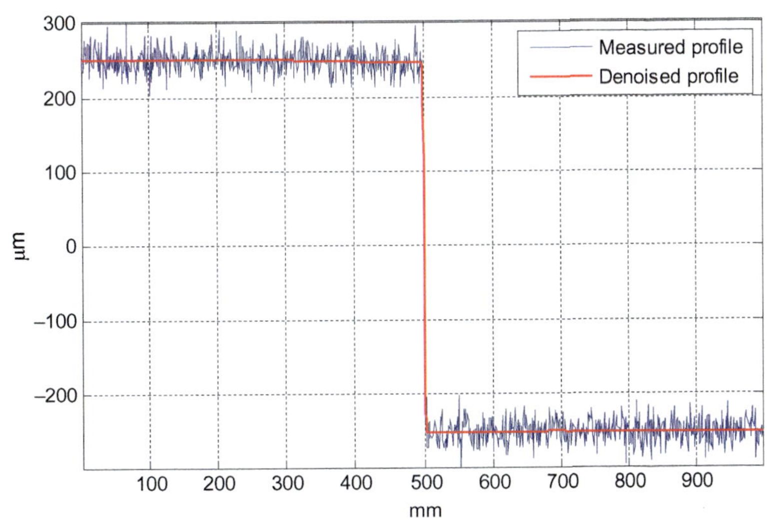

Fig. 11.9 Diffusion filter on a simulated single-sided step height.

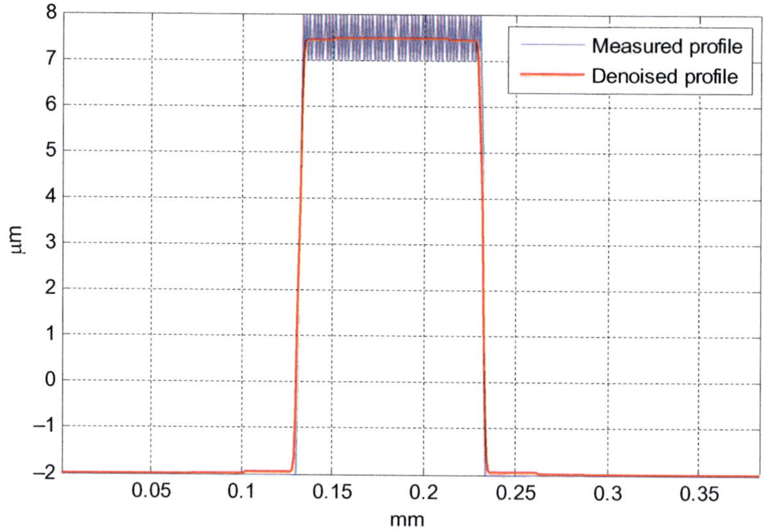

Fig. 11.10 Diffusion filter on a simulated double-sided step height.

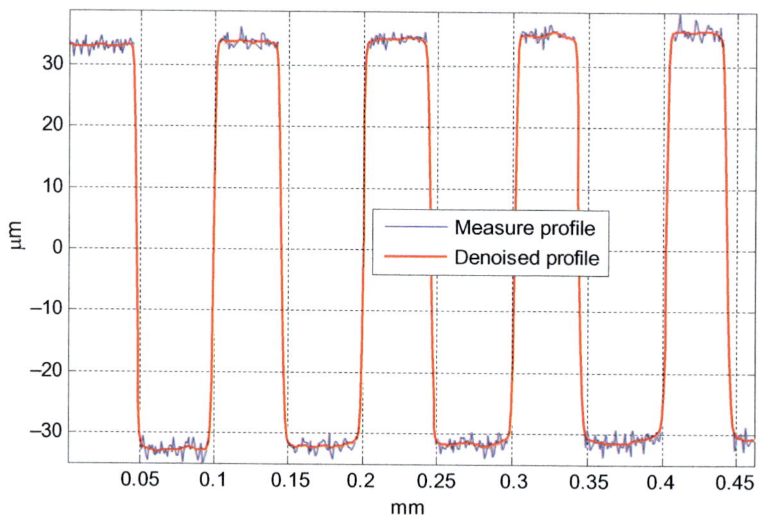

Fig. 11.11 Diffusion filter on a measured multistep height.

11.4.4.2 Areal surface analysis

For areal surface analysis, Let $z_{i,\ j}$ be the height of the measured surface data at position $(i\Delta x, j\Delta y)$; $u_{i,j}^{t}$ and $u_{i,j}^{t+1}$ be the filter surface data at time scale t and $t+1$, respectively; i,j be the index of the surface coordinate in the x and y directions; and $i=0,\ \ldots,\ M-1;\ j=0,\ \ldots,\ N-1$, M, N be the number of points in x and y directions, respectively, and the lateral

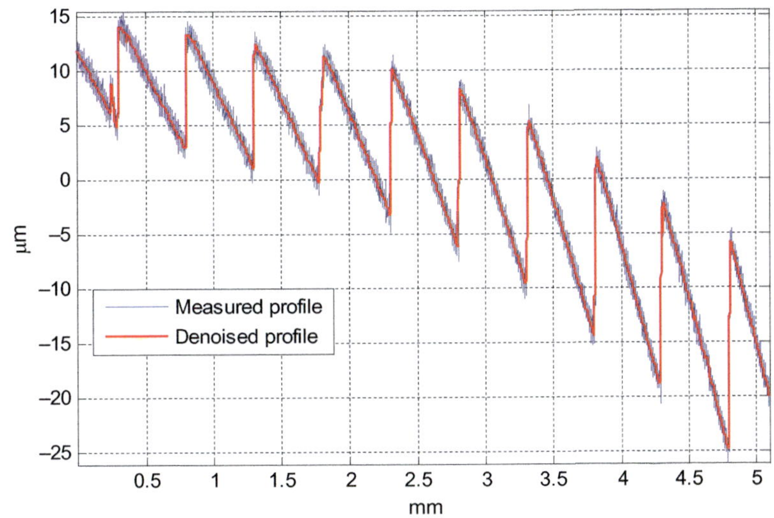

Fig. 11.12 Diffusion filter on an aspheric diffractive lens.

sample spacing are Δx, Δy. The model can be numerically described as the following by using a 4-neareset neighbor discretization scheme of the Laplacian operator [7, 13]:

$$\begin{aligned} u_{i,j}^{t+1} &= u_{i,j}^{t} + \tau\left[c_N \cdot \nabla_N u + c_S \cdot \nabla_S u + c_E \cdot \nabla_E u + c_W \cdot \nabla_W u\right]_{i,j}^{t} \\ u_{i,j}^{0} &= z_{i,j} \end{aligned} \tag{11.35}$$

The symbol ∇ indicates nearest neighbor difference:

$$\begin{aligned} \nabla_N u_{i,j} &= u_{i-1,j} - u_{i,j} \\ \nabla_S u_{i,j} &= u_{i+1,j} - u_{i,j} \\ \nabla_E u_{i,j} &= u_{i,j+1} - u_{i,j} \\ \nabla_W u_{i,j} &= u_{i,j-1} - u_{i,j} \end{aligned} \tag{11.36}$$

The conduction coefficients are updated at each iteration as a function of the gradient:

$$\begin{aligned} c_{N_{i,j}}^{t} &= g\left(\left|\nabla_N u_{i,j}^{t}\right|\right) \\ c_{S_{i,j}}^{t} &= g\left(\left|\nabla_S u_{i,j}^{t}\right|\right) \\ c_{E_{i,j}}^{t} &= g\left(\left|\nabla_E u_{i,j}^{t}\right|\right) \\ c_{W_{i,j}}^{t} &= g\left(\left|\nabla_W u_{i,j}^{t}\right|\right) \end{aligned} \tag{11.37}$$

Fig. 11.13 gives an example of the diffusion filter on a practical measure MEMS surface. To make it easy to see, Fig. 11.14 is the section profiles from Fig. 11.13. From

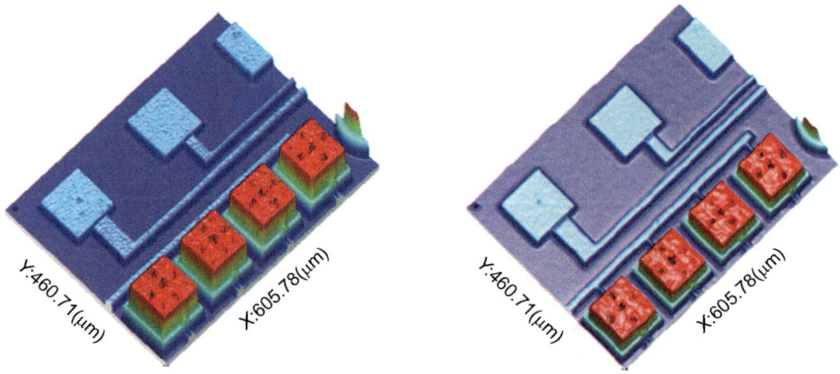

Fig. 11.13 Diffusion on a measured MEMS surface (up, measured, and down, denoised result).

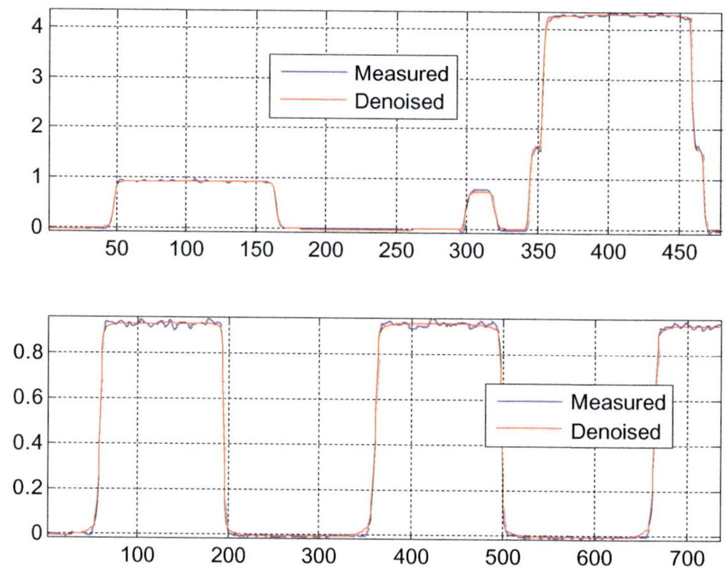

Fig. 11.14 Section profile form Fig. 11.13.

this example, one can clearly see that the most important feature on the surface, lines, steps, and even very small stairs are very well preserved.

11.5 Feature extraction

Identification of surface geometrical features on a free-form structured surface topography is more important than for general globe surface topography. It depends on a method to separate and quantify zones of differing planar height and then being able to analyze these zones both individually and in relation to each other.

11.5.1 Sobel edge operator

There are many edge detection methods by gradient operators such as the Sobel, Roberts, and Laplace operators, many of which are used to clarify the local transformation (e.g., sharp edges) in the optical metrology.

A Sobel edge detection operator consists of a pair of convolution kernels as shown in the succeeding text. The second kernel is simply a rotation of the first:

$$Sx = \begin{bmatrix} +1 & 0 & -1 \\ +2 & 0 & -2 \\ +1 & 0 & -1 \end{bmatrix}, \quad Sy = \begin{bmatrix} +1 & +2 & +1 \\ 0 & 0 & 0 \\ -1 & -2 & -1 \end{bmatrix} \tag{11.38}$$

These kernels are designed to respond maximally to edges running vertically and horizontally relative to the pixel grid, with one kernel for each perpendicular orientation. The kernels can be applied separately to surface measurement, to produce separate calculations of the gradient component in each orientation:

$$\begin{aligned} Gx &= Sx * Z \\ Gy &= Sy * Z \end{aligned} \tag{11.39}$$

These can then be combined together to find the absolute magnitude of the gradient at each point and the orientation of the gradient:

$$G = \sqrt{Gx^2 + Gy^2} \tag{11.40}$$

Other widely used edge operators include the Roberts cross operator. Similar with the Sobel operator, the Roberts cross operator can be defined by two convolution kernels as

$$Sx = \begin{bmatrix} +1 & 0 \\ 0 & -1 \end{bmatrix} \quad Sy = \begin{bmatrix} 0 & +1 \\ -1 & 0 \end{bmatrix} \tag{11.41}$$

The Prewitt operator can also be defined by a pair of convolution kernels as

$$Sx = \begin{bmatrix} -1 & 0 & +1 \\ -1 & 0 & +1 \\ -1 & 0 & +1 \end{bmatrix} \quad Sy = \begin{bmatrix} -1 & -1 & -1 \\ 0 & 0 & 0 \\ +1 & +1 & +1 \end{bmatrix} \tag{11.42}$$

For MEMS/MOEMS surface feature characterization (as shown in Fig. 11.3, concepts of MEMS/MOEMS surface characterization), the edge operator is only used here for initial region of the gradient as a preprocess for next pattern analysis/segmentation operation; in this case Sobel operator has been selected [5, 6].

11.5.2 Segmentation

Segmentation is the next step following on from the edge operator, and it aims to segment the surface topography into regions of interest, for example, to separate and delineate the step planes. These segmentation techniques developed by Scott [4] consider the surface as a continuous function and split the surface into segments consisting of hills and dales (Fig. 11.8). The techniques are based on work by Maxwell [15] who proposed dividing a landscape into regions consisting of hills and regions consisting of dales. A Maxwellian hill is an area from which maximum uphill paths lead to one particular peak, and a Maxwellian dale is an area from which maximum downhill paths lead to one particular pit. By definition the boundaries between hills are course lines (watercourses), and the boundaries between dales are ridge lines (watershed lines). Maxwell was able to demonstrate that ridge and course lines are maximum uphill and downhill paths emanating from saddle points and terminating at peaks and pits. Scott formulated this technique [4] for surface metrology and defined as texture primitives, the critical points and lines defining hills and dales on a surface representation as shown in Fig. 11.15.

A change tree [4] can be used to present the relationships between critical points on hills and dales and retain relevant information [4]. It represents the relationships between contour lines from a surface where the vertical direction represents height. In the change tree, adjacent contour lines form a line representing the contour lines continuously varying with height. Saddle points are represented by the merging of two or more of these lines into one line; peaks and pits are represented by the termination of lines, as

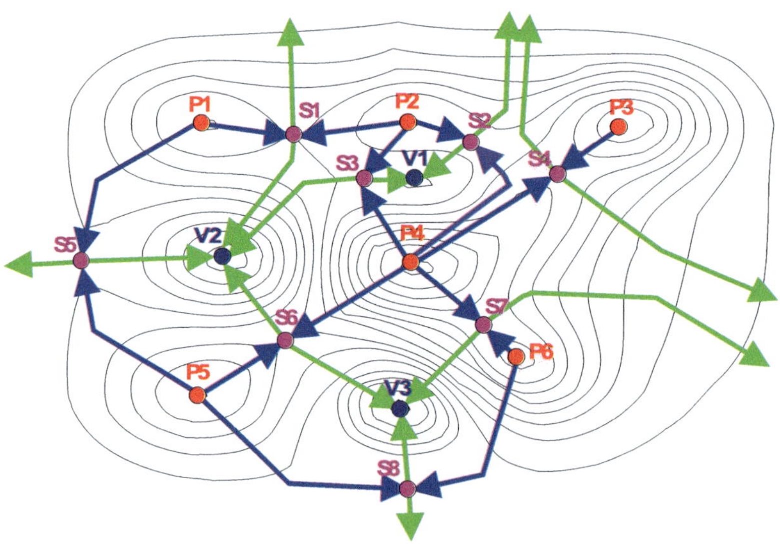

Fig. 11.15 Primitives—critical points, lines, and areas. Key: P = peak, V = pit, S = saddle point; ridgelines connect peaks to saddle points; course lines connect pits to saddle points.

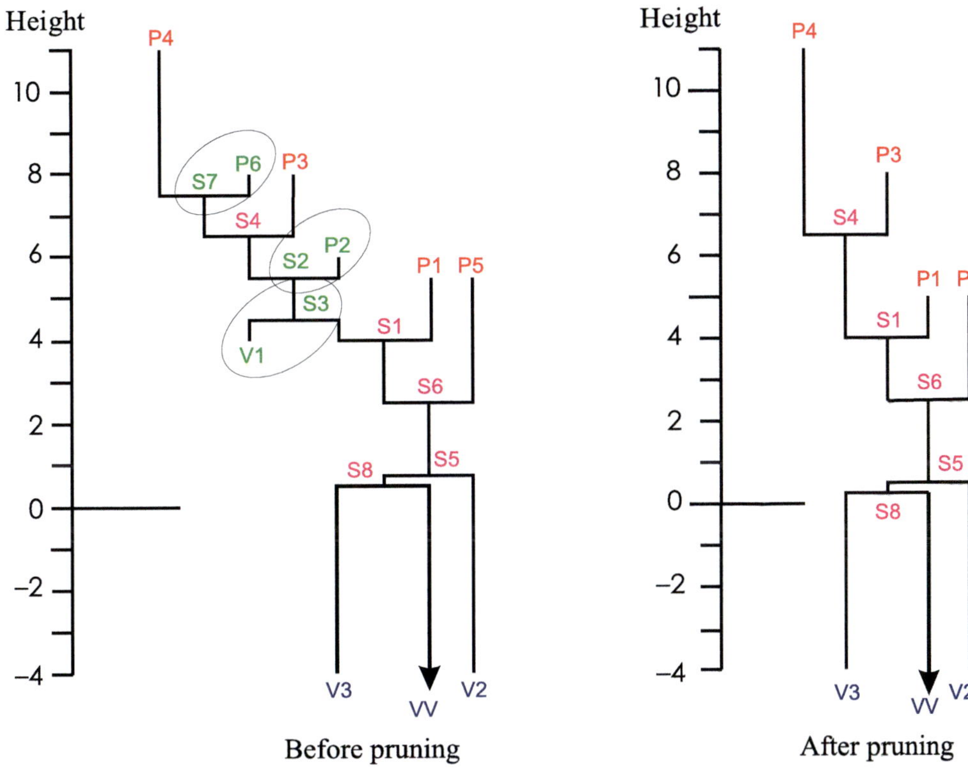

Fig. 11.16 Change trees.

shown in Fig. 11.15. A hill is an area from which maximum uphill paths lead to one particular peak, and a dale is an area from which maximum downhill paths lead to one particular valley. By definition the boundaries between hills are course lines, and the boundaries between dales are ridgelines. The topological relationship between critical points can be well encapsulated in a change tree graph as shown in Fig. 11.16.

11.5.3 Wolf pruning

To ensure the stability and robustness of the segmentation process, a set of necessary and sufficient criteria were identified [4]. The method of Wolf pruning satisfies these criteria. This pruning method consists of finding the peak or pit with smallest height difference to its adjacent saddle point on the change tree and then combining the peak or pit with the adjacent saddle point. The process is then repeated with that peak or pit with the smallest height difference to its adjacent saddle point on the change tree being eliminated until some height threshold is reached. For microcomponent surfaces, this threshold could be when all remaining height differences are above a fixed value. Normally, it is a value less than 10% of the local height.

11.6 Case studies for the characterization of MEMs surfaces

Most MEMS/MOEMS and semiconductor devices are manufactured by planar technologies such as lithography at micro-/nanoscale. In terms of geometry, these miniaturized components are a merging of microgeometry, surface geometrical features and texture. Their shape and size within specific planar areas and their texture become entities in their own right, having specific functional properties. As highlighted by in [16], this situation has not previously existed in traditional engineering, where the shape and texture are considered secondary to the size. This section focusses on surface features of a MEMS/MOEMS component and provide an unambiguous characterization for surface features of microcomponents, by using ISO 25178-2 [17] and ISO 25178-3 [18] together with MEMS filtering technology described previously.

As mentioned in the introduction, microcomponents in MEMS/MOEMS and semiconductor devices are primarily fabricated by planar technologies, such as physical or chemical deposition, patterning, lithography, and etching processes. By the very nature of these methods, most microcomponents have essentially a planar form over which is distributed surface geometrical features, whose primary geometrical constraints comprise feature width, height, and feature numbers and feature spatial relationships; for example, the "surface" of a microfluidic component is composed of zones of different planar heights and geometrically dispersed channels. Although functional features of MEMS/MOEMS surfaces have the same scale of heights as general surface texture, the characterization of MEMS/MOEMS surfaces is different from general engineering surfaces. The field parameters are of little use for these MEMS/MOEMS features.

The characterization system for MEMS/MOEMS surface must allow the segmentation of the surface into regions of interest followed by techniques to identify the basic geometrical features within these regions and then analyze their attributes before a departure from nominal as a function of their spatial arrangement. Based on these requirements the concepts for microcomponent surface characterization can be defined as shown in Fig. 11.17.

The only difference between Fig. 11.17 and previous figure is that the F-operator here use simple leveling method, due to the MEMS surface's property (Fig. 11.18).

An F-operator is necessary to generate a level planar datum or a reference surface of a measured MEMS surface topography. Since the setup errors will typically be very small angles, a simple way for a microcomponent surface to achieve this is to best fit a linear least squares plane (rather than best fit a total least squares plane) to a specified area, which can be an island region or a group of regions selected and treated as a composite region. Calculating the coefficients of the least squares plane is based on the specified region.

For a selected surface region $z_{\text{region}}(x, y)$, the plane is expressed as

$$f(x, y) = a_{00} + a_{10}x + a_{01}y \tag{11.43}$$

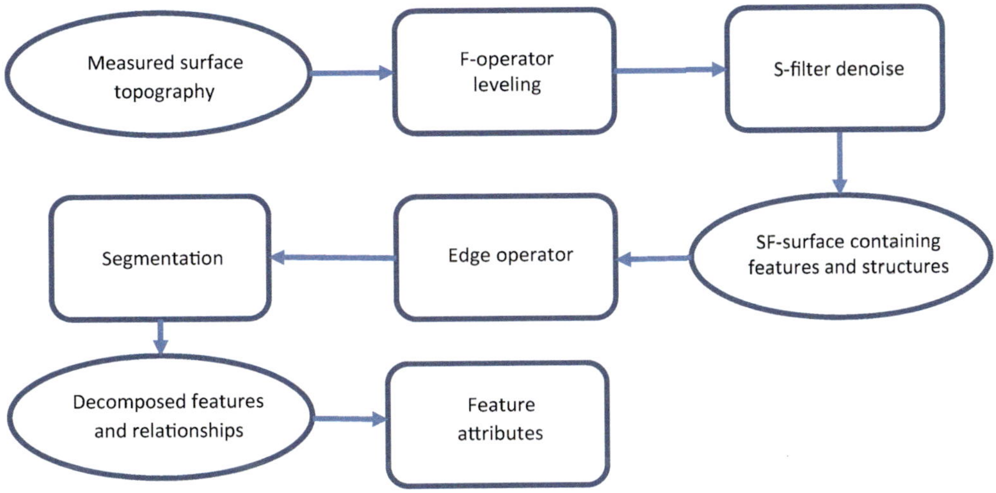

Fig. 11.17 Concepts of MEMS surface characterization.

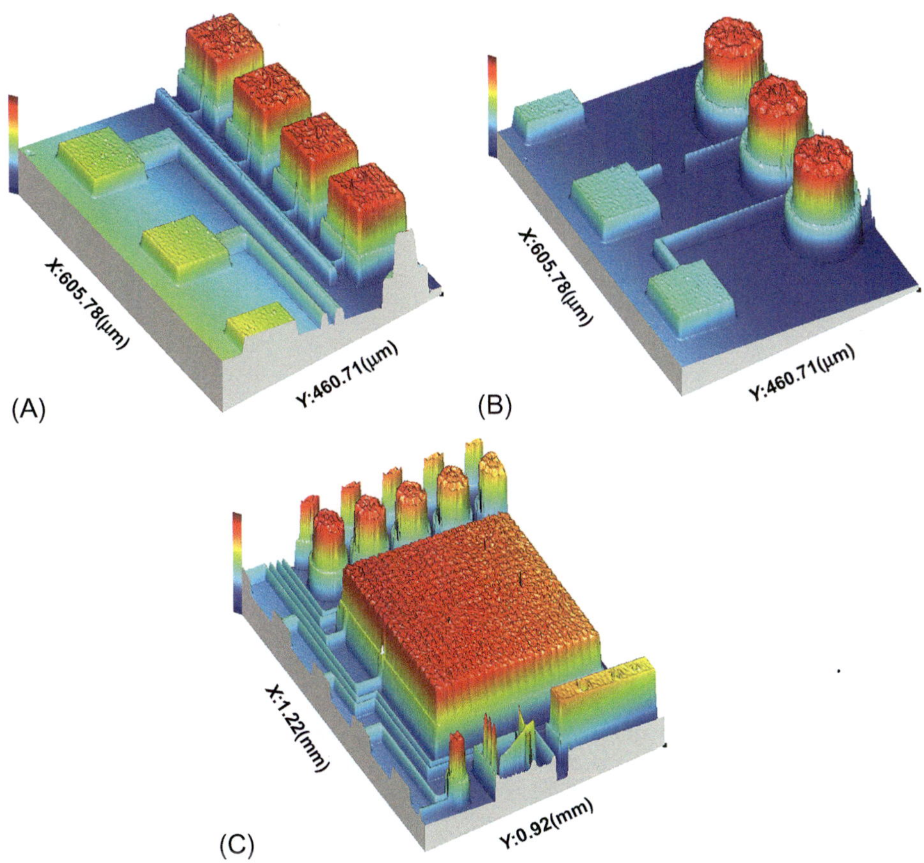

Fig. 11.18 Measured surface topographies of typical microcomponents. (A) Sample I, (B) Sample II, and (C) Sample III, all of them have clear slopes that cause by the measurement datum setup.

The solution of the least squared level surface can be solved by minimizing

$$\varepsilon^2 = \sum_{l=1}^{N} \sum_{k=1}^{M} \left(z_{\text{region}}(x_k, y_l) - f(x_k, y_l)\right)^2 = \sum_{l-1}^{N} \sum_{k=1}^{M} \left(z_{\text{region}}(x_k, y_l) - a_{00} + a_{10}x_k + a_{01}y_l\right)^2 \tag{11.44}$$

Fig. 11.19 shows the leveled measured surface topographies from Fig. 11.18 obtained by using this method.

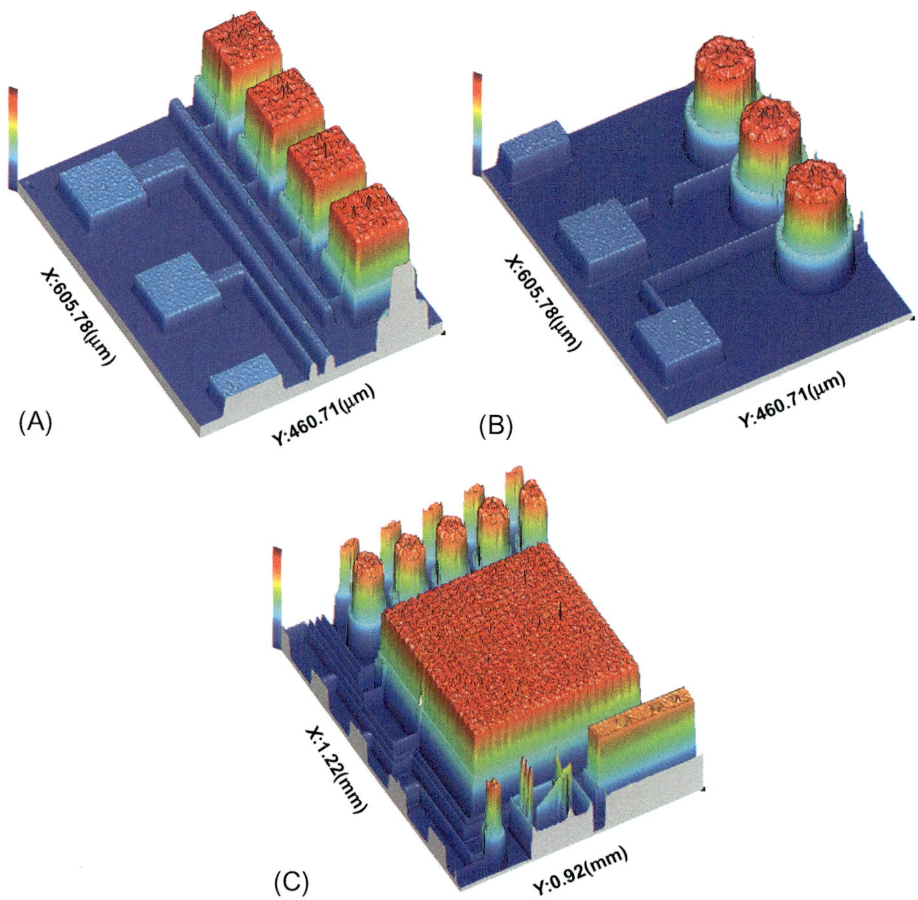

Fig. 11.19 Leveled surface topographies of the Fig 11.18 microcomponents. (A) Is the leveled surface topography from the Sample I (as shown Fig. 11.18A), (B) is the leveled surface topography from Sample II (as shown Fig. 11.18B), and (C) is the leveled surface topography from Sample III (as shown Fig. 11.18C).

After using F-operator to remove the form, the next step is to use S filter to remove the high-frequency/short-wavelength noise/roughness. Fig. 11.20 shows the results by using the previous presented nonlinear diffusion filter to remove noise from the surface data shown in Fig. 11.19.

Fig. 11.21 shows an example where the Sobel edge operator and segmentation have been applied to analyze the planar heights on a microfluidic device. In Fig. 11.21A the entrance to the reactor channels is shown at higher magnification. The channels have a close spacing of 3 μm; both plasma flow and consequent fluid separation are greatly influenced by the channel dimensions and any defect within the channels. The step planes can

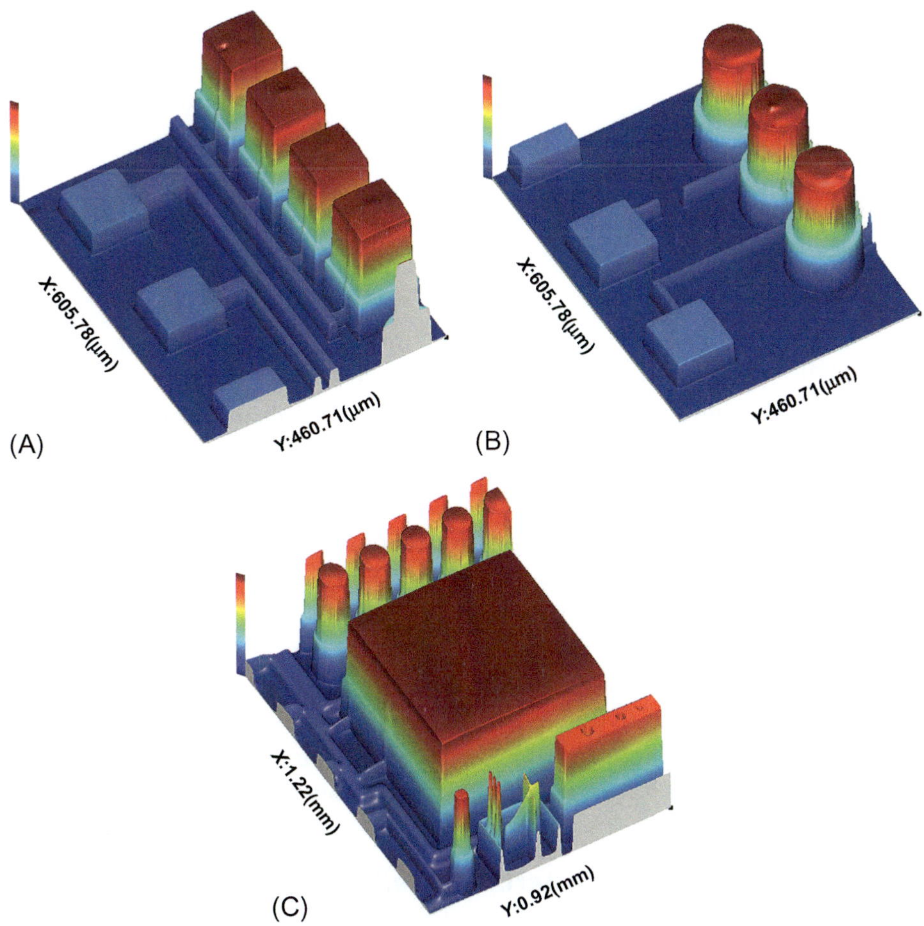

Fig. 11.20 Diffusion on the leveled MEMS surfaces as shown in Fig. 11.19. (A) Filtered surface topography of Sample I (as shown in Fig. 11.19A), (B) filtered surface topography of Sample II (as shown in Fig. 11.19B), and (C) filtered surface topography of Sample III (as shown Fig. 11.19C).

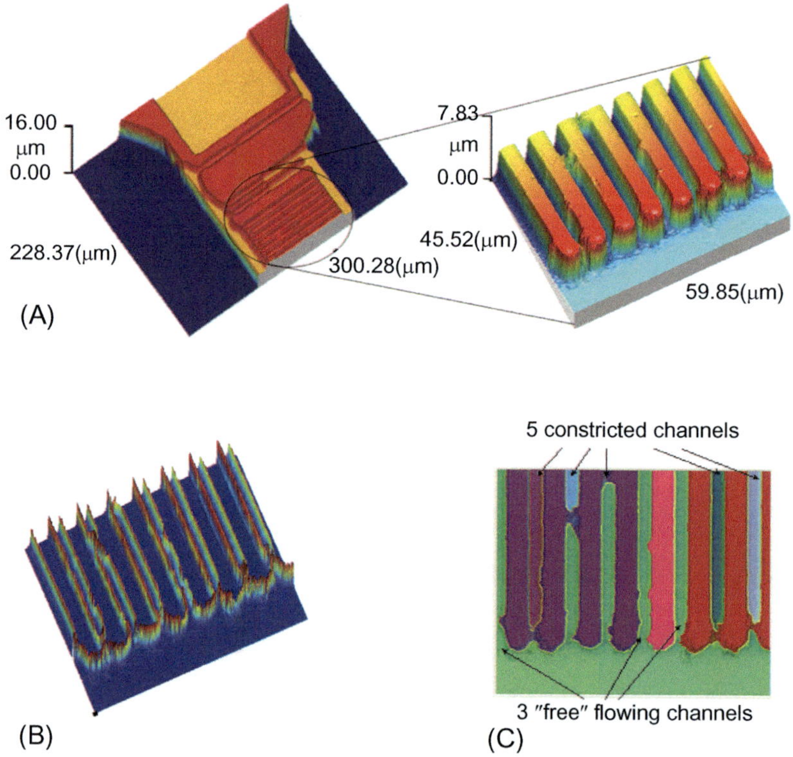

Fig. 11.21 Surface features of a microfluidic plasma device. (A) The entrance to the reactor channels is shown in higher magnification, (B) Sobel edge image of reactor channels, and (C) segmented image of device.

be distinguished from the base plane by using a Sobel edge operator; the edges build a "fence" to form the closed contours. In Fig. 11.21B the edges of the step planes form clear boundaries with the base plane, and positions of the step planes can be easily identified. Fig. 11.21C displays the result of the segmentation using Wolf pruning. The whole surface is segmented into nine sections. The edge lines are consistent with the edge of the channel walls. It is clear that five channels are partially blocked by defects and only three channels can be considered free flowing. By counting the free flowing channels, it is clear that this device would have suboptimal fluid flow through the reactor zone.

11.7 Methodology to characterize tessellation surfaces

There are two basic elements to define a tessellation surface: (1) primitive structures can also be cited as a single-tessellated tile, which is the basic structure (building blocks) that form the tessellation surface, and (2) the repeating rules that describe how the primitive structure will be repeated. Correspondingly, there are three basic objectives that need to be reached for the characterization of tessellated surfaces: (1) classification of different tessellation surfaces, (2) identification of the basic tessellated tile attributes, and (3) the definition of the relationships between tessellated tiles in a stable and robust way.

11.7.1 Methodology

To reach the previous objectives, two different types of methodology can be used. One is bottom-up method, and the other is top-down method. In the bottom-up method, primitive structures (single-tessellation tiles) of a tessellation are extracted first. The spatial relationships of these features are then analyzed and stored in a data structure according to their spatial relationship. Finally the periodicity is obtained by extracting the two independent translation vectors. As the bottom-up approach starts with feature extraction process, it is generally sensitive to distortions in the texture, and also the feature exaction methods (such as motif analysis, morphology analysis, and pattern analysis) are very complicated and time consuming. In contrast the top-down approaches extract the texture structure before the feature extraction process is performed. Fig. 11.22 shows the scheme structures of the two different methodologies. In this paper the author will use the top-down method.

As for the top-down approaches, Conners et al. [19] have used a parallelogram to describe the primitive structure of the tessellations. In their method the two sides of the parallelogram were used to represent the two translation vectors that describe repeating patterns of the tessellation [19]. Zuker and Terzopoulos [20] use a cooccurrence matrix (CM) method to extract the primitive structure [11]. However, in their method, only one translation vector can be extracted, which means the locations of the primitive structure cannot be located correctly. Kim and Park [21] improved the speed of the CM-based method by using the projection information. However, all CM-based methods are computationally expensive because the CM features are computed for all possible orientations and resolutions. The Fourier transform has been used to extract periodicity of a tessellation. In Fourier transform-based methods, the peaks in the Fourier spectrum are used to characterize the periodicity of tessellation. However, it is difficult to extract correct periodicity of a tessellation because peaks in the Fourier spectrum of the practical measured tessellation surface are usually not significant and thus difficult to find.

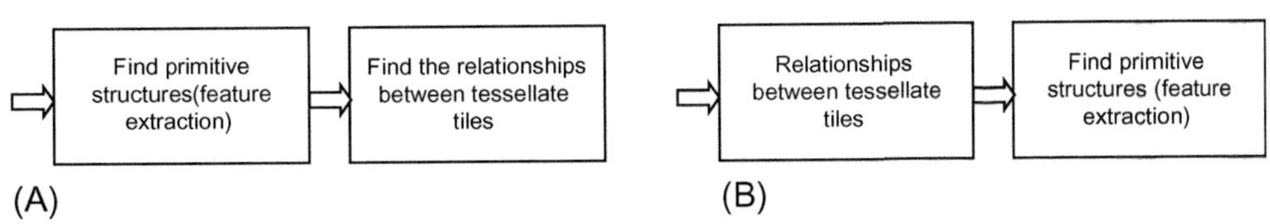

Fig. 11.22 Methods to characterize tessellation surfaces. (A) Bottom-up method. (B) Top-down method.

To overcome the previous weakness in top-down approach, Lin et al. [14] reported an autocorrelation function-based method to extracting periodicity of a regular texture [14]. In this method an autocorrelation function is used as the basis of a texture measure, and peaks in the autocorrelation function characterize the texture periodicity. Compared with the peaks in the Fourier spectrum, the peaks in the autocorrelation function are prominent and thus easier to find. However, their method has limited use as the peaks in the ACF will be dominated by the form and form error and the peaks that reflect the pattern of the tessellation will be weakened and not easy to find.

To characterize the tessellation surface, the authors use a lattice-based method combined with the spectral analysis to achieve characterization of the tessellation surface. Fig. 11.23 gives the structure of the proposed method. The basic procedure includes six steps. In the first step, preprocessing the original measured data is needed. The aim of this step is to remove the irrelevant information from the tessellations that are to be characterized. A regression filter technique is used for this purpose as this method can remove form very well while not changing the tessellations itself. In the second step the filtered data are converted into the transfer domain to make the relationship of the tessellated tile more significant. Our method here is the areal autocorrelation function (AACF) of the surface as it has been proved to be a good representation for the repetitions over planar space. This method can improve the significance of the peaks in the AACF that represent the periodicity of the tessellations; thus it can improve the accuracy of the extraction. In the third step the peak points of the AACF are linked, and then, by using a statistic histogram method, two main linearly independent translation vectors, which can attribute the relationship of the tessellation tiles, can be constructed. In the fourth step, linking the vectors a lattice grid can be built. And from the relationship of the length and the angles between the two sides of the lattice, the lattice can be classified as parallelogram, rectangular, rhombic, square, and hexagonal. Accordingly, tessellated surfaces can be categorized by the previous lattice types. In the fifth step the "mean tessellation" or the "reference tessellation" will be built according to the extracted primitive structures

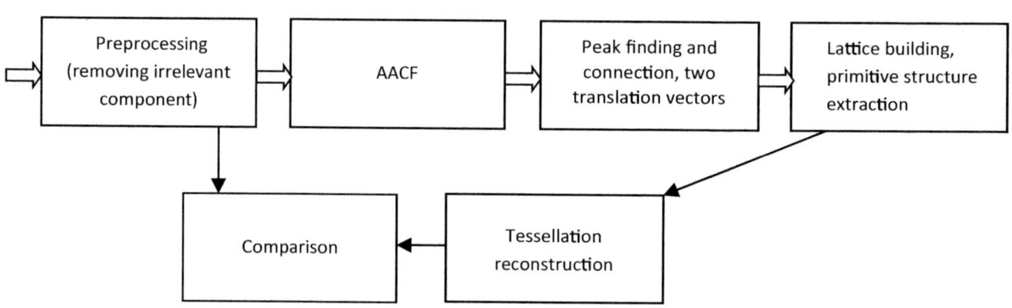

Fig. 11.23 Structure of the proposed method to characterization tessellation surface.

and the translation vectors. And finally, in the sixth step, compare the preprocessed tessellation with the reconstructed tessellations; the "roughness" parameters can be calculated.

11.7.2 AACF

The autocorrelation function is a very useful tool for processing random signals. It describes the general dependence of the topographical values at one position on the topographical values at another position. For areal surface evaluation, it can not only describe the spatial relation dependences of the surface topography but also describe the direction and periodicity of the surface texture. The AACF has been used to describe different machining methods as each has very different surface texture patterns and consequently very different patterns for their AACF spectra [22]. Furthermore the AACF spectra can reflect the texture periodicity and directionality more clearly than visualization of a surface topography map. Here the AACF is used to measure the repeating rules of the tessellation surfaces.

The AACF is defined as the average of a function multiplied with a shift translated version of itself, describing the interdependence of the surface heights at different positions. The AACF is defined mathematically as

$$R(\tau_x, \tau_y) = E[\eta(x, y)\eta(x+\tau_x, y+\tau_y)] = \lim_{l_x, l_y \to \infty} \frac{1}{4l_x l_y} \int_{-l_y}^{l_y} \int_{-l_x}^{l_x} \eta(x, y)\eta(x+\tau_x, y+\tau_y)\,dx\,dy \tag{11.45}$$

where $\eta(x_k, y_l)$ is a filtered surface data set after the form has been removed. The discrete AACF takes the form

$$R(\tau_i, \tau_j) = \frac{1}{(M-i)(N-j)} \sum_{l=1}^{N-j} \sum_{k=1}^{M-i} \eta(x_k, y_l)\eta(x_{k+i}, y_{l+j}) \tag{11.46}$$

$$i = 0, 1, \ldots, m < M;\ j = 0, 1, \ldots, n < N;\ \tau_i = i \cdot \Delta x;\ \tau_j = j \cdot \Delta y$$

The AACF of a surface signal has three properties: (1) symmetry, $R(\tau_i, \tau_j) = R(\tau_{-i}, \tau_{-j})$; (2) the maximum value is at the central point; and (3) similar pattern and periodicity as the surface texture.

A discrete Fourier transform (DFT)-based algorithm can be used to calculate the AACF very efficiently:

$$R(\tau_i, \tau_j) = \mathrm{IDFT}(\mathrm{DFT}(\eta(x_k, y_l))\mathrm{DFT}^*(\eta(x_k, y_l))) \tag{11.47}$$

where DFT represents the discrete Fourier transform, IDFT represents the inverse Fourier transform, and * represent the complex conjugate. Figs. 11.24A and 11.25A are two measured standard calibration block surfaces. For these surfaces the mean feature spacing (mean spacing between blocks) is one of the important parameters required to be

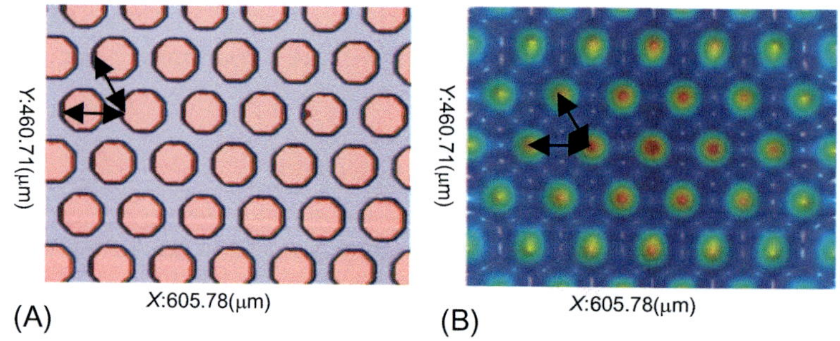

Fig. 11.24 (A) Measured calibration block and (B) AACF.

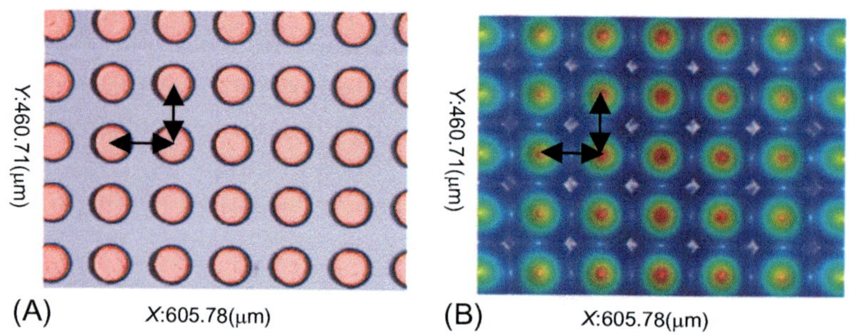

Fig. 11.25 (A) Measured calibration block and (B) AACF.

calculated. To calculate these parameters, the conventional way, each individual block needs to be separated firstly and then the centers of the blocks need to be found. However, as mentioned previously, separating the individual blocks is time consuming, and the calculated centers of these block, due to the noise and the irregularity of each block, are not very accurate. Figs. 11.24B and 11.25B are the AACFs of the measured surfaces. It is very clear that the AACF peaks can significantly represent the periodicity of the tessellations. One can obtain the interfeature distance by simply measuring the distance between the peaks in the AACF. The other advantage of this method is that the result is robust against noise and the shape deviation of individual features due to the AACF naturally being an "average" effect.

11.7.3 Lattice building

Four basic planar translational symmetries can be used to categorize the tessellation surface: (1) Translations, mean one can slide a pattern along a certain direction with certain distance, and will fall back upon itself with all the patterns exactly matching. (2) Rotations fix one point in the place and turn the rest of it by some angle around that point. The angle of rotation can only be 180, 120, 90, and 60 degrees. (3) Reflection fixes one line in the plane, called the axis of reflection, and exchanges points on one side of the axis with point on the other side of the axis at the same distance from the axis. (4) Glide reflection is composed of a reflection across an axis and a translation along the axis. According to these four basic planar translational symmetries, the wallpaper group method has been used to classify the tessellations. However, this method is very complex and trivial for engineering surfaces as some different types are actually very similar.

Here the simplified method, called the lattice method, is used to characterize the tessellations of engineering surfaces. For any point on the tessellation, the collection of translates, by translation symmetries of a pattern, forms a lattice. There are five different kinds of lattice:

(1) Parallelogram: if a lattice has a parallelogram as a fundamental region, it is parallelogram lattice. A parallelogram lattice's symmetry group has translations and half-turns, but there are no reflections nor glide reflections.

(2) Rectangle: if a lattice has a rectangle as a fundamental region, it is rectangle lattice. Rectangular lattice is a special parallelogram lattice (90 degrees parallelogram). A rectangular lattice's symmetry group has translation, half-turns, and reflections.

(3) Rhomboid: if a lattice has a rhombic as a fundamental region, it is a rhombic lattice. (A rhombus is a parallelogram with equal sides.) A rhombic lattice's symmetry group has translation, half-turns, reflections, and glide reflections.

(4) Square: if a lattice has a square fundamental region, it is called a square lattice. Square lattice is very special rhombic lattice (90 degrees rhombus). A square lattice's symmetry group has translation, rotations of 90 and 180 degrees, reflections, and glide reflections.

(5) Hexagon: if a lattice has a 60 degrees rhombus as a fundamental region, it is called a hexagonal lattice. That's because, in that case, the points in the lattice nearest any one point in the lattice are the vertices of a regular hexagon. Hexagonal lattice is a very special rhombic lattice (60 degrees rhombus). A hexagonal lattice's symmetry group has translation; rotations of 60, 120, and 180 degrees; reflections; and glide reflections.

The fundamental region can be spanned by two displacement vectors. The generalized Hough transform (GHT) has been used to find two displaced vectors from the peaks of the AACF. This pair of displacement vectors is then used to construct the primitive structure (fundamental region) of the tessellations, and from the relationship of the length and the angles between the two displaced vectors, the type of the tessellations can be classified accordingly.

11.7.4 Case studies

To test effectiveness and efficiency of the lattice method, a number of different types of simulated and measured tessellation surface have been processed. Some typical data are selected in this paper to demonstrate the results. Fig. 11.26A is the measured EBT surface.

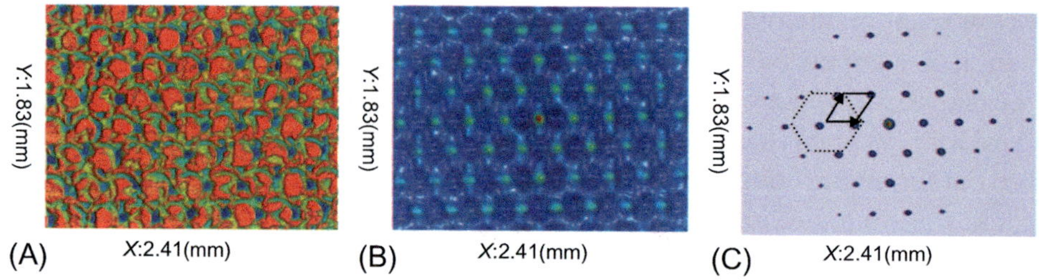

Fig. 11.26 (A) EBT surface, (B) AACF, and (C) translation vector.

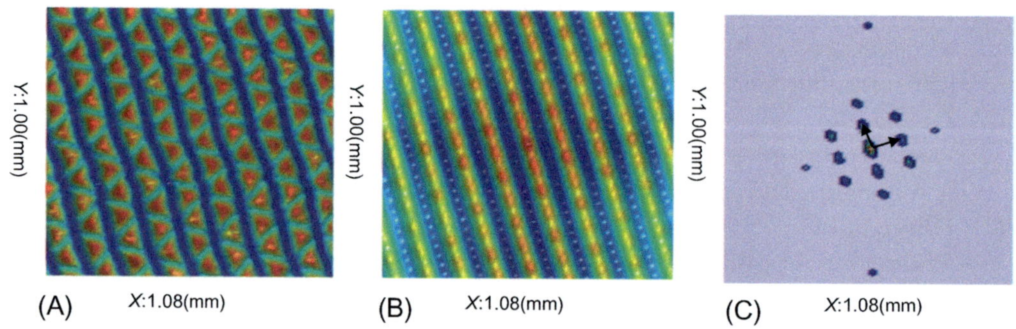

Fig. 11.27 (A) 3M TRIZAC abrasive paper surface, (B) AACF, and (C) translation vector.

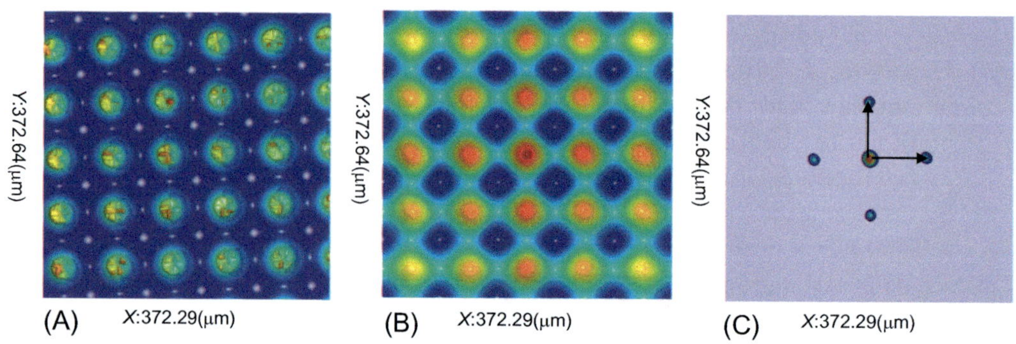

Fig. 11.28 (A) Microneedles, (B) AACF, and (C) translation vector.

One can intuitively see that tessellations do exist in the surfaces, but it is still difficult to quantify what the primitive structure is and how is it repeated from the original surface. By using the lattice method, one can clearly see, in Fig. 11.26C, that the peaks in the AACF can represent the periodicity of the tessellations and direction and length of two translation vectors that show how the primitive structure is repeated. Using the lattice method the EBT surface can also be classified as hexagon type tessellation surface.

Figs. 11.27 and 11.28 depict 3M TRIZAC abrasive paper surface and microneedles array surface, respectively. By using the presented method, they can be classified as rectangle tessellations, and the primitive structure and the repeating rules can be recognized by the two displacement vectors.

11.8 Summary

In this chapter a general framework for characterization of free-form structured surface is proposed, which includes four steps:

(1) form/form error separation to find a reference surface and the form-removed structured feature,

(2) denoising/filtration to reduce the effect of instrument noise or separate the features from surface texture,

(3) feature extraction to separate feature regions following by identification of metrological characteristics of surface features and relationships, and

(4) feature characterization to quantitative assessment of feature attributes.

For step (1), two methods can be used. The first uses a higher-order robust Gaussian regression filter, and the second is the robust spline filter.

For the step (2) a partial differential equation (PDE)-based adaptive nonlinear diffusion filter is proposed. The proposed diffusion filter is based on the PDE method and can be seen as a nonlinear heat equation, which describes the distribution of heat (or variation in temperature) in a given region overtime. The diffusivity function based on the gradient of the surface can help to separate the internal region area of boundary area of the measured feature. In this way the diffusion process will take place mainly in the interior regions (line, step, etc.) of the surface, and it will not affect the region boundaries where the magnitude of gradient is large.

For the step (3), data segmentation method based on edge operators and pattern analysis is proposed.

For the step (4), feature parameters that defined in ISO25178-2 can be used to quantitatively assess the geometrical size, individual zones of surface features, relationship to each other, and surface feature parameters.

Furthermore, for the tessellated surface, two routes can be followed: one is bottom-up method that is same as the previous framework, and the other is the top-down method. The top-down method to characterize the tessellated surface is based on lattice building combined with the spectral analysis for the characterization of tessellation surfaces. The structure of the proposed method basically includes six steps: firstly, preprocessing the measured data to remove the irrelevant information; secondly, converting the filtered data to transfer domain (AACF) to make the relationship of the tessellated tile more significant; thirdly, finding peaks in AACF and obtain the translation vectors; fourthly, building the lattice and extracting the primitive structure; and the last two steps are tessellation reconstruction and comparison.

References

[1] Jiang X, Scott P, Whitehouse D, Blunt L. Paradigm shifts in surface metrology part I: historical philosophy. Proc R Soc A 2007;463:2049–70.
[2] Zeng W, Jiang X, Scott P. Fast algorithm of the robust Gaussian regression filter for areal surface analysis. Meas Sci Technol 2010;21:055108.
[3] Goldstein T, Osher S. The split Bregman method for L1-regularized problems. SIAM J Imaging Sci 2009;2(2):323–43.
[4] Scott PJ. Pattern analysis and metrology: the extraction of stable features from observable measurements. Proc R Soc Lond A 2004;460:2845–64.
[5] Blunt L, Xiao S. The use of surface segmentation methods to characterise laser zone surface structure on hard disc drives. Wear 2011;271(3–4):604–9.
[6] Xiao S, Xie F, Blunt L, Scott P, Jiang X. Feature extraction for structured surface based on surface networks and edge detection. Mater Sci Semicond Process 2006;9:210–4.
[7] Perona P, Malik J. Scale-space and edge detection using anisotropic diffusion. IEEE Trans Pattern Anal Mach Intell 1990;12(7):629–39.
[8] Goto T, Miyakura J, Umeda K, Kadowaki S, Yanagi K. A robust spline filter on the basis of L2-norm. Precis Eng 2005;29:157–61.
[9] ISO/DTS 16610-32:2002. Geometrical product speciation—filtration, part 32: robust profile filters: spline filters; 2002.
[10] Krystek M. Form filtering by splines. Measurement 1996;18:9–15.
[11] Zeng W, Jiang X, Scott P. A generalised linear and nonlinear spline filter. Wear 2011;271(3/4):544–7.
[12] Seewig J. Linear and robust Gaussian regression filters. J Phys Conf Ser 2005;13:254–7.
[13] Weickert J, et al. Efficient and reliable schemes for nonlinear diffusion filtering. IEEE Trans Image Process 1998;7:398–410.
[14] Lin HC, Wang L, Yang S-N. Extraction periodicity of a regular texture based on autocorrelation functions. Pattern Recogn Lett 1997;18:433–43.
[15] Maxwell JC. On hills and dales. Philos Mag 1870;40:421–7.
[16] Jiang X, Scott P, Whitehouse D, Blunt L. Paradigm shifts in surface metrology part II: the current shift. Proc R Soc A 2007;463:2071–99.
[17] ISO 25178-2:2012. Geometrical product specification (GPS)—surface texture: areal—part 2: terms, definitions and surface texture parameters. Geneva: International Organization for Standardization; 2012.
[18] ISO 25178-3:2012. Geometrical product specification (GPS)—surface texture: areal—part 3: operators. Geneva: International Organization for Standardization; 2012.
[19] Conners RW, Harlow CA. Towards a structural textural anlyzer based on statistical methods. Comput Graph Image process 1980;12(3):224–56.
[20] Zuker SW, Terzopoulos D. Finding structure in coocurrence matrices for texture analysis. Comput Graph Image Process 1980;12:286–308.
[21] Kim HB, Park SN. Extracting spatial arrangement of structural textures using projection information. Pattern Recogn 1992;25(3):237–47.
[22] Zeng W, Jiang X, Blunt L. Surface characterisation-based tool wear monitoring in peripheral milling. Int J Adv Manuf Technol 2009;40:226–33.

Further reading

Brinkmann S, Bodschwinna H. Accessing roughness in three-dimensions using Gaussian regression filter. Int J Mach Tools Manuf 2001;41:2153–61.
Jiang X, Blunt L. Third generation wavelet for the extraction of morphological features from micro and nano scalar surfaces. Wear 2004;257:1235–40.

Jiang X, Zeng W, Scott P, Ma J. Linear feature extraction based on complex ridgelet transform. Wear 2008;264(5–6):428–33.

Jiang X, Blunt L, Stout KJ. Three-dimensional surface characterization for orthopaedic joint prostheses. Proc Inst Mech Eng H 1999;213:49–68.

Jiang X, Blunt L. Morphological assessment of in vivo wear of orthopaedic implants using multi-scalar wavelet. Wear 2001;250:217–21.

Jiang XJ, Whitehouse DJ. Technological shifts in surface metrology. CIRP Ann Manuf Technol 2012.

Stout KJ, Sullivan PJ, et al. The development of methods for the characterisation of roughness in three dimensions. London: Penton Press; 2000.

CHAPTER 12

Smart surface design and metrology platform

12.1 Introduction

Smart manufacturing urges machines to interpret and exchange large amounts of data in an accurate and efficient manner, "smart" in that different types of machines can work harmoniously and collaboratively with little human intervention. Current machines use a digitalized symbolic language to represent and then exchange the data. They are, however, often incapable of directly interpreting the meaning of the data [1], that is, the semantics of the data. As a result, information loss and incorrect interpretation can often happen during the communication. To improve machine intelligence the specifications of a product, such as geometric shapes and geometrical variability, cannot merely be graphic symbols with text, but its semantics have to be explicitly represented in the machine [2, 3], such that its meaning can be unambiguously interpreted to aid decision-making prior to the manufacturing and measurement processes.

To accomplish this the first step is to make a specification machine readable. To do this a machine-readable language must first be adopted [4]. There are a number of direct machine-readable models to date, including EXPRESS model [5], an extensive markup language (XML)-based model [6], ontology-based (OWL) model [7, 8], and category theory-based model [9–12]. These models can unambiguously represent the syntax of a specification. However, not all of them represent a specification at a semantic level where the models should have a rigorous structure and be able to be reasoned and interrogated; that is, information is managed in a structural way; therefore the end users can ask (query) specific or conceptual questions to the system.

The second step is to make the machine-readable information also machine interpretable, that is, enabling knowledge interrogation operations such as query and reasoning to deduce the requested detail. Among the aforementioned models, only ontology-based language OWL is directly machine interpretable, as OWL has formal semantics (description logics, DLs) as its foundation [13]. DL-based ontologies and the tools developed to support them have rapidly become de facto standards for ontology development and deployment. OWL-based technologies have exerted a major influence on the development of information technology. DLs are based on set theory and are best suited to represent relationships between sets. They are therefore limited in extent and cannot guarantee the direct merger of two different ontologies or construct complex

Advanced Metrology
https://doi.org/10.1016/B978-0-12-821815-0.00012-5

relationships among ontologies. DLs are also not equipped with the apparatus to represent many types of mathematical operations and constraints that would turn it into a more generic formalism [14]. It therefore cannot provide adequate expressive power to fully build semantic knowledge models for smart manufacturing where extensive mathematical theories, operations, and analysis are involved.

A category-based approach is currently being developed into a knowledge representation language equipped with syntax, semantics, and reasoning rules based on categorical laws and concepts [15]. It is named category semantic language (CSL), a smart language that uses category theory as its foundation. In particular, it uses category theory for deduction rather than DLs as in ontology. Thus CSL is substantially distinct from ontology. One of the distinguishing features over ontology is that CSL naturally supports multilevel representation. At the basic level, objects and the relationships between objects are represented and grouped (level of set). At a higher level, relationships between the groups are structured (sets of sets). At another level, complex relationships between the relationships of groups are also structured. The language is gifted with strong expressive power as representations of many other mathematical theories beyond sets, such as topological space, graphs, groups, rings, and monoids, all of which are supported. Moreover, CSL provides a direct solution for the structure-preserving mappings between the two totally ordered sets of operations (i.e., specification and verification operators), namely, the duality principle [16] in the ISO GPS system. The mapping is defined as a *functor*, and the dual mappings between the specification and verification operators in particular take the structure of an *adjoint functor*, which is seen as central to category theory.

To this end a smart semantic model, based on CSL, is used to decode the geometrical requirements (i.e., derive the semantics of a symbolic specification) into a machine-readable and machine-interpretable format. The smart model implements the duality principle to deduce corresponding verification operations in detail from the specification. Based on the knowledge model, a smart platform with user-friendly interfaces, named CatSurf, is developed to facilitate decision-making for end users in the design and measurement of surface texture.

12.2 Semantics of surface design and metrology

12.2.1 Semantic flows among the four conceptual realms

In Chapter 1 an analogy between the chains of four realms was used to illustrate the basic idea of specification, manufacture, verification, and comparison. It implies that there is a rich flow of semantics among the four realms. Here, semantic flows of geometric variability among the four realms are explained in a higher order; see Fig. 12.1.

In the realm of design, the designer translates functional requirements into geometrical requirements (specification), where technical drawing is used to represent the semantics of the specification. In the realm of manufacture, the semantics of the drawings

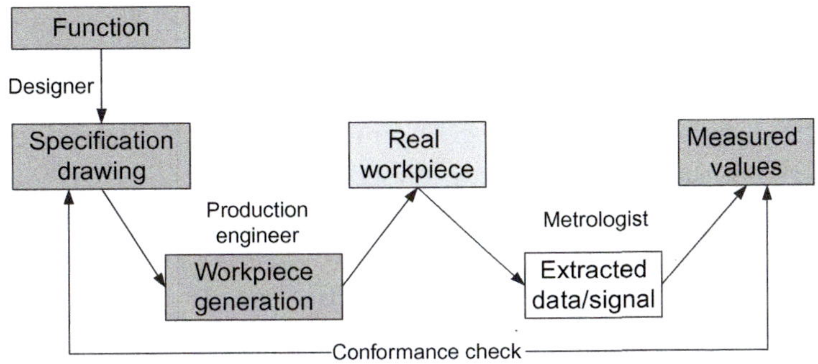

Fig. 12.1 Semantic flows among the four conceptual realms.

should be understood by the production engineer, to allow them to design, analyze, and manage the planning of the manufacturing process. In the realm of verification, prior to the measurement, the metrologist should also understand the drawing to develop a reasonable measurement plan. The finished workpiece is then measured according to the plan, and data are extracted and then analyzed under a set of planed operations. Obtained measured values can then be compared with the specification in which the conformance rule is used as the comparison criteria. In the realm of product life, the workpiece may be required for further measurement and diagnosis and, whenever necessary, subject to further manufacturing processes and subsequent measurements to extend its functionality.

It is rather clear that the semantics of a specification is of utmost importance among the four realms. The semantics need to be understood and translated into different perspectives (manufacture and verification). Working toward smart manufacturing and to improve machine intelligence, the machines should be equipped with apparatus to interpret the semantics in a robust and stable manner.

The dual semantic mappings between the specification and verification are an adjoint functor, as shown in Fig. 12.2. Among concepts in category theory, an adjoint functor is seen as central to category theory. Some category theorists consider adjoint functors as

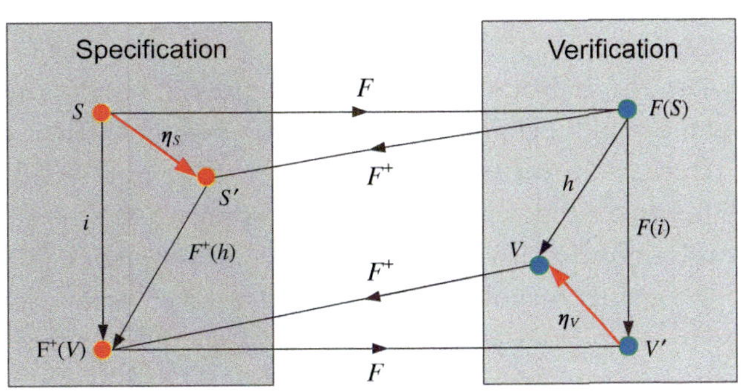

Fig. 12.2 An adjoint functor between specification and verification.

dictionaries that translate back and forth between categories [17]. If the two categories are two languages (say English and Chinese) that are equally expressive, then a good dictionary will be an explicit exchange of ideas. However, in category theory, we often have two categories that are not in the same conceptual world, and the adjoint functors connect two different structures by structure-preserving mapping. That is why adjoint functors often come in the form of "free" and "forgetful." One particular example is a forgetful functor that is defined from a category of algebraic structures (group or vector spaces) to the category of sets. The forgetful functor forgets the arrows, remembering only the underlying set regardless of their algebraic properties.

Here a functor F represents a structure-preserving mapping from the specification to the verification, that is, carrying out verification operations according to the specification operations. For example, object S in specification is mapped to object $F(S)$ in verification, by the F functor. Another functor F^+ represents the conformance check, in which verification operations are mapped to specification. For example, object S' in specification is mapped from object $F(S)$ in verification, by the F^+ functor. The object S' is clearly related to the object S, as it is mapped from S and then mapped back and $S' = F^+ F(S)$.

Using the translation example between English and Chinese, let S be an English sentence; F translates S into a Chinese sentence $F(S)$. F^+ then translates the Chinese sentence $F(S)$ back to an English sentence S'. There exists an arrow/relationship η_S from the English sentence S to the S', and it represents the difference, or accuracy, or resolution of the dual translations F and F^+. Similarly, in specification, η_S represents a unique relationship from a specified operation to the resulted operation after measurement process. The property of η_S decides the robustness of the semantic flow between specification and verification. If S is equivalent to S', the dual translations are flawless. If S is bijective (exchangeable) to S', the dual mappings are considered robust. If S is injective (one to one) to S', the dual translations generate more information, and there may exist redundancy. If S is surjective (onto) to S', the dual translations simplify S.

12.2.2 Representing the semantics

12.2.2.1 CSL semantics

CSL uses basic syntax and an intuitive interface to represent semantics. It starts with a set of basic boxes and meaningful arrows to connect one box to another, in such a way that more and more boxes can form complex structures that can be used to assemble with other modules to become a bigger model. In this case the basic boxes are *objects*; the various forms of arrows connecting two objects together are *morphisms*. Different numbers of morphisms binding together that obey categorical laws form various *morphism structures*, in such a way that more complex structures can be constructed. The set of boxes and arrows, which has a specific functionality, is a *category*.

Basic semantics

A category is a collection of objects with arrows between them. We do not need to know the details about the object; all we need to know is the arrows (morphisms). Arrows acting like functions (mappings between sets) have some function-like properties, for example, composability, which is associative. An arrow f represents a relationship from object A to B in a category $\mathbf{C}$, written as f: A (*notion*) $\rightarrow B$. f may also have one of six special properties and different types of arrows representing different properties. An arrow can be epic (denote as $\twoheadrightarrow$), monic ($\rightarrowtail$), isomorphic ($\leftrightarrow$), retraction ($\bullet\!\!\rightarrow$), section ($\blacksquare\!\!\rightarrow$), and both epic and monic but not isomorphic ($\rightarrowtail\!\!\!\!\twoheadrightarrow$). For more details regarding the properties of morphism, see Appendix and Fig. A.2. A notion of an arrow often represents the meaning of the relationship between the domain and codomain objects, and it could be started with characters such as "is," "has," "with," "applied to," and "assigned to."

Take areal surface texture as an example; we can construct a category to represent a collection of areal surface texture parameters. Here, objects can be the name, definition, type, or unit of a parameter, as shown in Fig. 12.3, where four objects are as follows:

[Para Type]: the type of a parameter, such as height parameters, spatial parameters, and feature parameters;

[Para Name]: the name of the defined parameter, for example, *Sq*, *Sal*, *Str*, *Vvv*, and *Spd*;

[Para Unit]: the unit of a parameter; and

[Para Def]: the definition of a parameter,

and three arrows are defined as follows:

The arrow s_1: [Para Name] (belongs to) $\twoheadrightarrow$ [Para Type] states that every parameter belongs to a parameter type; for example, the parameter *Str* (texture aspect ratio) is classified as spatial parameters as listed in Table 12.1. Arrow s_1 is epic (onto) as all parameters are defined into different types, for example, height, spatial, and feature parameter.

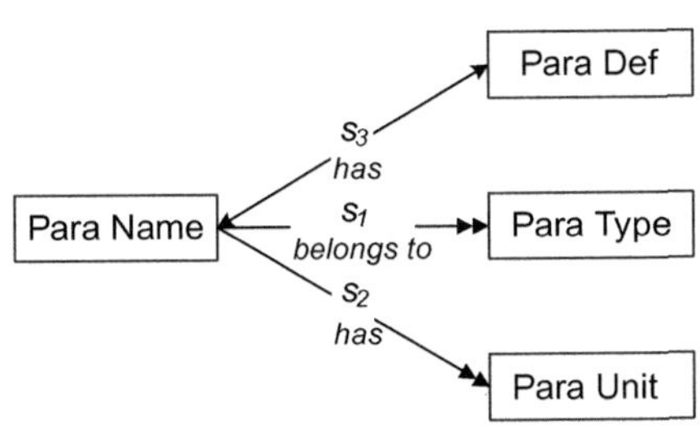

Fig. 12.3 Some objects and arrows in a category of "areal surface texture parameters."

Table 12.1 Data examples for characteristic of areal surface texture parameters (ISO 25178-3, 2010).

Parameter type	Para	Parameter name	Default unit
Height parameters	*Sq*	Root-mean-square height	μm
	Ssk	Skewness	Unitless
	Sa	Arithmetical mean height	μm
Spatial parameters	*Sal*	Autocorrelation length	μm
	Str	Texture aspect ratio	Unitless
Functions and related parameters	*Vvv*	Dale void volume	mL/m^2
	Vvc	Core void volume	mL/m^2
	Vmp	Peak material volume	mL/m^2
	Vmc	Core material volume	mL/m^2
	Sxp	Peak extreme height	μm
Feature parameters	*Spd*	Density of peaks	$1/mm^2$
	Spc	Arithmetic mean peak curvature	1/mm
	S5p	Five-point peak height	μm
	S5v	Five-point pit height	μm

The arrow s_2: [Para Name] (has) $\twoheadrightarrow$ [Para Unit] shows that every parameter has a related unit, and it is also epic.
The arrow s_3: [Para Name] (has) $\leftrightarrow$ [Para Def] indicates that every parameter has a unique parameter definition, and then it is isomorphism (bijective).

Enriched semantics

While objects and arrows form the basic semantic blocks, the semantics can also be enriched in various ways. One of the ways is by constructing bigger objects from the existing ones, which can be carried out using product and coproduct. A *product* of two objects A and B is denoted as $A \times B$, as shown in Fig. 12.4A; A can be an axis, B can be a hexagon, and the product of the two is a uniform hexagonal prism.

A *coproduct* of two objects A and B is denoted as $A \sqcap B$, as shown in Fig. 12.4B; let A be a motif filter; B be a sphere segment filter; and the coproduct of A and B is the disjoint

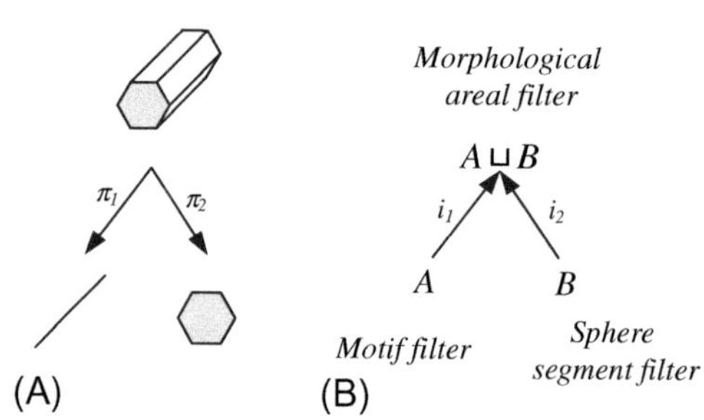

Fig. 12.4 Enriched semantics using product and coproduct. (A) A product: a hexagon surface. (B) A coproduct: morphological areal filter.

union of the two filters, which is a morphological areal filter, as both motif filter and sphere segment filter are morphological areal filters.

Another way of enriching the semantics is by combining arrows together to form a more complicated semantic web. As an arrow has a predefined path, different numbers of arrows binding together (which obey categorical laws) form various structures to enrich the semantics. The most basic structure is a triangle △, where two arrows are commutative and the third arrow is the composition of the two. Fig. 12.5 shows a typical triangle structure, where the triangle starts from the object [Para Def], the arrow s_2: [Para Def] (has) $\leftrightarrow$ [Para Name], and s_3: [Para Name] (has) $\twoheadrightarrow$ [Para Unit] commute; the third arrow s_4: [Para Def] (decides) $\twoheadrightarrow$ [Para Unit] is the composition of s_2 and s_3. The arrow s_4 represents the most basic semantic composition, that is, given two arrows f: $A \rightarrow B$ and g: $B \rightarrow C$, the composition $g \circ f$ represents relationship from A to C. In Fig. 12.5, as the definition of a parameter decides its unique name and the name must have a unit with it, then the composition means that the definition of the parameter also decides its unit.

If two triangles share the same composition arrow, a rectangle □ can then be constructed, as shown in Fig. 12.6. Here the arrow s_4: [Para Def] (decides) $\twoheadrightarrow$ [Para Unit] represents the composition of s_2 and s_3 and the composition of s_5: [Para Def] (decides) $\twoheadrightarrow$ [Para Value] and s_6: [Para Value] (with) $\twoheadrightarrow$ [Para Unit]. The semantics behind the triangle is that the relationship from the definition of a parameter to its unit is either decided using its unique name or via its value. This means that a rectangle represents a binary relationship between two objects, using two different routes.

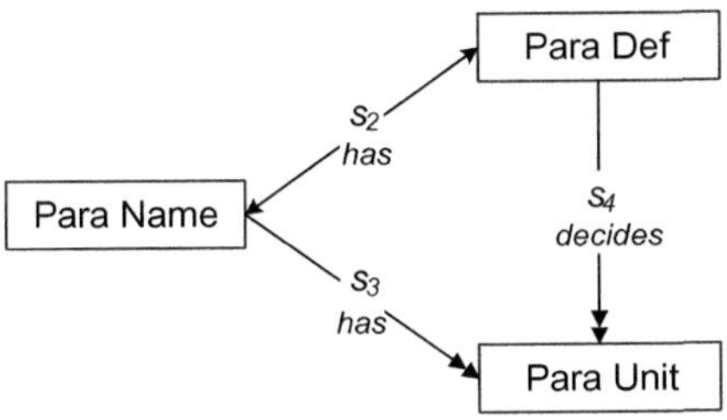

Fig. 12.5 Enriched semantics using triangle structure.

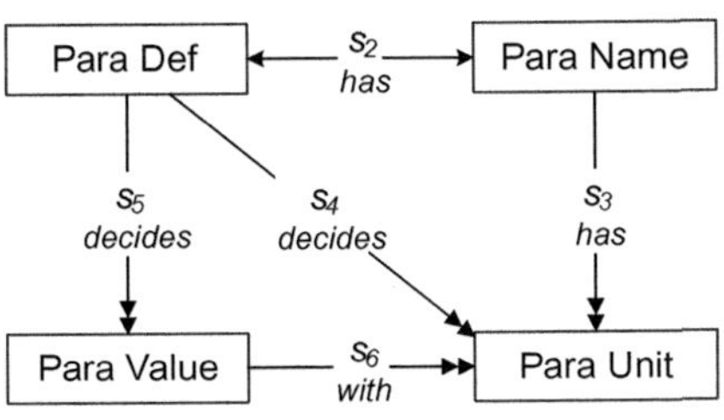

Fig. 12.6 Enriched semantics using rectangle structure.

We can also combine the aforementioned product and coproduct structures with other arrows to form a rectangle. For example, if we have two arrows as shown in Fig. 12.7A, we can then construct a product to connect with the two arrows, such that a rectangle as indicated in Fig. 12.7B can be constructed. The semantics of such a rectangle indicates that, as shown in Table 12.2, the name of a parameter together with its value is always with a unit.

Hierarchical semantics

As category theory in nature is hierarchical, at a higher level, whenever one category is related with another category, there usually is a mapping between the two. For this mapping to be meaningful, it should preserve the structure of the category. Therefore every object from one category has to be mapped into an object from another category, and all morphisms must be mapped correctly. Such mapping is called a *functor*; see Appendix in Fig. A.4 for more detail.

As we have mentioned in the previous section, the dual semantic mapping between the specification and verification is an adjoint functor, which is a pair of functors. As shown in Fig. 12.8, the forward functor F represents the mapping from the specification to the verification. Every object in the category of specification is mapped to an object in the category of verification; for example, objects S_1, S_2, and S_3 in the specification are mapped to objects $F(S_1)$, $F(S_2)$, and $F(S_3)$ in the verification, respectively. Also, morphisms i, j, and l in the specification are mapped to morphisms $F(i)$, $F(j)$, and $F(l)$ in the verification, and the composition between i, j, and l is preserved.

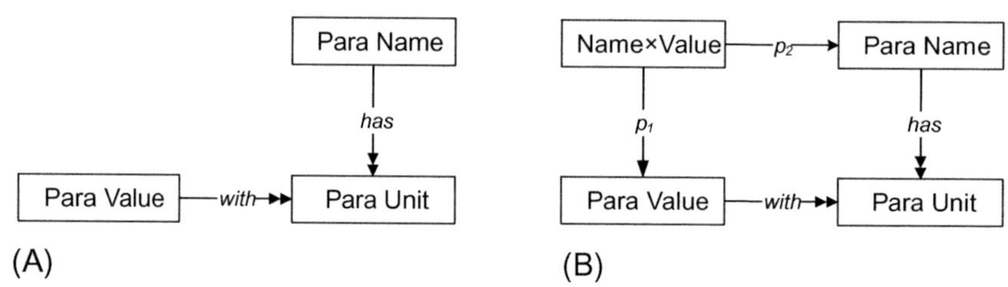

Fig. 12.7 A product forms a rectangle structure. (A) Two arrows with the same codomain. (B) A product to form a rectangle.

Table 12.2 Data example for the rectangle structure in Fig. 12.6B.

Name × value	Para value	Para name	Para unit
(*Ra*, 0.1)	0.1	*Ra*	μm
(*Ra*, 0.8)	0.8	*Ra*	μm
(*Rz*, 12)	12	*Rz*	μm
(*Rsm*, 0.04)	0.04	*Rsm*	μm

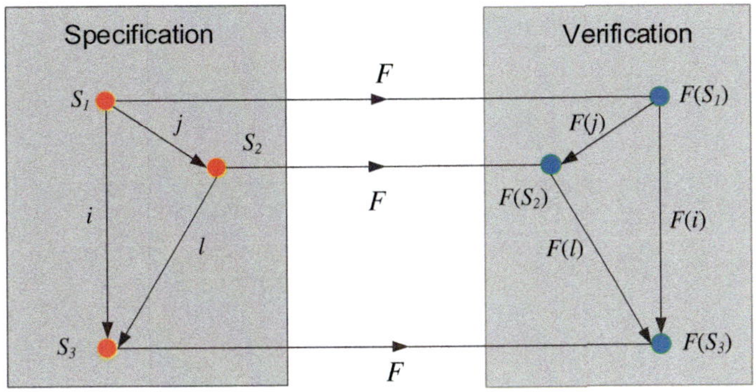

Fig. 12.8 A functor between specification and verification.

The mappings between the objects and morphisms may have certain properties, and these properties decide the properties of a functor. Let F^M be the mapping between the two morphism sets and F^O be the mapping between the two morphism sets, and $F^{\rightarrow}$ represents the mapping from the set of morphisms from A to B in the specification to the set of morphisms from $F(A)$ to $F(B)$ in the verification. A functor may also have a special property. As shown in Table A.1 in Appendix, nine properties are listed. They are full (denote as $\bullet\!\Rightarrow$), faithful ($\blacktriangleright\!\Rightarrow$), full faithful ($\Leftrightarrow$), injective on objects ($\blacktriangleright\!\xrightarrow{O}$), injective on morphisms ($\blacktriangleright\!\xrightarrow{M}$), surjective on objects ($\overset{O}{\twoheadrightarrow}$), surjective on morphisms ($\overset{M}{\twoheadrightarrow}$), bijective on objects ($\overset{O}{\leftrightarrow}$), and bijective on morphisms ($\overset{M}{\leftrightarrow}$).

Different numbers of functors binding together (which obey categorical laws) form various functor structures (adjoint functor is one of them), in such a way that a higher level of complex structures can be established.

12.2.2.2 CSL reasoning

Reasoning is the key to generating new knowledge and new structures and to further enrich the semantics. CSL reasoning mechanisms are designed naturally from categorical concepts and laws. One of the most useful categorical laws is the composition operation, where two commutative morphisms $f: A \rightarrow B$ and $g: B \rightarrow C$ can form a third morphism $g \circ f: A \rightarrow C$ and the three morphisms can then form a triangle structure $\triangle(\{A, B, C\}, \{f, g, g \circ f\})$. Generating a new morphism is only an initial step; configuration of the properties of the generated morphism is of importance for further reasoning and generating a mapping pair (data) of the new morphism. The properties of the third morphism can be established via deduction of the two known properties of f and g. For example, if f is epic, g is isomorphic, then $g \circ f$ is epic.

Proof. Let morphisms $h, h': C \rightarrow E$. If $h \circ (g \circ f) = h' \circ (g \circ f)$, from the associate law $(h \circ g) \circ f = (h' \circ g) \circ f$. As f is epic, this implies $h \circ g = h' \circ g$. As g is also epic, this implies $h = h'$; therefore $g \circ f$ is epic.

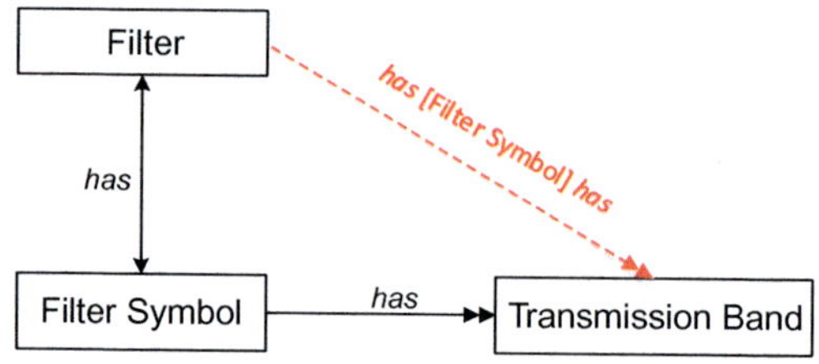

Fig. 12.9 An example of composition rule.

An example of this type of composition is shown in Fig. 12.9, in which the composition of an isomorphism [Filter] (has) ↔ [Filter Symbol] and an epic morphism [Filter Symbol] (has) ↠ [Transmission and] generates a new epic morphism [Filter] (has [Filter Symbol] has) ↠ [Transmission Band].

The morphism structures and functor structures are established not only to enrich the semantics and query efficiency but also to provide the foundations of the reasoning rules. As triangle structures can be established by the comparison rule, then two triangle structures that share the same composition morphism form a rectangle structure. Semantics of a rectangle structure can also be further enriched by pullback and pushout rules, if its morphisms have certain properties (monic morphism preserves pullbacks, and epic morphism preserves pushout). Therefore pullback structures and pushout structures can be deduced or established from a rectangle, such that more objects and morphisms can be generated.

As shown in Fig. 12.10A, applying the pullback rule based on a rectangle structure $\square_1$ will generate a product object ($O_2 \times O_3$), two projection morphisms (p_1: $O_2 \times O_3 \rightarrow O_2$ and p_2: $O_2 \times O_3 \rightarrow O_3$), and one unique morphism from the product object toward the start object of the based rectangle structure (u_2: $O_2 \times O_3 \rightarrow O_1$). Also generated are a set of new rectangle structures and triangle structures. A rectangle structure of a model of work

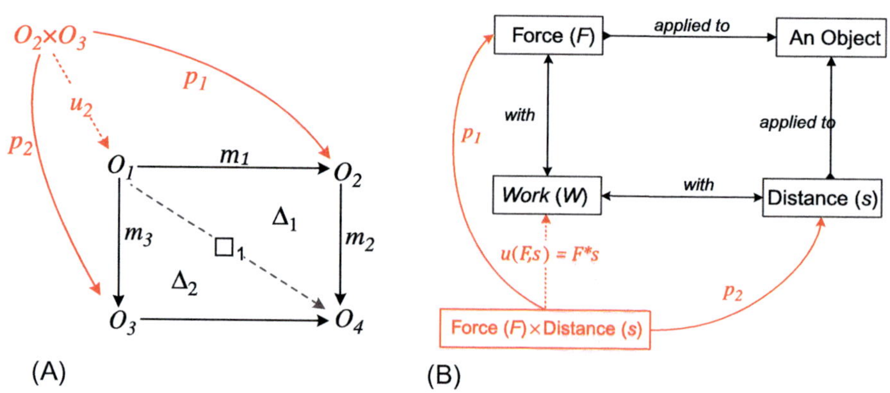

Fig. 12.10 Pullback rule and an example. (A) Pullback rule. (B) An example of applying pullback rule.

as illustrated in Fig. 12.10B satisfies the prerequisite of the pullback rule where morphism between objects [Force (*F*)] and [An object] and morphism between objects [Distance (s)] and [An object] are monic. Applying the pullback rule to the model produces a new object of [Force (*F*) × Distance (*s*)]; two projection morphisms p_1 and p_2; and a unique morphism *u*, which is a function of work $u\ (F,\ s) = F^*s$.

A set of reasoning rules of CSL in the category level can then be developed to construct triangle structures, rectangle structures, and pullback and pushout structures. The rules can be classified into two groups: The first is processing rules that generate new morphism structures without consideration of their data, and the second is a set of instance rules for generating new data. Because the reasoning rules are naturally designed from the CSL structures, one of the bonuses is that the reasoning efficiency is greatly enhanced as reasoning process uses/generates the same structure as in the CSL model.

12.3 Semantics of surface texture specification

Modeling a set of information is the first step to enabling the constructed model to be interrogated. In this section, using the CSL semantic representation method, an example of areal surface texture specification will be applied to demonstrate how to decode tolerancing information into a form of CSL.

As shown in Fig. 12.11, an areal surface texture specification symbol is illustrated in ISO 25178-1:2016 [18] to explain the meaning of each control elements of a specification symbol.

To start with a set of objects need to be identified firstly from a set of specification elements of the symbol. Then, relationships (morphisms) among the objects should be identified. If possible the nature of the relationship (properties of the morphism) should also be identified.

A specification defines a set of specification operations, which are defined as "specific tools required to obtain features or values of characteristics, their nominal value and their limit(s)" [16]. The set of specification operations is called a specification operator. Eight feature operations are defined in the ISO GPS system. They are "partition," "extraction," "filtration," "association," "collection," "construction," "reconstruction," and "evaluation," where

partition is to identify bounded features, for example, points, straight lines, or planes from nonideal surface features;

extraction is used to identify a finite number of points from a feature with specific rules;

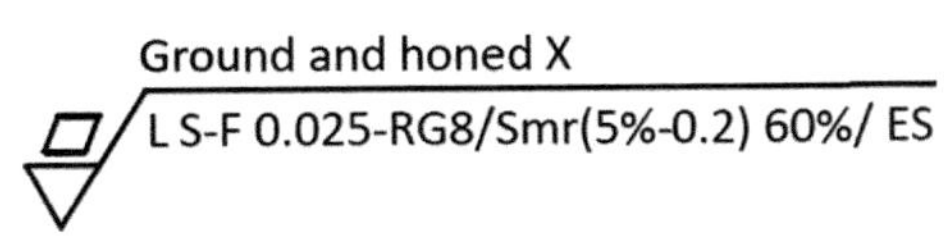

Fig. 12.11 An areal surface texture specification symbol.

filtration is used to distinguish the feature in different scales, for example, in profile surface texture, and is also used to distinguish roughness, waviness, and form;
association is to fit ideal features to nonideal features according to specific criteria that give an objective for a characteristic and set constraints;
collection is to identify and consider some features that together play a functional role;
construction is to build ideal features from other features;
reconstruction is to reconstruct a continuous feature from a finite number of points and is the inverse of extraction; and
evaluation is to identify either the value of a characteristic or its nominal value and its limit(s).

A category $\mathbf{S}_{\mathrm{AST}}$ can then be established to represent the operator. For surface texture, four ordered feature operations are defined; they are "partition," "extraction," "filtration," and "evaluation," each of which has a group of objects and morphisms as shown in Fig. 12.12.

The *partition* operation encloses a single object [Orientation] with element {crossed}, which represents the symbol "X" in Fig. 12.11.

The *filtration* operation consists of five objects and five morphisms. Object [Filter Symbol (S, F)] with element {(G, RG)} indicates a pair of filter symbols for the S filter and F operator, [Filter (S, F)] {(Gaussian, Robust Gaussian)} represents the meaning of the filter symbols of "G" and "RG," [Nesting Index (S, F)] {(0.025, 8)} is a pair of nesting indices for the S filter and F operator, and two objects [F Nesting Index (mm)] and [S Nesting Index (mm)]{0.025} represent nesting indices for S filter and F operator. In this operation, there is a product structure $\times_1$([F Nesting Index], [S Nesting Index], p_0, p_1); an isomorphism between object [Filter]{(Gaussian, Robust Gaussian)} and [Filter Symbol] {(G, RG)}; an epic morphism m_2: [S Nesting Index] (decides) ↠ [F Nesting Index]; and a general morphism m_1: [Filter (S, F)] (has) → [Nesting Index].

The *evaluation* operation encloses five objects, including [Comparison Rule] with element {max}, which is the default comparison rule, and [Specification type] with

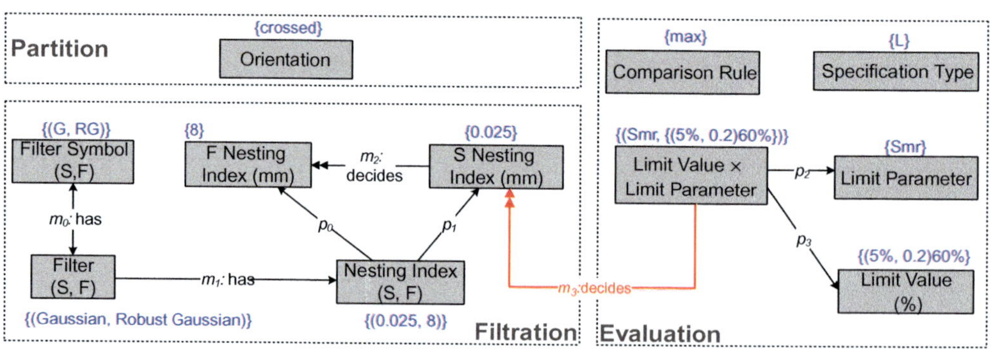

Fig. 12.12 A category $\mathbf{S}_{\mathrm{AST}}$ that consists of 11 objects and 8 morphisms.

element {L}, which is indicated in the symbol as shown in Fig. 12.11, and two objects [Limit Parameter] {Smr} and [Limit Value]{(5%, 0.2) 60%} form a product structure $\times_2$([Limit Parameter], [Limit Value], p_2, p_3).

An epic morphism connecting the filtration and evaluation operations, m_3: [Limit Value × Limit Parameter] (decides) ↠ [S Nesting Index], is to represent that limit value and limit parameter together decide the value of S nesting index.

12.3.1 Hierarchical operation model

The duality principle, as defined in ISO 8015:2011 [19], states that a specification operator is independent of any measurement characteristics, and the specification operator is physically realized in a verification operator, which is intended to mirror the specification operator but independent of such. To translate the duality principle, a set of specification operations will be mapped fully to a set of verification operations.

The specification category $\mathbf{S}_{AST}$ will be mapped fully to obtain a partial set of verification operations. To construct a full functor F: $\mathbf{S}_{AST} \Leftrightarrow \mathbf{M}_{AST}$, all objects and morphisms in the category $\mathbf{S}_{AST}$ are mapped isomorphically to that of measurement category $\mathbf{M}_{AST}$, as shown in Fig. 12.13.

As a measurement process requires more measurement procedures and details, additional measurement objects and morphisms (which obey GPS concepts and philosophy)

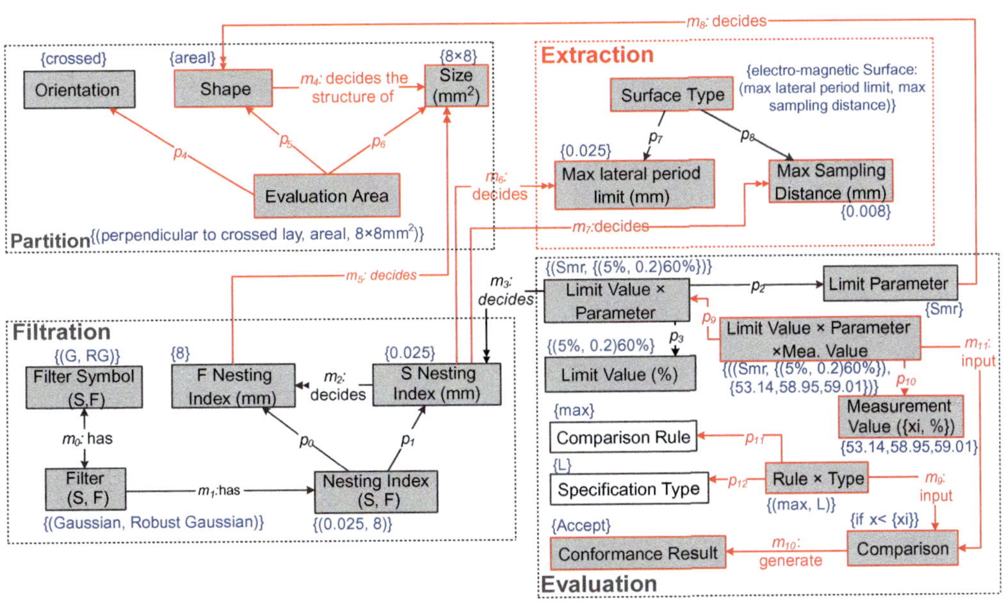

Fig. 12.13 The implementation of duality principle: A verification category $\mathbf{M}_{AST}$ encloses a full set of verification operations.

can then be added into the mapped verification category $\mathbf{M}_{\text{AST}}$. The actual verification operator for areal surface texture is an ordered set of operations of partition, extraction, filtration, and evaluation. Therefore an extraction operation including one product structure $\times_5$(Max Lateral Period, Max Sampling Distance, p_7, p_8) is added into the resulting structures, and object [Surface Type] is the product object of $\times_3$.

In the *partition* operation, two objects [Shape]{areal} and [Size] were added to construct a nesting of product structure. Therefore, in this operation, two product structures, which are $\times_3$([Shape], [Size], p_5, p_6) and $\times_4$(Orientation, $\times_1$, p_4, $\{p_5, p_6\}$); three projection morphisms p_4, p_5, and p_6, which represent that [Evaluation Area] is a nesting product of [Orientation], [Shape], and [Size]; and one morphism m_4: [Shape] (decides the structure of) $\rightarrow$ [Size] that represents the shape of the evaluation area and decides the structure of the size are generated.

In the *evaluation* operation, two more product structures $\times_6$([Comparison Rule], [Specification Type], p_{11}, p_{12}) and $\times_7$($\times_2$, [Measurement Value], p_9, p_{10}) are generated. Apart from the two new product objects, three new objects [Measurement Value], [Comparison], and [Conformance Result] are also added. Three morphisms are then identified as follows:

m_9: [Rule $\times$ Type] (input) $\rightarrow$ [Comparison] indicates that the comparison rule and specification type are the input of the comparison process.

m_{10}: [Comparison] (generate) $\rightarrow$ [Conformance Result] represents that the comparison process will generate a conformance result.

m_{11}: [Limit Value $\times$ Parameter $\times$ Mea. Value] (input) $\rightarrow$ [Comparison] represents that the limit value, parameter, and measurement values are all input by the comparison process.

Four epic morphisms are then established to connect the four operations, where

m_5: [F Nesting Index] (decides) $\twoheadrightarrow$ [Size] is to represent that the value of F nesting index decides the size;

m_6: [S Nesting Index] (decides) $\twoheadrightarrow$ [Max Lateral Period Limit] indicates that the value of S nesting index decides the value of the max lateral period limit;

m_7: [S Nesting Index] (decides) $\twoheadrightarrow$ [Max Sampling Distance] denotes that the tolerance type decides the type of parameter; for example, roundness tolerance should be assigned with roundness parameter only; and

m_8: [Limit Parameter] (decides) $\twoheadrightarrow$ [Shape] is to represent that the type of limit parameter decides the shape of the evaluation area.

A full set of ordered verification operations can then be established. After the known input elements in $\mathbf{S}_{\text{AST}}$ were mapped into $\mathbf{M}_{\text{AST}}$, and applying the reasoning rules, all elements of the set of objects in $\mathbf{M}_{\text{AST}}$ are deduced in the modeling process. A measurement guidance can then be generated, such that the duality principle is ensured to reduce specification uncertainty and method uncertainty.

12.3.2 Reasoning on the semantics

CSL reasoning can also be used to check redundancy and incompleteness of a specification using specific rules. The set of specification operations has to be simplified to obtain a specification structure with only necessary details. To ensure the stability of a design, when there are two or more specification elements in the solution, they must be independent. Those independent specification elements can then be mapped back to form a complete specification operator and then construct a complete verification operator using the forward mappings *F*. Therefore it can ensure that the forward mapping from the simplified specification to verification is without loss of information. Simplifying the objects in a specification is decided by the special type of properties of its objects and morphisms and decided by the types of morphism structures [15].

For example, given a surface texture specification as shown in Fig. 12.14A, to detect any redundancy or incompleteness of such a symbol in an automatic way, the first step is to model the specification based on four totally ordered operations, which are partition, extraction, filtration, and evaluation, as shown in Fig. 12.15. After the simplification process the result specification is indicated in Fig. 12.14B.

Similar processes can also be applied to enrich an incomplete specification. Fig. 12.16A shows a typical cylindricity symbol. And five totally ordered operations can be constructed as shown in Fig. 12.17. After the simplification process, it can identify that objects [Limit Parameter], [Limit Value], and [Filter Symbol] are necessary objects. To complete the specification the default settings in ISO GPS system should also be considered. The [Filter] object can use default Gaussian filter, but in the [Nesting Index], the filter undulation per revolution (UPR) has no default values, and for a cylindricity specification UPR, both generatrix and redial section have to be defined. After the incompleteness check an example of a complete cylindricity specification is indicated in Fig. 12.16B.

12.4 CatSurf: An integrated surface texture information system

After the semantics has been represented into a machine-readable and machine-interpretable format, the semantic model can then be developed into a system with

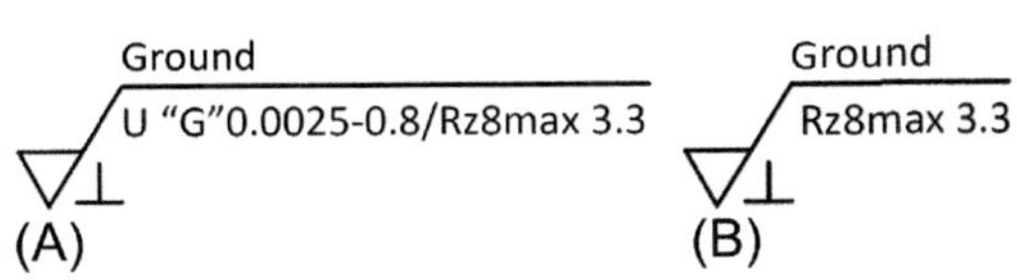

Fig. 12.14 A simplification example of a specification. (A) A specification with redundancy. (B) Result after the simplification process.

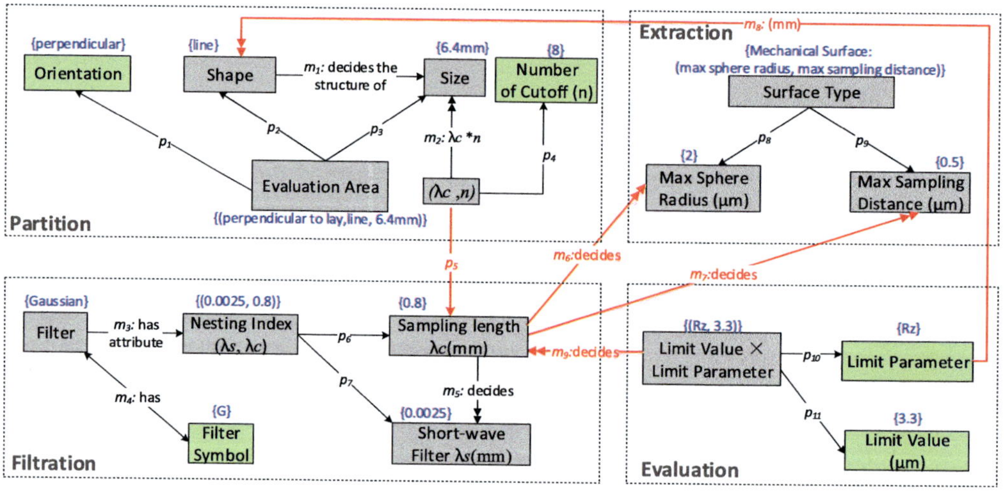

Fig. 12.15 CSL model of a profile surface texture specification.

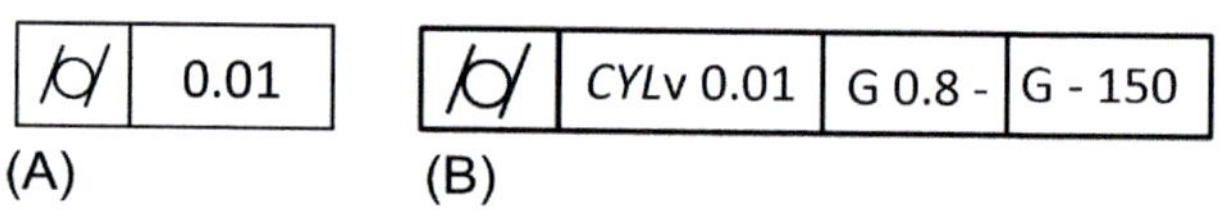

Fig. 12.16 Specification examples of cylindricity. (A) An incomplete specification. (B) The proposed cylindricity specification.

user-friendly interfaces to facilitate end users to interrogate the system with minimum amount of input requirement. An intelligent information system, named "CatSurf" [11], is developed for surface texture with two general modules to support design and measurement of profile and areal surface texture, respectively.

12.4.1 CatSurf: System architecture and components

The architecture of the CatSurf system is designed in accordance with the product chain from ISO GPS in which surface texture is defined. As shown in Fig. 12.18, the main components of the CatSurf system are presented with one knowledge management model and two modules (for profile and areal surface texture) each with five components, which are *function*, *manufacture*, *specification*, *verification*, and *help*, as shown in Fig. 12.19. The five components are designed to provide both designers and metrologists with related information. The first three components are part of the design phase; the *verification* component is designed for surface texture measurement; the *help* component is developed to provide all the help information for the system. A categorical knowledge management model is developed to support all the knowledge, manipulation, querying, and reasoning within the two modules.

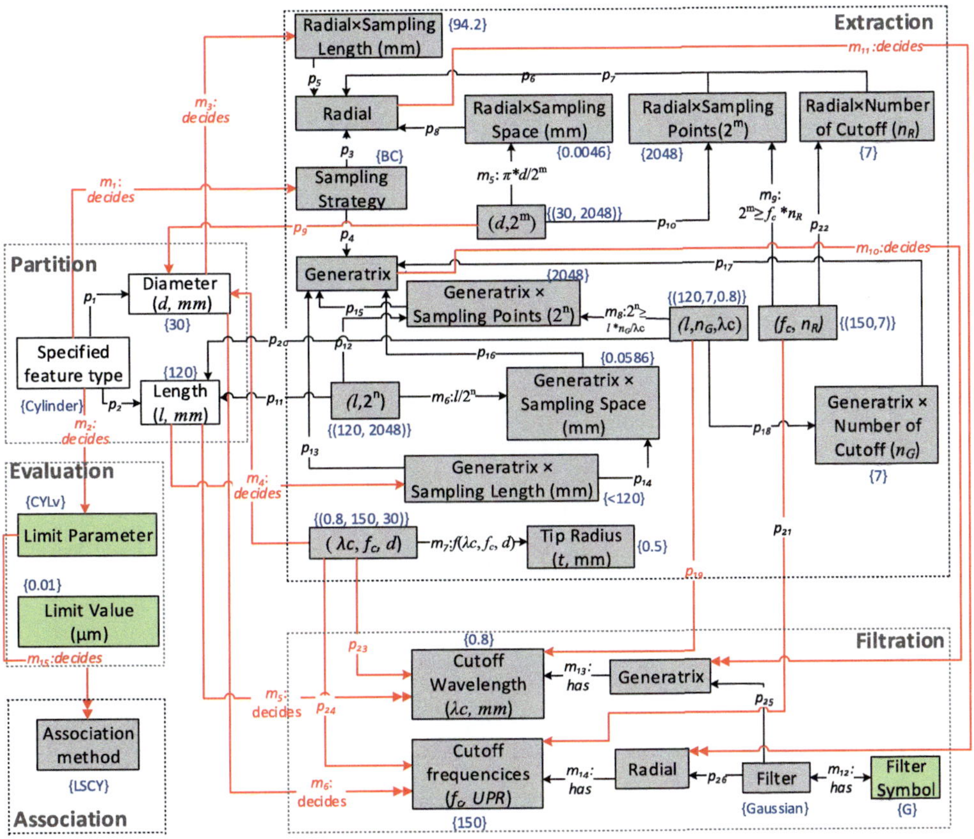

Fig. 12.17 CSL model of a cylindricity specification.

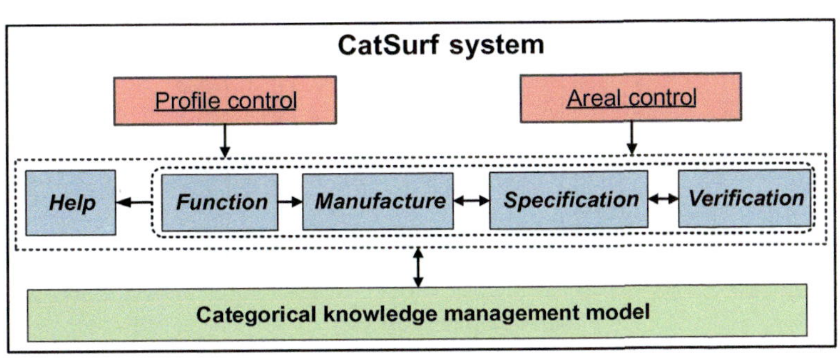

Fig. 12.18 Main components in the CatSurf system.

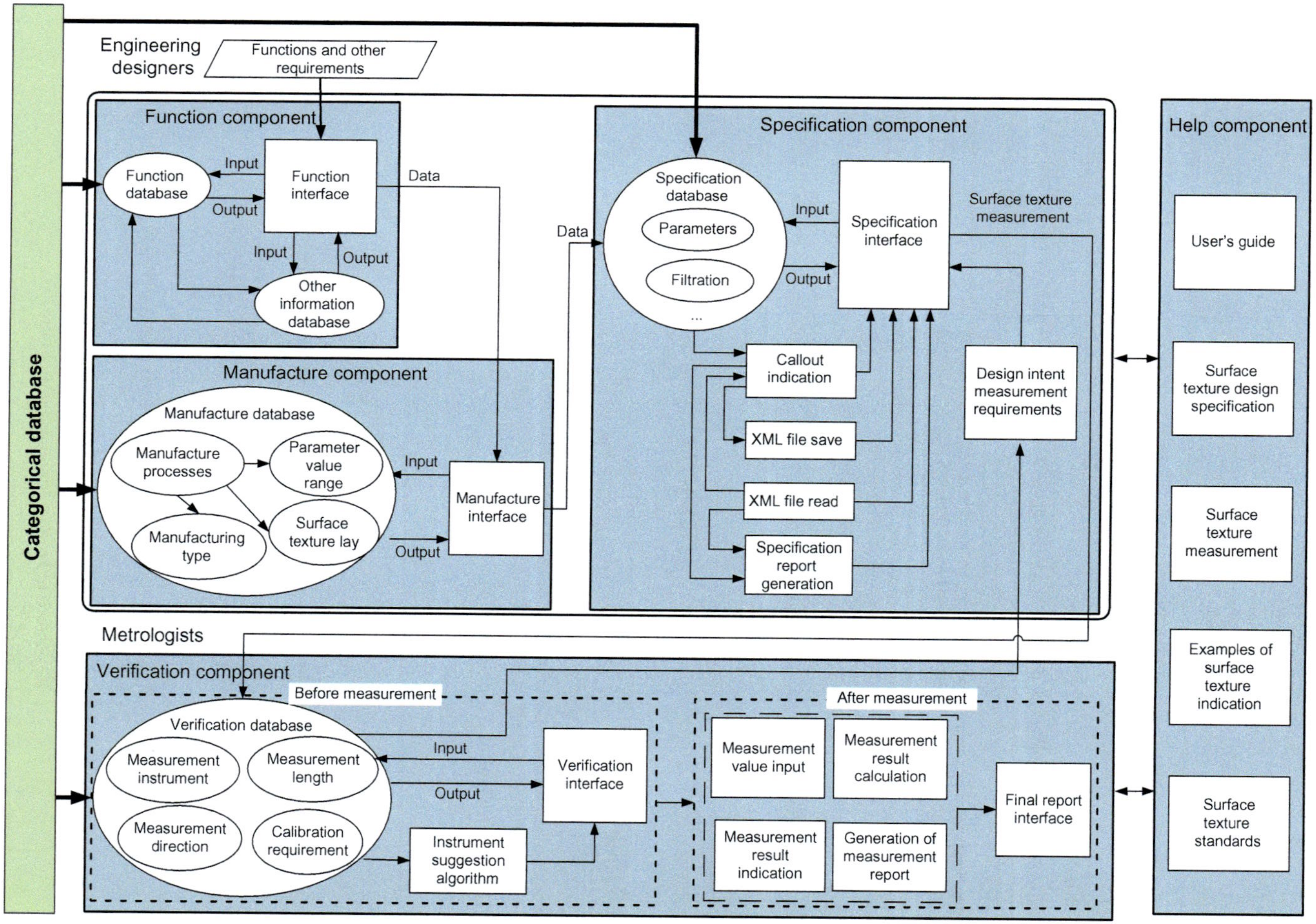

Fig. 12.19 The architecture of the CatSurf system.

12.4.1.1 Design your specifications with functionality and production capability

Two components (*function* and *manufacture*) as shown in Fig. 12.20 provide all relevant information for the engineered artifact surface before the assignment of a specification, to ensure the unambiguity and functionality of the assigned specification.

12.4.1.2 Intelligent specification generation

A *specification* component, as indicated in Fig. 12.21, provides optimized and complete surface texture specifications for designers with the least amount of input information. It is possible to avoid the indiscriminate use of surface texture values that result in impractical and costly production requirements. It can generate a complete specification based on the information gained in previous components, and specification data and symbols can also be generated and saved for future use. Designers can also revise the generated specification in accordance with their specialized requirements. A specification report can also be generated with a full interpretation of the semantics of the specification and basic measurement information based on such semantics.

12.4.1.3 Guidance and analysis of measurement

A *verification* component provides detailed measurement parameters such as the measurement environment, measurement direction, and length and calibration requirements. It also provides a suggested instrument according to the specification and then generates a measurement strategy (see Fig. 12.22A).

It is also designed to record the details of the measurement environment such as measurement time, humidity, and operators, to calculate the number of measurements, to estimate the measurement uncertainty, to indicate the measurement result, and to

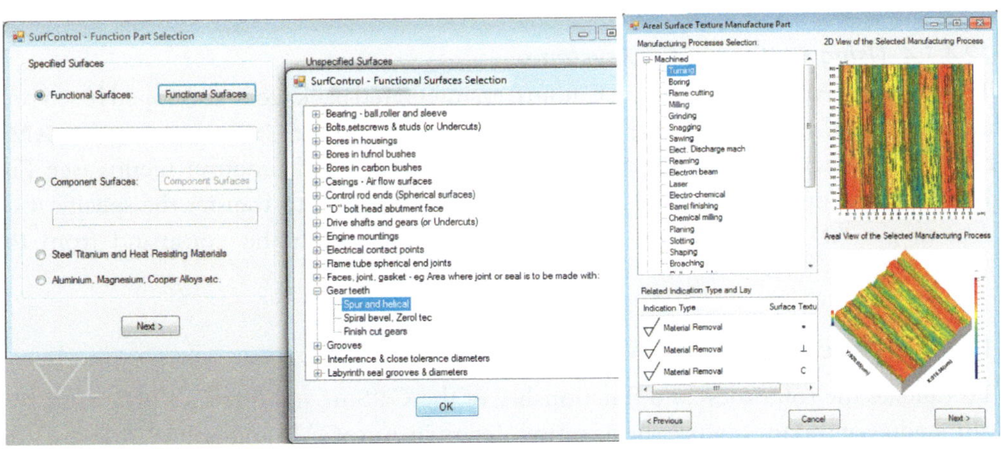

Fig. 12.20 The selection of function requirements (left) and manufacturing process (right).

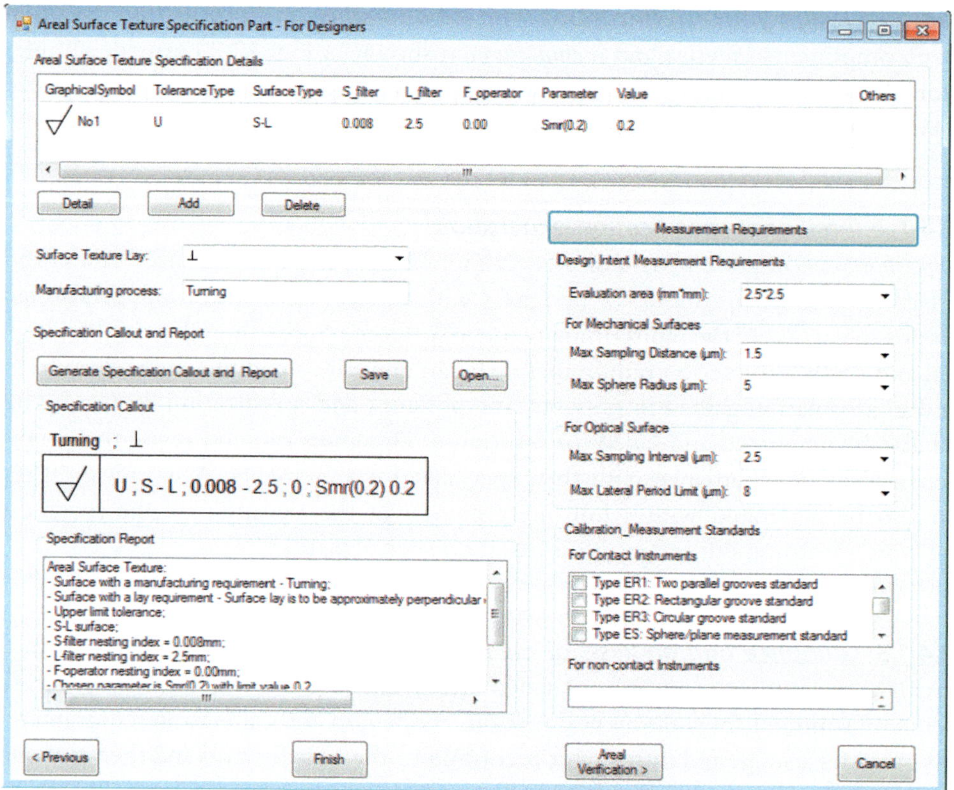

Fig. 12.21 The generation of the specification for profile surface texture.

provide a conformance zone to make a measurement result decision with reference to the specification and uncertainty (see Fig. 12.22B).

12.4.1.4 Using CatSurf in CAD systems

To facilitate CAD end users to use CatSurf, an integrated package is developed based on a universal XML approach. As the designed specifications in the CatSurf are saved to XML files, an interface application program with two embedded function menus (see Fig. 12.23) is developed to read the XML files (see Fig. 12.24), to transfer the specification data to a CAD database (see Fig. 12.25) and to execute the command from the interface in the CAD.

12.4.2 Test cases

To validate the robustness and functionality of the CatSurf and interface programs, two case studies of surface texture specification design in AutoCAD and SolidWorks are presented. The first test case is the design of the profile surface texture specifications in

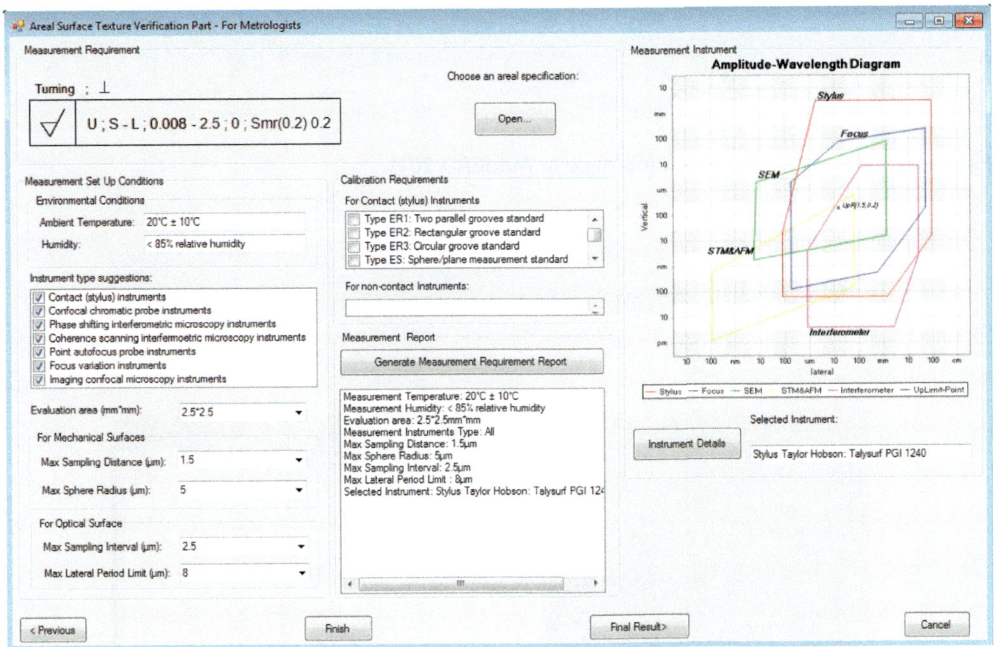

(A)

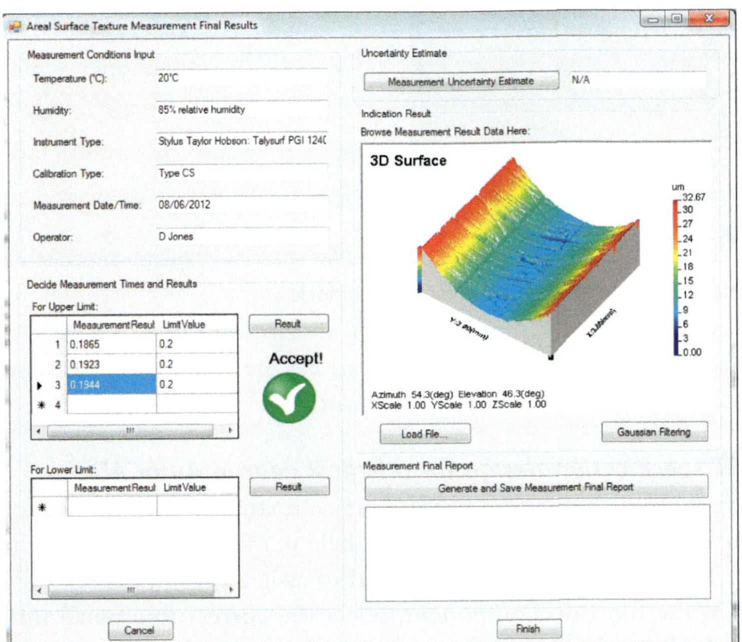

(B)

Fig. 12.22 The interfaces of verification component for areal surface texture module. (A) Measurement guidance interface. (B) Postmeasurement analysis interface.

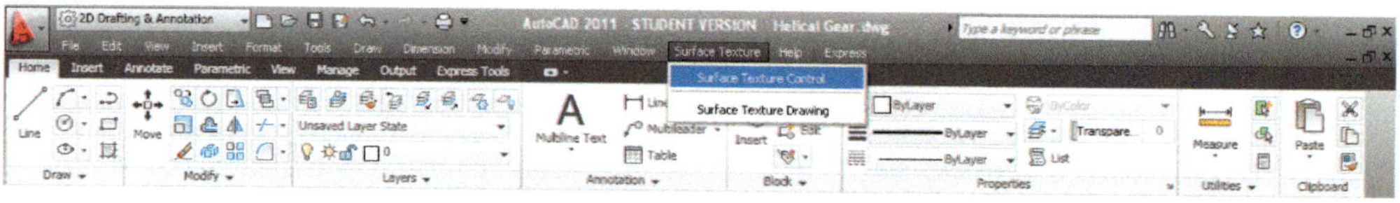

Fig. 12.23 The embedded menu interface in AutoCAD 2011.

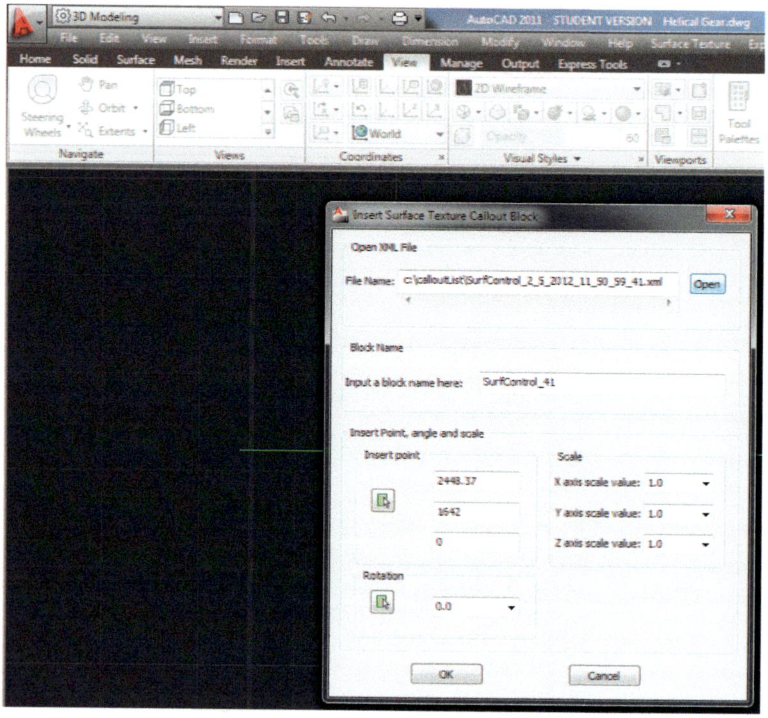

Fig. 12.24 Open XML file and insert the indication block.

AutoCAD for a helical gear. The second test case is the design of the areal surface texture specifications in SolidWorks for a stepped shaft.

12.4.2.1 PST specification design for a helical gear in AutoCAD

The first case study aims to assign profile surface texture specifications for a helical gear that is shown in Fig. 12.26. The case study is held in the *ProfileControl* module and AutoCAD 2011. There are three steps in CatSurf to assign a specification.

Step 1: In the function component, select the correct functional surface type and material. As shown in Fig. 12.27, the selected functional surface is "spur and helical" under the category of "gear teeth," and the selected material is "steel titanium and heat-resisting materials."

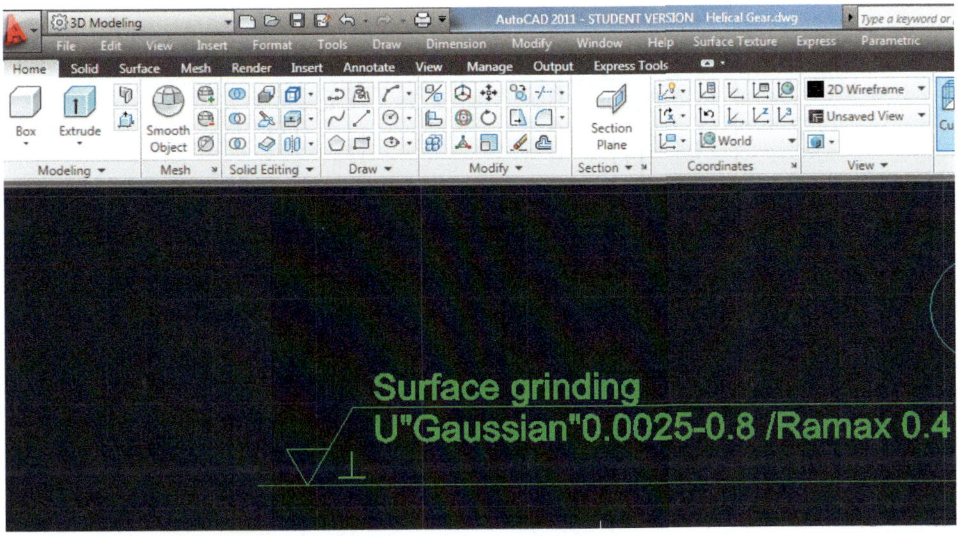

Fig. 12.25 The inserted profile surface texture specification.

Step 2: In the manufacture component the manufacturing process of "surface grinding" is selected automatically by the system as the default manufacturing process for helical gear teeth. Accordingly the related *Ra* value range is 0.1–0.8 μm, and lays are "=," "⊥," and "R." The lay "⊥" is then selected.

Step 3: In the specification component the details of the specification are generated automatically. The indication and XML file are saved as shown in Fig. 12.28.

Returning to the AutoCAD 2011 environment, there are three steps to insert the designed specification.

Step 4: Click "surface texture drawing" menu and open the "insert surface texture callout block" interface as shown in Fig. 12.29. In the interface, open the saved XML file.

Step 5: Change the name of the block and select the insertion point, scale, and rotation. Insert the block in the drawing (as shown in Fig. 12.30).

Step 6: Repeat steps 1–5 to design more specifications for a different surface in the helical gear. Alternatively, it is possible to insert the saved blocks for the surfaces with the same requirements. The finished surface texture specifications are shown in Fig. 12.31.

12.4.2.2 Areal specification design for a stepped shaft in solidworks

The second case study aims to assign areal specifications for a stepped shaft that is shown in Fig. 12.32. According to the functional requirements, the shaft is divided into six segments:

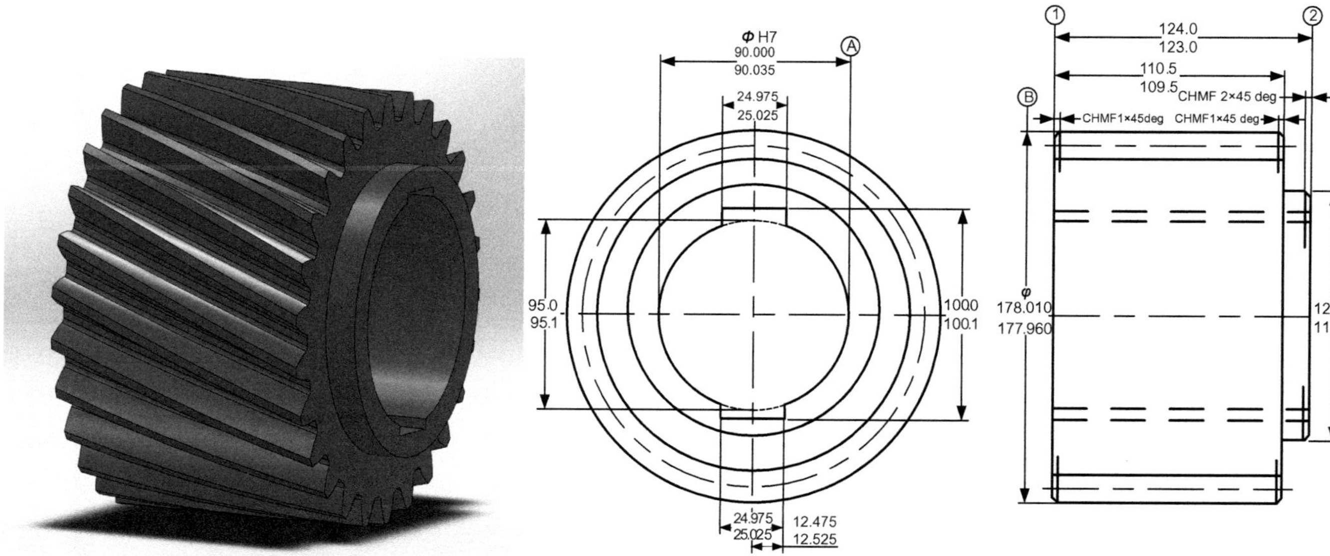

Fig. 12.26 The design of a helical gear.

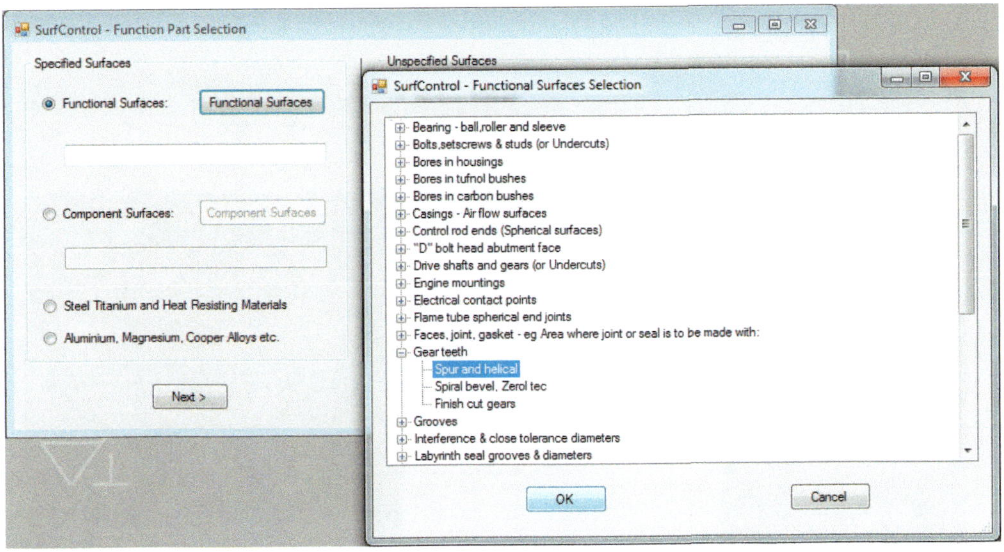

Fig. 12.27 The selection of function requirement in the function component.

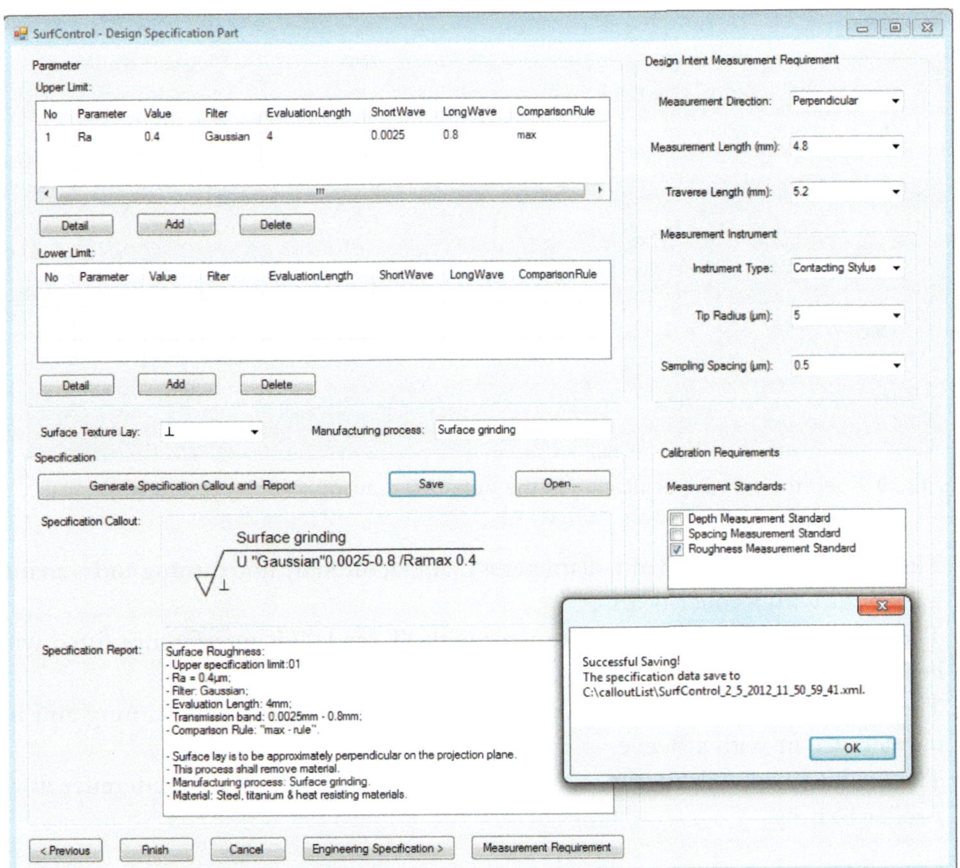

Fig. 12.28 The generation of the specification in the specification component.

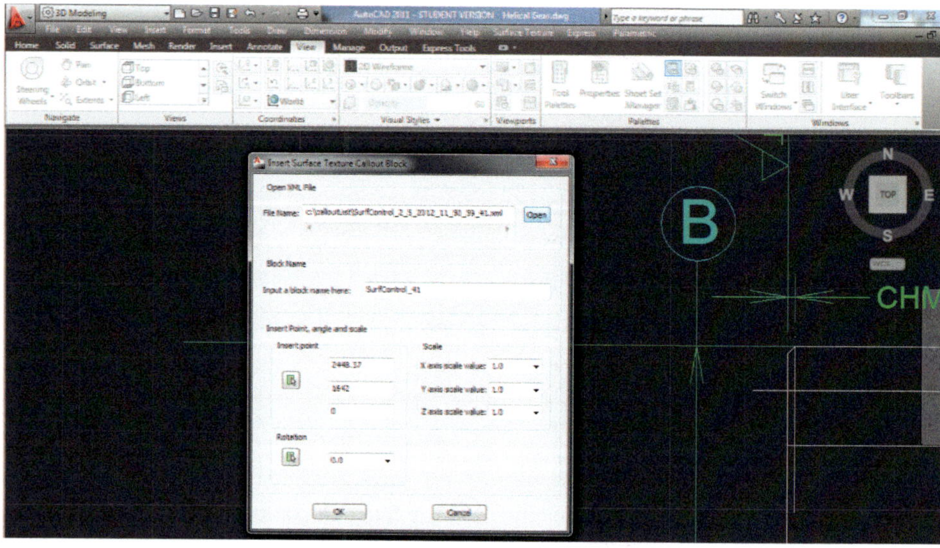

Fig. 12.29 Open XML file and insert the indication block.

Fig. 12.30 Insert the saved specification in the AutoCAD drawing.

- The shaft segment 1 of 55 mm diameter is manufactured by fine turning and is an interference fit with a roller bearing.
- The shaft segment 2 of 58 mm diameter with IT grade 7 is interference fitted with a helical gear.
- The shaft segment 3 of 55 mm diameter is manufactured by fine turning and is an interference fit with a sleeve.
- The shaft segment 4 shares the same shaft with segment 3 and is an interference fit with a roller bearing.

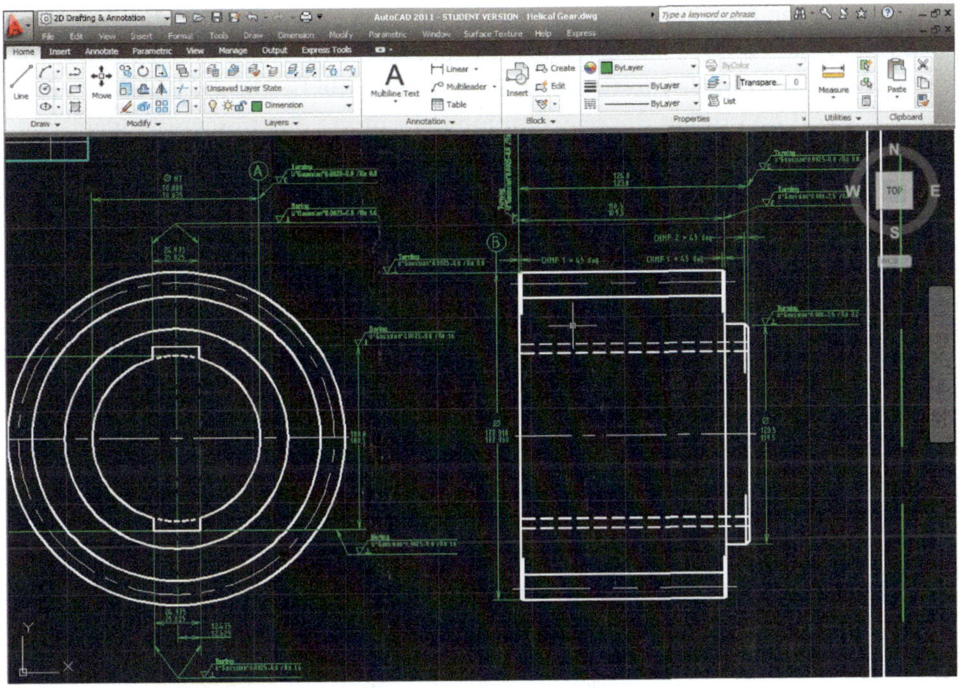

Fig. 12.31 The completed PST specifications design for a helical gear.

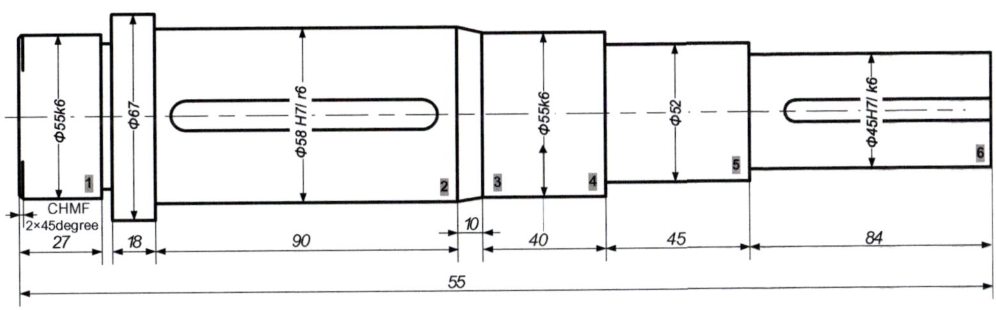

Fig. 12.32 The design of a stepped shaft.

- The shaft segment 5 of 55 mm is manufactured by turning and is a sealing fit with an end plate.
- The segment 6 with IT grade 7 is an interference fit with a flat key.

By accessing the CatSurf system in SolidWorks, the *ArealControl* module is applied to carry out the specification assignment. Taking the shaft segment 1 as an example, there are three steps in the specification assignment in CatSurf.

Step 1: In the function component, select functional surfaces "shaft fit with rolling bearing"; although the normal chosen parameter for turning surfaces is *Ra*, for the purpose of functionality testing, the *Sa* of 0.4 μm will be chosen here as a substitute of *Ra*. Fig. 12.33 shows the selection interface of the function component.

Step 2: In the manufacture component, fine turning is selected (with lay "⊥").

Step 3: In the specification component the details of areal specification are generated automatically. The indication and XML file are saved as shown in Fig. 12.34.

Returning to the SolidWorks 2009 environment, there are three steps to insert the saved specification in the drawing.

Step 4: Click "insert block" menu and open the "insert surface texture callout block" interface as shown in Fig. 12.35. In the interface, open the saved XML file.

Step 5: Change the name of the block; select insert point, scale, and rotation. Insert the block in the drawing (as shown in Fig. 12.36).

Step 6: Repeat steps 1–5 to design specifications for segment 2–6. The suggested parameter for segment 2 is *Sa* of 0.8 μm, for segment 3 is *Sa* of 0.8 μm, for segment 4 is *Sa* of 0.4 μm, for segment 5 is *Sa* of 0.6 μm, and for segment 6 is *Sa* of 1.6 μm. The finished surface texture specifications are shown in Fig. 12.37.

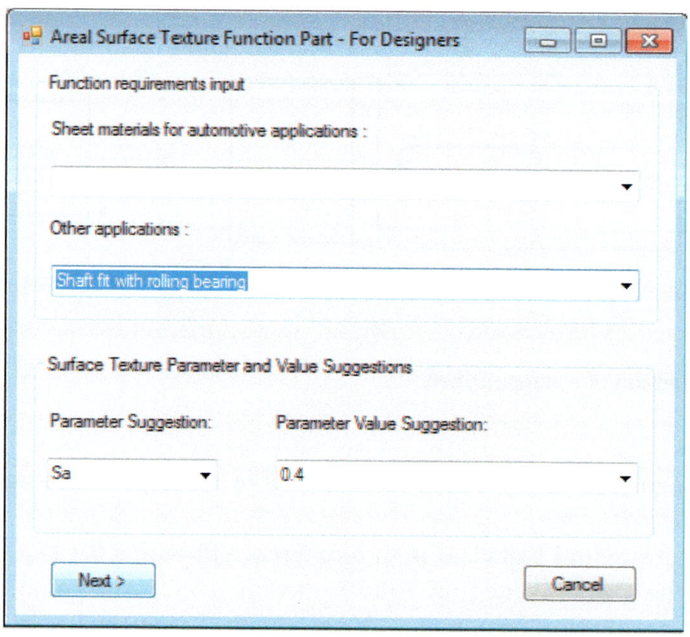

Fig. 12.33 The selection of function requirements in the function component.

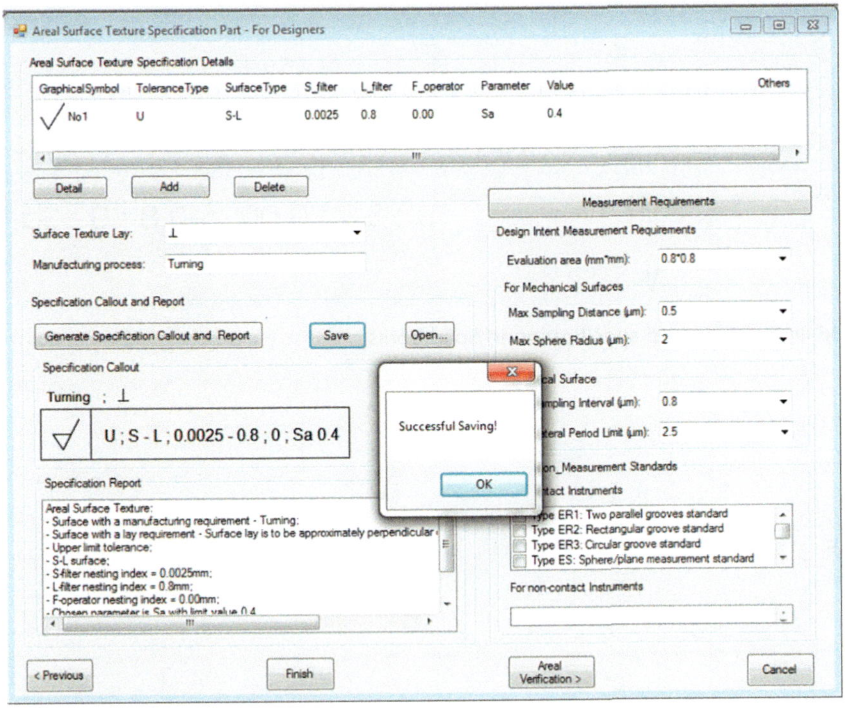

Fig. 12.34 The generation of specification in specification component.

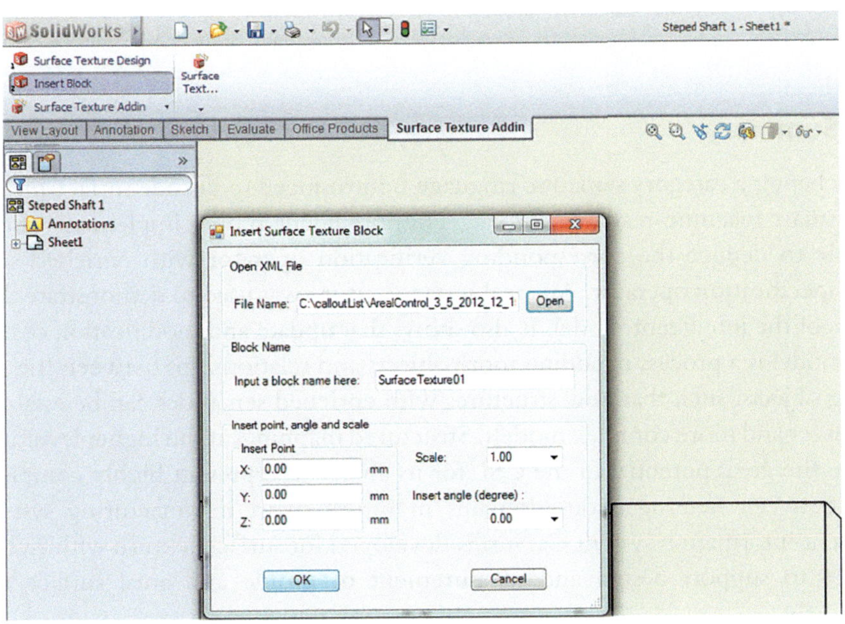

Fig. 12.35 Open the saved XML file.

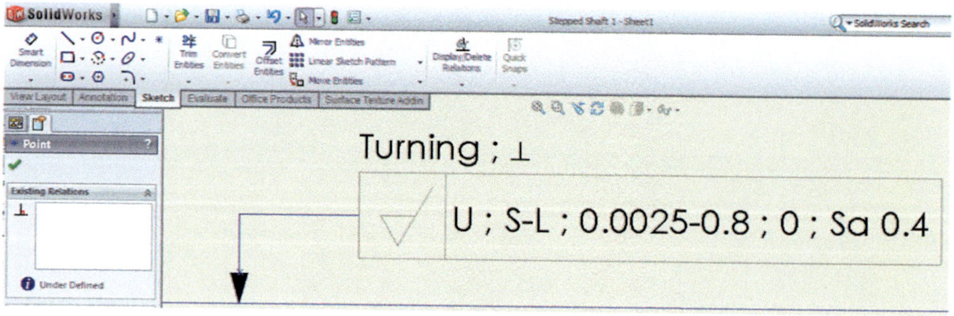

Fig. 12.36 Insert the saved specification in SolidWorks.

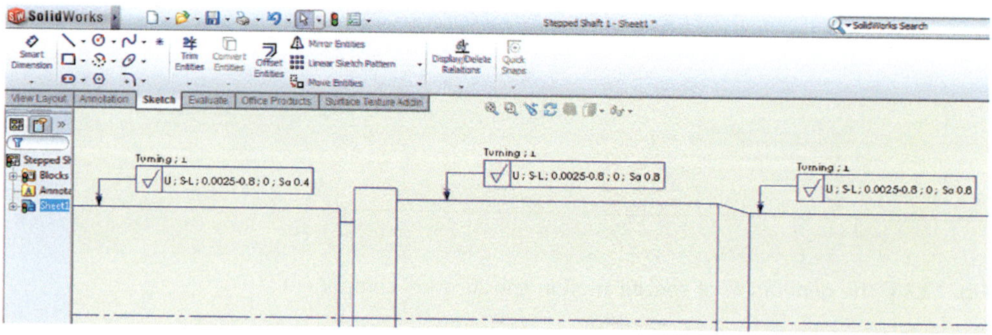

Fig. 12.37 The completed areal surface texture specification design for the stepped shaft.

12.5 Summary

In this chapter a category semantic language is introduced to decode surface information into a smart machine-readable format. The intelligent model implements the duality principle to deduce the corresponding verification operator with enriched semantics from a specification operator. An areal surface texture was used to demonstrate the effectiveness of the intelligent model. It also shows that update and modification of the intelligent model is a process of adding more objects and relationships between the new and existing objects, such that new structures with enriched semantics can be established to build larger and more complex models. Structured mappings at the higher level also demonstrate the great potential of the CSL for its ability to represent highly complex operations between heterogeneous domains in future smart manufacturing systems. An intelligent information system CatSurf is developed for surface texture with two general modules to support design and measurement of profile and areal surface textures, respectively.

Probably the biggest challenge for smart specification and verification lies in the manual work required for semantic representation. This is a challenge that cannot be fully solved with current technology, as no methods are available to date to derive and to structure the information in an automatic manner. A number of approaches can be applied in the future to reduce manual input to an extent. One approach can be the so-called "Lego" approach. Basic operation models for geometrical variability can be constructed as standard Lego blocks. For specific applications, necessary operation blocks can be directly adopted to assemble a specific model and later to add more objects to it. Another approach is to develop a set of algorithms to discover automatically the nature of a relationship, therefore enriching the semantics structure and facilitating further reasoning.

Appendix

A.1 Basic category concepts

In category theory a category is constructed as a collection of *objects* and relationships from one object to another. The relationship is called *morphism* as shown in Fig. A.1A and satisfies the following conditions:

- For each arrow f, there are given objects: $dom(f)$ and $cod(f)$, called the domain and codomain of f, respectively. We write $f: A \rightarrow B$ to indicate that $A = dom(f)$ and $B = cod(f)$.
- Given morphisms $f: A \rightarrow B$ and $g: B \rightarrow C$, there is a morphism $g \circ f: A \rightarrow C$, called the composite of f and g, and $\circ$ is the composition operation, as shown in Fig. A.1B.
- For each object A, there is an identify morphism $id_A: A \rightarrow A$ that satisfies the identity law, for any morphism $f: A \rightarrow B$, $id_B \circ f = f$ and $f \circ id_A = f$, as shown in Fig. A.1A.

The collection of all morphisms from A to B in category C is denoted as $Hom_C(A, B)$ and called the hom-set between A and B (the collection of morphisms is not required to be a set).

There are many types of morphisms as shown in Fig. A.2. A morphism $f: A \rightarrow B$ is *monic* (Fig. A.2A; definition refer to [20] p. 29); if given any $g, h: C \rightarrow A, f \circ g = f \circ h$ implies $g = h$; f is *epic* (Fig. A.2B; definition refer to [20] p. 31); if given any $g, h: B \rightarrow C, g \circ f = h \circ f$ implies $g = h$; f is *isomorphic* (Fig. A.2C; definition refer to [20] p. 12); there exists a morphism $g: B \rightarrow A$, such that $g \circ f = id_A$, *and* $f \circ g = id_B$; f is *retraction* (Fig. A.2D; definition refer to [21] p. 32); there exists a morphism $g: B \rightarrow A$, such that $g \circ f = id_A$; f is *section* (Fig. A.2E; definition refer

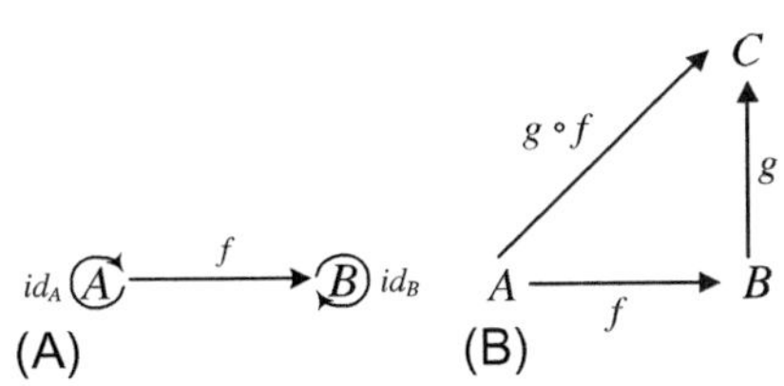

Fig. A.1 Basic concepts of category theory. (A) A morphism $f: A \rightarrow B$. (B) Composite of f and g.

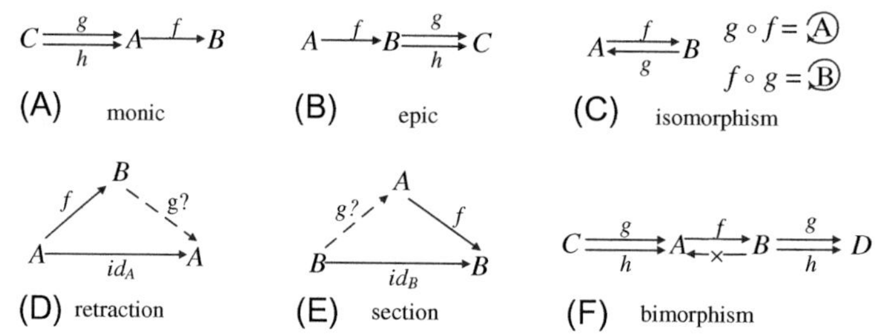

Fig. A.2 Different types of morphisms. (A) monic. (B) epic. (C) isomorphism. (D) retraction. (E) section. (F) bimophism.

to [21] p. 32); there exists a morphism g: $B \rightarrow A$, such that $f \circ g = id_B$; f is *bimorphism* (both *monic and epic* but not *isomorphic*; Fig. A.2F) whenever there is no inverse morphism of f; note that for category of sets, if f is both monic and epic, then f is isomorphic.

The sets of objects and morphisms, which have a specific functionality, form a *category*.

There are some of the fundamental operations of objects in a category, such as product, coproduct, and pullback. A *product* of two objects A and B, denoted $A \times B$, if both A and B are sets, $A \times B$ is defined as the ordered pairs (a, b) where $a \in A$ and $b \in B$, and there are two natural projection functions, π_1: $A \times B \rightarrow A$ and π_2: $A \times B \rightarrow B$.

A *coproduct* of two objects A and B, denoted $A \sqcap B$, if both A and B are sets, $A \sqcap B$ is defined as the disjoint union of A and B; that is, $A \sqcap B$ includes nonoverlapping copies of A and B, and there are two natural inclusion functions i_1: $A \rightarrow A \sqcap B$ and i_2: $B \rightarrow A \sqcap B$.

Given two morphisms f and g, with a same codomain, that is, $cod(f) = cod(g)$, the *pullback* of f and g consists of two morphisms p_1 and p_2 such that $f \circ p_1 = g \circ p_2$ and given any object E with two morphisms that are directed to A and B, respectively, $f \circ q_1 = g \circ q_2$; there exists a unique morphism u: $E \rightarrow D$ with $q_1 = p_1 \circ u$ and $q_2 = p_2 \circ u$, as shown in Fig. A.3.

A.2 High-level categorical concepts

At a higher level a relationship from on category to another is called a *functor*, denoted as F:$\mathbf{C} \rightarrow \mathbf{D}$ as shown in Fig. A.4, and F satisfies the following conditions:

- For each morphism f: $A \rightarrow B$ in $\mathbf{C}$, there is a morphism $F(f)$: $F(A) \rightarrow F(B)$ in $\mathbf{D}$.
- For each object A in $\mathbf{C}$, $F(id_A) = id_{FA}$ holds in $\mathbf{D}$.

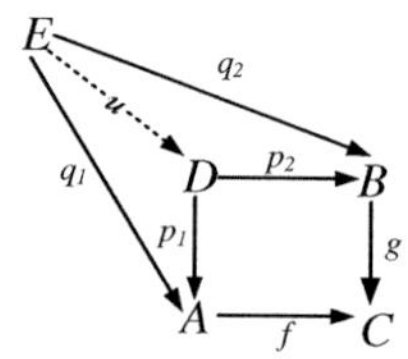

Fig. A.3 A pullback in category theory.

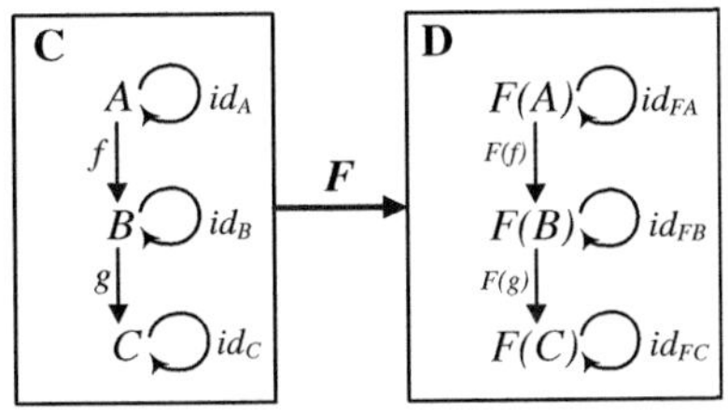

Fig. A.4 A functor F:**C** → **D**.

- For each object pair of morphisms f: $A \rightarrow B$ and g: $B \rightarrow C$ in **C**, $F(g \circ f) = F(g) \circ F(f)$ holds in **D**.

A functor F is constructed by a tuple (F^O, F^M) where F^O is the mapping between the two object sets $O_\mathbf{C}$ and $O_\mathbf{D}$ and F^M is the mapping between the two morphism sets $M_\mathbf{C}$ and $M_\mathbf{D}$, in such a way that for every pair of objects A, $B \in O_\mathbf{C}$, the functor F represent the mapping from the set of morphism from A to B in **C** to the set of morphisms from $F(A)$ to $F(B)$ in **D**, written as $F^{\rightarrow}$: $Hom_\mathbf{C}(A, B) \rightarrow Hom_\mathbf{D}(F(A), F(B))$. A functor may also have a special property as shown in Table A.1, where nine properties are listed. There are more than nine properties in category theory, but only nine of them will be discussed here.

A functor is full (denote as $\bullet\!\Rightarrow$) when $F^{\rightarrow}$ is injective, is faithful ($\blacktriangleright\!\Rightarrow$) when $F^{\rightarrow}$ is surjective, is full faithful ($\Leftrightarrow$) when $F^{\rightarrow}$ is bijective, is injective on objects ($^{\blacktriangleright O}\!\!\rightarrow$) when F^O is injective, is injective on morphisms ($^{\blacktriangleright M}\!\!\rightarrow$) when F^M is injective, is surjective on objects ($^O\!\twoheadrightarrow$) when F^O is surjective, is surjective on morphisms ($^M\!\twoheadrightarrow$) when F^M is surjective, is bijective on objects ($^O\!\leftrightarrow$) when F^O is bijective, and is bijective on morphisms ($^M\!\leftrightarrow$) when F^M is bijective (Fig. A.4).

There are also morphisms between functors, and they are called *natural transformations*, and if there is a functor that has an inverse functor, the pair is called *adjoint functor.* A *natural transformation* is a mapping of one functor to another. If F and E are two functors from **C** to **D**, as shown in Fig. A.5, a natural transformation η from F to E associates to

Table A.1 Special properties of a functor F:**C** → **D**.

Properties	Symbol	Semantics
Faithful	$\blacktriangleright\!\Rightarrow$	$F^{\rightarrow}$ is surjective
Full	$\bullet\!\Rightarrow$	$F^{\rightarrow}$ is injective
Full faithful	$\Leftrightarrow$	$F^{\rightarrow}$ is bijective
Injective on objects	$^{\blacktriangleright O}\!\!\rightarrow$	F^O is injective
Injective on morphisms	$^{\blacktriangleright M}\!\!\rightarrow$	F^M is injective
Surjective on objects	$^O\!\twoheadrightarrow$	F^O is surjective
Surjective on morphisms	$^M\!\twoheadrightarrow$	F^M is surjective
Bijective on objects	$^O\!\leftrightarrow$	F^O is bijective
Bijective on morphisms	$^M\!\leftrightarrow$	F^M is bijective

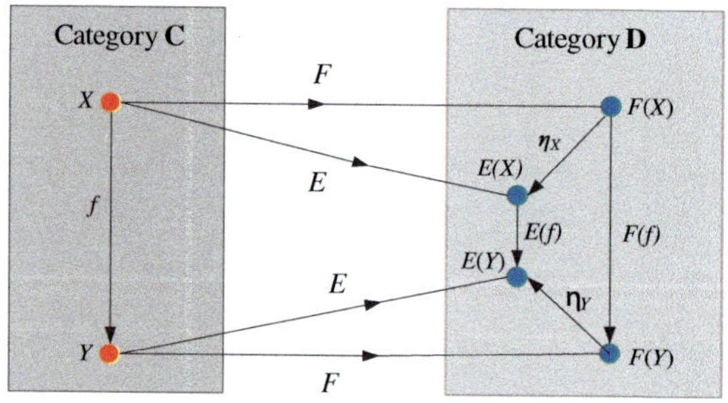

Fig. A.5 Natural transformation.

every object X in **C**, which includes a morphism η_X: $F(X) \rightarrow E(X)$ between objects of **D**, called the component of η at X, such that for every morphism f: $X \rightarrow Y$ in **C** we have: $\eta_Y \circ F(f) = E(f) \circ \eta_X$.

References

[1] Chandrasegaran SK, Ramani K, Sriram RD, Horváth I, Bernard A, Harik RF, et al. The evolution, challenges, and future of knowledge representation in product design systems. Comput Aided Des 2013;45(2):204–28.

[2] Verhagen WJ, Bermell-Garcia P, van Dijk RE, Curran R. A critical review of knowledge-based engineering: an identification of research challenges. Adv Eng Inform 2012;26(1):5–15.

[3] Qin Y, Qi Q, Lu W, Liu X, Scott PJ, Jiang X. A review of representation models of tolerance information. Int J Adv Manuf Technol 2018;95(5–8):2193–206.

[4] Negri E, Fumagalli L, Garetti M, Tanca L. Requirements and languages for the semantic representation of manufacturing systems. Comput Ind 2016;81:55–66.

[5] ISO 10303-11. Industrial automation systems and integration—product data representation and exchange—part 11: description methods: the EXPRESS language reference manual. Geneva: International Organisation for Standardisation; 2004.

[6] Zhao YF, Horst JA, Kramer TR, Rippey W, Brown RJ. Quality information framework–integrating metrology processes. IFAC Proc Vol 2012;45(6):1301–8.

[7] Lu W, Qin Y, Liu X, Huang M, Zhou L, Jiang X. Enriching the semantics of variational geometric constraint data with ontology. Comput Aided Des 2015;63:72–85.

[8] Qin Y, Lu W, Qi Q, Li T, Huang M, Scott PJ, et al. Explicitly representing the semantics of composite positional tolerance for patterns of holes. Int J Adv Manuf Technol 2017;90(5-8):2121–37.

[9] Lu W, Jiang X, Liu X, Qi Q, Scott P. Modeling the integration between specifications and verification for cylindricity based on category theory. Meas Sci Technol 2010;21(11):115107.

[10] Qi Q, Jiang X, Scott PJ. Knowledge modeling for specifications and verification in areal surface texture. Precis Eng 2012;36(2):322–33.

[11] Qi Q, Scott PJ, Jiang X, Lu W. Design and implementation of an integrated surface texture information system for design, manufacture and measurement. Comput Aided Des 2014;57:41–53.

[12] Xu Y, Xu Z, Jiang X, Scott P. Developing a knowledge-based system for complex geometrical product specification (GPS) data manipulation. Knowl-Based Syst 2011;24(1):10–22.

[13] Baader F, Calvanese D, McGuinness D, Patel-Schneider P, Nardi D. The description logic handbook: theory, implementation and applications. Cambridge University Press; 2003.

[14] Freitas F, Lins F, editors. The limitations of description logic for mathematical ontologies: an example on neural networks; ONTOBRAS-MOST. 2012.

[15] Qi Q, Pagani L, Scott PJ, Jiang X. Enabling metrology-oriented specification of geometrical variability—a categorical approach. Adv Eng Inform 2019. ADVEI873.

[16] ISO 17450-1:2011. Geometrical product specifications (GPS)—general concepts—part 1: model for geometrical specification and verification. Geneva, CH: International Organization for Standardization; 2011.
[17] Spivak DI. Category theory for scientists. Cambridge, MA: The MIT Press; 2014.
[18] ISO 25178-1:2016. Geometrical product specifications (GPS)—surface texture: areal—part 1: indication of surface texture. Geneva, CH: International Organization for Standardization; 2016.
[19] ISO 8015:2011. Geometrical product specifications (GPS)—fundamentals—concepts, principles and rules. Geneva, CH: International Organization for Standardization; 2011.
[20] Awodey S. Category theory. New York: Oxford University Press Inc.; 2010.
[21] Lawvere FW, Schanuel SH. Conceptual mathematics: a first introduction to categories. Cambridge University Press; 2009.

Index

Note: Page numbers followed by *f* indicate figures, *t* indicate tables and *b* indicate boxes.

G

T

U

V

W

Z

9780128218150